SCIENCE AND SCIENTISTS

SCIENCE AND SCIENTISTS

Essays by
Biochemists, Biologists and Chemists

Edited by
M. KAGEYAMA, K. NAKAMURA,
T. OSHIMA and T. UCHIDA

JAPAN SCIENTIFIC SOCIETIES PRESS, Tokyo
D. REIDEL PUBLISHING COMPANY, Dordrecht Boston London

Library of Congress Cataloging in Publication Data

DATA APPEAR ON SEPARATE CARD

ISBN-13: 978-94-009-7757-0   e-ISBN-13:978-94-009-7755-6
DOI: 10.1007/978-94-009-7755-6

Published by JAPAN SCIENTIFIC SOCIETIES PRESS, Tokyo, in co-publication with
D. REIDEL PUBLISHING COMPANY, P.O. Box 17, 3300 AA Dordrecht, Holland.

Sold and distributed in Japan, China, Korea, Taiwan, Indonesia, Cambodia,
Laos, Malaysia, Philippines, Thailand, Vietnam, Burma, Pakistan, India, Bangla
Desh, Sri Lanka, by JAPAN SCIENTIFIC SOCIETIES PRESS, 6-2-10 Hongo, Bunkyo-ku,
Tokyo 113, Japan.

Sold and distributed in the U.S.A. and Canada by KLUWER BOSTON INC., 190
Old Derby Street, Hingham, MA 02043, U.S.A.

Sold and distributed in all other countries by KLUWER ACADEMIC PUBLISHERS
GROUP, P.O. Box 322, 3300 AH Dordrecht, Holland.

Preface

In this volume are collected essays contributed by biochemists, biologists and chemists from all over the world who are prominent in their respective domains. The themes dealt with in these essays are related to different aspects of today's respective sciences including reminiscence into the past, problems we are facing today and the outlook for the future. This collection itself will undoubtedly provide a source of enlightenment not only for accomplished researchers but also to those young students who are about to embark upon their research activities.

Originally, this project was initiated to commemorate the 70th birthday of Dr. F. Egami, who headed the laboratories to which all of the members of this Editorial Committee once belonged, and the essays contributed by his friends and associates were forged into this one volume. It is expected that this publication will receive widespread scholastic attention because of its instructive substance covering diversified areas of science.

Editorial Committee

Contents

Invited Contributors and Co-contributors*

BECK, M.T.   Institute of Physical Chemistry, Kossuth Lajos University, H-4010 Debrecen, Hungary

BENSON, A.A.   Scripps Institution of Oceanography, University of California, La Jolla, California 92093, USA

BRAUNSTEIN, A.E.   Institute of Molecular Biology, Academy of Sciences of the USSR, 117984 Moscow B334, USSR

BRODA, E.   Institut für Physikalische Chemie der Universität Wien, A-1090 Wien, Austria

CALVIN, M.   Department of Chemistry, University of California, Berkeley, California 94720, USA

CAMPBELL, P.N.   Courtauld Institute of Biochemistry, The Middlesex Hospital Medical School, London W1P 7PN, UK

CHAPEVILLE, F.   Institut de Recherche en Biologie Moléculaire, Université Paris VII et C.N.R.S., 75221 Paris Cedex 05, France

CHARGAFF, E.   350 Central Park West, New York, New York 10025, USA

CHENG, D.M.*   Department of Biochemical and Biophysical Sciences, School of Hygiene and Public Health, The Johns Hopkins University, Baltimore, Maryland 21205, USA

COHEN, S.S.   Department of Pharmacological Sciences, State University of New York at Stony Brook, Long Island, New York 11794, USA

CRAMER, F.   Max-Planck-Institut für Experimentelle Medizin, Abteilung Chemie, 3400 Göttingen, FRG

DESNUELLE, P.   Centre de Biochimie et de Biologie Moléculaire, C.N.R.S., 13009 Marseille Cedex 2, France

DODGSON, K.S.   Department of Biochemistry, University College, Cardiff CF1 1XL, Wales UK

DORFMAN, A.   Department of Pediatrics and Biochemistry, University of Chicago, Chicago, Illinois 60637, USA

EBERT, J.D.   Carnegie Institution of Washington, Northwest, Washington D.C. 20005, USA

EDSALL, J.T.   Department of Biochemistry and Molecular Biology, Harvard University, Cambridge, Massachusetts 02138, USA

ESSEN-MÖLLER, J.*   Institute of Molecular Cytogenetics, University of Lund, S-223 63 Lund, Sweden

FOX, S.W.   Institute for Molecular and Cellular Evolution, University of Miami, Coral Gables, Florida 33134, USA

FRIEBELE, E.*   Laboratory of Chemical Evolution, Department of Chemistry, University of Maryland, College Park, Maryland 20742, USA

FRUTON, J.S.   Department of Molecular Biophysics and Biochemistry, Yale University, New Haven, Connecticut 06520, USA

GEST, H.   Department of Biology, Indiana University, Bloomington, Indiana 47405, USA

HAKOMORI, S.   Division of Biochemical Oncology, Fred Hutchinson Cancer Research Center, Seattle, Washington 98104, USA

HOFFMANN-OSTENHOF, O.   Institut für Allgemeine Biochemie der Universität Wien, A-1090 Wien, Austria

HÖGBERG, B.   Aktiebolaget Leo Research Laboratories, S-25109 Helsingborg, Sweden

HORECKER, B.L.   Roche Institute of Molecular Biology, Nutley, New Jersey 07110, USA

ISAKSSON, M.*   Institute of Molecular Cytogenetics, University of Lund, S-223 63 Lund, Sweden

JAYARAMAN, K.*   Department of Biochemical and Biophysical Sciences, School of Hygiene and Public Health, The Johns Hopkins University, Baltimore, Maryland 21205, USA

JORDAN, D.K.*   Scripps Institution of Oceanography, University of California, San Diego, La Jolla, California 92093, USA

KAMEN, M.D.   Department of Chemistry, University of California, San Diego, La Jolla, California 92093, USA

KAN, L.-S.* Department of Biochemical and Biophysical Sciences, School of Hygiene and Public Health, The Johns Hopkins University, Baltimore, Maryland 21205, USA

KATCHALSKI-KATZIR, E. Department of Biophysics, The Weizmann Institute of Science, Rehovot 76100, Israel

KRASNOVSKY, A.A. A.N. Bakh Institute of Biochemistry, Academy of Sciences of the USSR, 117071 Moscow B-71, USSR

LEDERER, E. Laboratoire de Biochimie, C.N.R.S., 91190 Gif-Sur-Yvette, France

LEUTZINGER, E.* Department of Biochemical and Biophysical Sciences, School of Hygiene and Public Health, The Johns Hopkins University, Baltimore, Maryland 21205, USA

LEWIN, R.A. Scripps Institution of Oceanography, University of California, San Diego, La Jolla, California 92093, USA

LIMA-DE-FARIA, A. Institute of Molecular Cytogenetics, University of Lund, S-223 63 Lund, Sweden

LJUNGDAHL, L.G. Department of Biochemistry, University of Georgia, Athens, Georgia 30602, USA

MACGREGOR, J.S.* Roche Institute of Molecular Biology, Nutley, New Jersey 07110, USA

MARK, H. Polytechnic Institute of New York, New York, New York 11201, USA

MEYER, T.E.* Department of Chemistry, University of California, San Diego, La Jolla, California 92093, USA

MILLER, P.S.* Department of Biochemical and Biophysical Sciences, School of Hygiene and Public Health, The Johns Hopkins University, Baltimore, Maryland 21205, USA

MOTHES, K. Deutsche Akademie der Naturforscher Leopoldina, 4010 Halle (Saale), DDR

NAKANISHI, K. Department of Chemistry, Columbia University in the City of New York, New York, New York 10027, USA

NEERGAARD, P. Danish Government Institute of Seed Pathology for Developing Countries, DK-2900 Hellerup, Copenhagen, Denmark

NEILANDS, J.B. Department of Biochemistry, University of California, Berkeley, California 94720, USA

NICHOLAS, D.J.D. Department of Agricultural Biochemistry, Waite Agricultural Research Institute, University of Adelaide, Glen Osmond, South Australia 5064, Australia

NOMURA, M. Department of Genetics and Biochemistry, Institute for Enzyme Research, University of Wisconsin, Madison, Wisconsin 53706, USA

OLSSON, E.*   Institute of Molecular Cytogenetics, University of Lund, S-223 63 Lund, Sweden

ORÓ, J.   Laboratory of Biomolecular Analysis, Department of Biophysical Sciences, University of Houston, Houston, Texas 77004, USA

OVCHINNIKOV, YU.A.   Shemyakin Institute of Bioorganic Chemistry, Academy of Sciences of the USSR, Moscow 117334, USSR

PONNAMPERUMA, C.   Laboratory of Chemical Evolution, Department of Chemistry, University of Maryland, College Park, Maryland 20742, USA

PONTREMOLI, S.*   Institute of Biological Chemistry, University of Genoa, Genoa, Italy

PULLMAN, B.   Institut de Biologie Physico-chimique (Fondation Edmond de Rothschild), 75005 Paris, France

QUASTEL, J.H.   Kinsmen Laboratory of Neurological Research, Department of Psychiatry, Faculty of Medicine, University of British Columbia, Vancouver, British Columbia, V6T 1W5, Canada

ROBERTS, E.   Division of Neurosciences, City of Hope Research Institute, Duarte, California 91010, USA

ROCHE, J.   Biochimie Générale, Collège de France, 75005 Paris, France

SCHIDLOWSKI, M.   Max-Planck-Institut für Chemie (Otto-Hahn-Institut) Abteilung Chemie der Atmosphäre, D-6500 Mainz, FRG

SHUGAR, D.   Institute of Biochemistry and Biophysics, Polish Academy of Sciences, 02-532 Warszawa, Poland

SHIMOYAMA, A.*   Laboratory of Chemical Evolution, Department of Chemistry, University of Maryland, College Park, Maryland 20742, USA

SIBATANI, A.   CSIRO Molecular and Cellular Biology Unit, North Ryde, N.S.W., 2113, Australia

SLATER, E.C.   Laboratory of Biochemistry, B.C.P. Jansen Institute, University of Amsterdam, 1018TV Amsterdam, The Netherlands

SUEOKA, N.   Department of Molecular, Cellular and Developmental Biology, University of Colorado, Boulder, Colorado 80309, USA

TODD, LORD A.R.   Christ's College, Cambridge, CB2 3BU, UK

Ts'o, P.O.P.   Division of Biophysics, School of Hygiene and Public Health, The Johns Hopkins University, Baltimore, Maryland 21205, USA

Tsou, C.L. — Institute of Biophysics, Academia Sinica, Peking, People's Republic of China

Tsutsui, M. — Department of Chemistry, Texas A&M University, College Station, Texas 77843, USA

Udenfriend, S. — Roche Institute of Molecular Biology, Nutley, New Jersey 07110, USA

Wang, Y.-L. — Shanghai Institute of Biochemistry, Academia Sinica, Shanghai, People's Republic of China

Witkop, B. — National Institute of Arthritis, Metabolism and Digestive Diseases, National Institutes of Health, Bethesda, Maryland 20014, USA

Yamada, T. — Institut Suisse de Recherches Expérimentales sur le Cancer, CH-1066 Épalinges, Switzerland

Yuasa, S.* — Department of Biology, College of General Education, University of Osaka, Toyonaka, Osaka 560, Japan

The Long Journey

MELVIN CALVIN

M. CALVIN, born in Minnesota in 1911, was graduated from Michigan College of Mining & Technology in 1931 and received Ph.D. degree in 1935 from University of Minnesota in the field of chemistry. He was a postdoctoral fellow at the University of Manchester from 1935 to 1937 and joined the staff of the Department of Chemistry, University of California, Berkeley in 1937. He became a Professor in 1947 and university professor in 1971. He was the Director of the Laboratory of Chemical Biodynamics from 1960 to 1980 as well as Director of the Chemical Biodynamics Division of the Lawrence Berkeley Laboratory (1945–1980) and associate director of the Lawrence Berkeley Laboratory (1967–1980). He is a Nobel Prize winner in chemistry, 1961. He also received many prizes and medals including the Priestley Medal and Gibbs Medal from the American Chemical Society in 1977 and 1978, respectively, the Gold Medal of the American Institute of Chemists in 1979, the Hales Award from the American Society of Plant Physiology in 1956, and the Davy Medal from The Royal Society in 1964. He was President of the American Chemical Society in 1971 and the American Society of Plant Physiology in 1963–1964.

Let me describe for you the genesis of this contribution. Recently, new developments have occurred which indicate the success which several research groups have attained in what has been called "genetic engineering." This term refers to the ability of men to introduce genetic information into a microbial cell from some source other than another microbe, or even from another microbe, in order to allow the manufacture of useful materials which the microbe normally would not produce and which could not be obtained in required amounts from those organisms which did produce it. I am referring, of course, to the successful transplantation of the gene for the production of protein materials, interferon, on the one hand and insulin, on the other, neither of which can be made in the laboratory as yet. They are both clinically useful materials.

It was with this idea in the back of my mind that I began to think about bringing you up to date in the developments which are taking place concerning the generation of fuel and materials by biological means to replace our dwindling fossil hydrocarbon supplies, the products of ancient photosynthesis. I would like to take you through a bit of the history of our ideas of chemical and molecular evolution and genetic modification and bring these concepts into the energy context with which we are all now concerned.

The fundamental problem is a very large one, with which all mankind has been concerned since man began to think about his nature and his origins. What I am referring to is the question of the origin of mankind and the origin of life itself. There have been many answers proposed to these questions, but I will address only those which are based on some kind of experimental observables and try to relate how the answers to that question(s) are relevant to the very practical problems of energy and materials with which we are faced today.

The modern history of this discussion begins a little over one hundred years ago, in the period just before Darwin. The earth had long since been removed from the center of the universe and placed in one of many solar systems amongst many galaxies. The idea that life on earth was an actual consequence of some greater law had already been bruited about, but it was not until the 19th century that the notion that life had a long history on the earth's surface, from very primitive and even prebiotic origins leading all the way to man, was postulated and described. The carefully documented journals of Charles Darwin on his voyage on the Beagle (1) in the 1830s during which he examined the variety of life on the east and west coasts of South America and in the Pacific, as well as the geological evidences of life's ancient remains, provided one of the earliest substantial documentations for this idea.

This is "The Long Journey" to which the title of this presentation refers which was plagiarised from a fictionalized history of mankind written by Johannes Jensen some years ago (2). In a sense, this "long journey" is beginning a new phase, in which man can have direct influence on its course in a way that no conscious activity heretofore has had.

Darwin not only recognized the close relation of all forms of life that existed on the earth during his time, but its relation to much earlier forms whose relics he was able to observe (3). But, later, following the voyage in the Beagle, in letters written to Alfred Wallace in 1871 and 1882 Darwin suggested the basic notion that life arose on the earth's surface as a consequence of the interaction of the materials that were present on the inanimate earth with various sources of energy, mostly in the form of radiation (4).

In 1925 this idea was more explicitly stated by another Englishman, J.B.S. Haldane (5), and a Russian biochemist, who has died just this year, by the name of A.I. Oparin (6). It was Oparin who undertook to do experimental tests of the notion that the impingement of high energy radiations, either from outside the solar system, or the sun itself, or perhaps even the energy generated by weather (such as lightning), might indeed have given rise to the molecules of which living things are made from the pre-existing atmosphere of the inanimate earth (7).

In order to do such experiments in a reasonable fashion, one had to have some notion of what the inanimate atmosphere of the prebiotic earth was like, and the first guess as to its composition postulated that it was mostly carbon dioxide, hydrogen and water vapor, together with traces of the metals which were dissolved in the sea (8). In this connection the pioneering concepts of Dr. Fujio Egami made much later are relevant: his hypothesis that the composition of contemporary sea water reflected that of primeval sea water and that chemical evolution might actually have proceeded in the primeval sea have been confirmed in the laboratory and have opened up still another avenue for research in chemical evolution (9, 10).

The first actual experiment to determine whether or not anything "useful" could be formed from such simple initial materials that were present in the primitive atmosphere was performed in the sixty inch cyclotron at the University of California in Berkeley in 1951 (11). That experiment led indeed to the formation of a large number of more complex materials from the simple molecules of hydrogen, carbon dioxide, and water. What was lacking in that initial experiment was the presence of reduced nitrogen, or nitrogen of any kind, in the initial simulated atmosphere. In 1953 that reduced atmosphere containing nitrogen was subjected to various kinds of radiation by experiments done by Urey and Miller at the University of Chicago in which, indeed, the very components of living things were generated, primarily amino acids which give rise to proteins (12, 13). Since that time an enormous number of small molecules, which are the constituent parts of today's living organisms, have been generated in this simple or random fashion.

The next step in this "chemical evolution" (14) as we have called it was to find ways in which these small units might be hooked together in very long chains which would then give rise to the proteins that are necessary for the construction of living things, as well as the very long chain materials in which the genetic information can ultimately be stored. All of this has now been achieved experimentally in the last thirty years, so there is no longer any question of the possible generation not only of small units, of which these giant molecules are composed, but actually the stringing together of the small units into long chains to give rise to the major biological giant molecules—carbohydrates, proteins and, finally, the nucleic acids in which the genetic information is stored.

During the past twenty years it became evident that there should be some evidence in the ancient geological formations of this early form of life. In order to search these things out it was necessary to have a geological time scale (Fig. 1). You can see that the age of the solid earth is set at somewhere about 4.5 billion years. The earliest microfossils that could be identified in the rocks, that is, forms whose physical shape could be recognized, are microorganism residues such as the blue-green algae (15). These have been found in rocks as old as 3.5 billion years; a

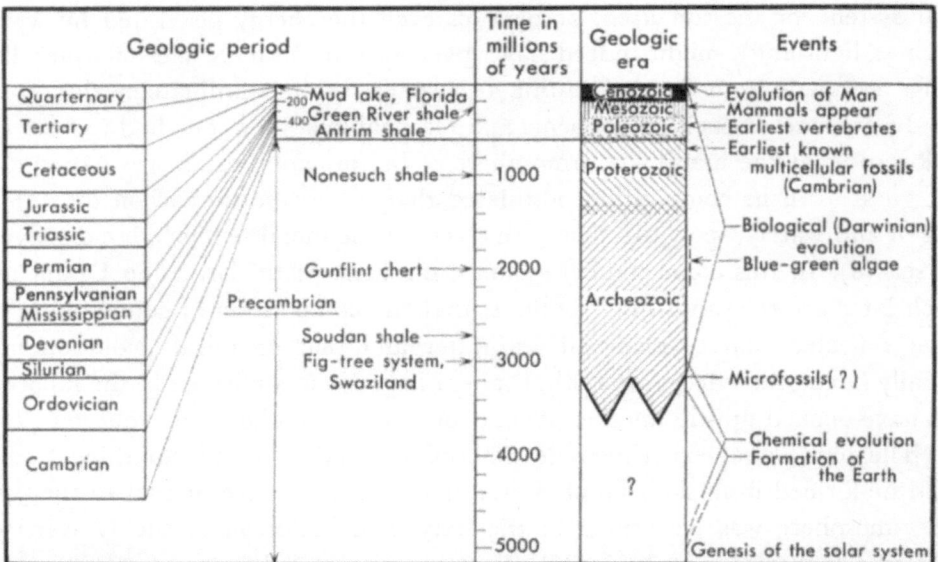

Geologic period		Time in millions of years	Geologic era	Events
Quarternary	—200 Mud lake, Florida —400 Green River shale Antrim shale		Cenozoic Mesozoic Paleozoic	Evolution of Man Mammals appear Earliest vertebrates
Tertiary				
Cretaceous	Nonesuch shale →	1000	Proterozoic	Earliest known multicellular fossils (Cambrian)
Jurassic				
Triassic				Biological (Darwinian) evolution
Permian	Gunflint chert →	2000		Blue-green algae
Pennsylvanian	Precambrian		Archeozoic	
Mississippian				
Devonian	Soudan shale →			
Silurian	Fig-tree system, → Swaziland	3000		Microfossils(?)
Ordovician				
Cambrian		4000	?	Chemical evolution Formation of the Earth
		5000		Genesis of the solar system

FIG. 1.   Time scale for evolution.

modern photograph of such organisms which still persist is shown in Fig. 2. Presumably before that time there must have been organizations of materials which had no way of leaving a formed physical residue which could be recognized in a microscope, that is, the initial life forms were all microbial in nature and some of them could indeed do ordered synthesis, but would have no way of leaving a fossilized, that is, a rock-formed, residue which could be recognized. Therefore, we examined still older rocks, of the order of 4 billion years, to see if we could not find "molecular fossils" (*16*), that is, arrangements of atoms which were not by their very organization randomly formed. We have found such molecules in rocks as old as 4 billion years, suggesting that there were organized synthetic sequences occurring as early as 4 billion years ago, a mere half a billion years after the formation of the solid earth.

The next difficult question, which has not yet been experimentally surrounded, so to speak, is the question of how the information for the construction of such organized units could be stored in a form that could be transmitted from one entity to another, that is, from one generation to the next. This requires the evolution of some kind of "molecular code". Here we are still at a loss, although we believe that this relationship of molecular structure to information storage has its origin in the same qualities of the atoms of which the initial molecules are made; it has yet to be demonstrated how indeed that relationship was first generated.

Having generated that relationship in the initial phases of the earth's history, it is now relatively simple, at least conceptually, to see how that information can be modified as it passes through successive generations by virtue of its ability to respond to the environment in which it is situated. This is simply Darwin's theory of natural selection, carried out on the molecular level and, eventually, through the microorganism to the multicellular organisms that we know today.

The process of selection was what Darwin called a "natural" one, that is, the environment would allow those organisms best suited to the environment in which they lived to reproduce and survive. Those which had less well adapted information stored in them would be superceded by the more adaptive ones. This system gave rise to the enormous variety of living things we have on the surface of the earth today, beginning (in the fossil record) with what we now recognized to be the kind of one-celled plant called blue-green algae (Fig. 2) and reaching to man. With our present knowledge of how the system works, mankind can now deliberately rearrange the genetic material and thereby alter the method of selection.

Very early in the history of life on the earth the green plants developed, and several hundred million years ago the earth was covered with a heavy vegetation which left its residues in the bottoms of lakes, ponds and oceans to be covered with mud, and eventually converted into what we call today our fossil fuels—coal, oil, gas. The increasing difficulty of finding oil is objective evidence of the exhaustion of what must be a limited supply of the products of ancient photosynthesis. We must now find annually renewable sources of energy and materials. There are families of green plants that produce material very closely related to oil, that is, hydrocarbons in the form of latex. From this plant latex it is possible to extract a material which very closely resembles crude oil (*17*).

Experiments conducted in our laboratory in Berkeley and elsewhere in the world using *Euphorbia lathyris* (Fig. 3) indicate that it is possible to produce approximately 10 barrels of oil-equivalent/acre/year with a minimum of water requirement. This yield is with seeds from wild plants in their first stages of domestication and without any agronomic improvement (*18*). We have also found that in addition to the oil-equivalent product from *E. lathyris* that there is another, equally useful product in the form of carbohydrate fermentable to ethanol, making the petroleum plantation an energy positive operation (*19*).

The next step, after further refinement of the extraction process of the material from *Euphorbias* and other oil producing plants, would be to improve yield by genetic selection, that is, insert genetic material into the plants that will create more hydrocarbons of the desired composition, for example, sesquiterpenes.

Here we come to the very last stage of our evolutionary sequence. We can now select from natural collections of genes the ones best suited to our needs. Also, we are now on the threshhold of capability of introducing the genes we want in the plant not only to make the desired hydrocarbons but the amount we want as well. This concept of so-called "genetic engineering" that I mentioned earlier is shown in Fig. 4.

We have thus passed on our "long journey" from the time when the evolution of life was determined solely by environmental natural selection into a period when the evolution of living things, such as plants, may actually be directed to man's

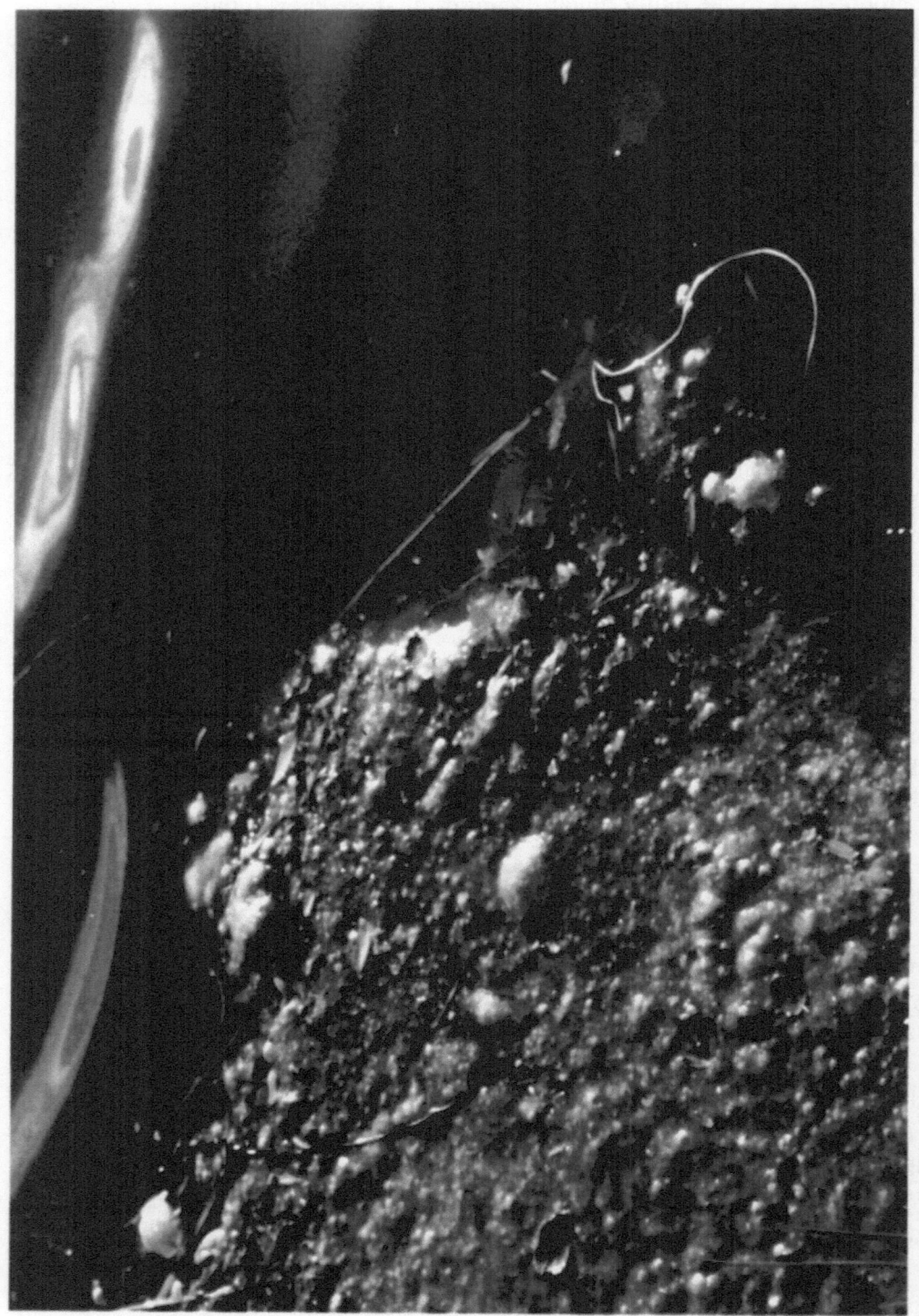

Fig. 2. Blue-green algae. (Photo by Gene Elle Calvin).

FIG. 3.   *Euphorbia lathyris,* Southern California. (Photo by Gene Elle Calvin).

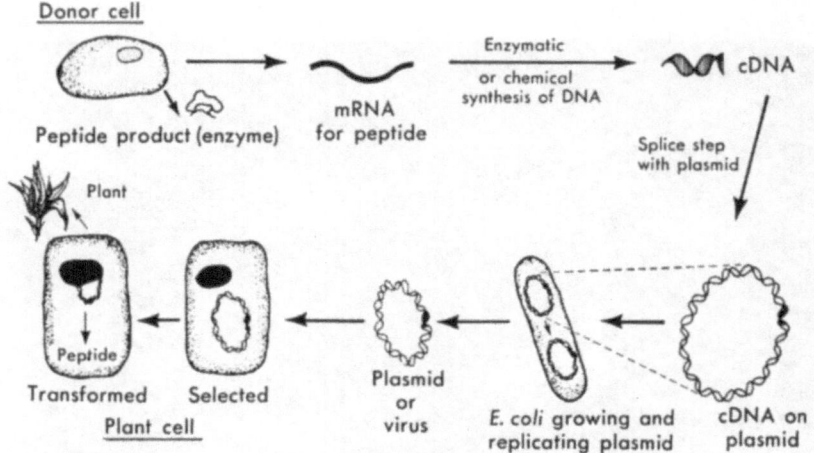

FIG. 4.   Genetic engineering (man-made evolution).

purpose. What this leads to ultimately is a very difficult question to answer, but obviously one which we can no longer avoid.

REFERENCES

1.   Darwin, C.: 1845. *The Voyage of the Beagle.*
2.   Jensen, J.: 1933. *The Long Journey*, Alfred P. Knopf Co., New York.
3.   Darwin, C.: 1859. *On the Origin of the Species.* 1st ed.
4.   Darwin, C.: 1871 (see Ref. *13*, pages 4–5).
5.   Haldane, J.B.S.: 1929. *Rationalist Annual.*
6.   Oparin, E.I.: 1924. Moskovski, R.; 1938, *The Origin of Life* (trans. by S. Margulis), Macmillan, New York.
7.   Oparin, A.I.: 1959. *The Origin of Life on the Earth*, Pergamon Press, London.
8.   Urey, H.C.: 1952. *The Planets*, Univ. Chicago Press, Chicago.
9.   Egami, F.: 1974. *J. Mol. Evol.* 4, 113–117.
10.   Ochiai, T., Hatanaka, H., Ventilla, M., Yanagawa, H., Ogawa, Y., and Egami, F.: 1978. *The Origin of Life* (Noda, H., ed.) pp. 135–139, Japan Sci. Soc. Press, Tokyo.
11.   Garrison, W.M., Morrison, D.C., Hamilton, J.G., Benson, A.A., and Calvin, M.: 1951. *Science* 114, 416–418.
12.   Miller, S.L.: 1953. *Science* 117, 528–530.
13.   Miller, S.L.: 1955. *J. Am. Chem. Soc.* 77, 2351–2360.
14.   (a) Calvin, M.: 1969. *Chemical Evolution: Molecular Evolution Towards the Origin of Living Systems on the Earth and Elsewhere*, Oxford Univ. Press, Oxford.
       (b) Egami, F.: translator of Ref. *14*(a).
15.   Lowe, D.R.: 1980. *Nature* 284, 441–443; and references cited therein.
16.   Calvin, M.: 1969. *Persp. Biol. Med.* 13, 45–62.
17.   Calvin, M.: 1980. *Naturwissenschaften* 67, 525–533.
18.   Calvin, M.: 1979. *BioScience* 533–538.
19.   Nemethy, E.K., Otvos, J.W., and Calvin, M.: 1980. *Pure Appl. Chem.*, in press.

Adventures in Coordination Chemistry

M. T. BECK

M. T. BECK was born November 14, 1929 in Szöreg, Hungary and educated at University of Szeged. First he dealt with analytical applications of complexes. His main field of interest is coordination chemistry in general, thermodynamics, and kinetics involving metal complexes in particular. A connected area is the thermodynamic and kinetic aspects of chemical feedback. Since 1968 he is Professor of Physical Chemistry at Kossuth Lajos University, Debrecen, Hungary. He is a member of the Hungarian Academy of Sciences, the Chairman of the Chemistry Section of this Academy since 1976, and the author of the books "Chemistry of Complex Equilibria," "Science and Pseudoscience," and about 200 papers; editor-in-chief of Scientometrics and Kémiai Közlemények.

E very scientist dreams of making great discoveries, but only a very few of these dreams are fulfilled. There are two ways to avoid a feeling of frustration. The first is a self-deceptive overestimation of the importance of one's own results, which necessarily implies a nasty underestimation of the achievements of others. The second is to realize that good research is an *adventure* leading to unknown fields, and besides new findings it allows us to explore the basic laws of epistemology and of the psychology of research, valid independent of the importance of the results. I would not like to give the impression that a feeling of the relevance of the work is not important; on the contrary, it is quite obvious that the hope of making a major contribution to the building of the stronghold of existing knowledge is a key element in realizing that hope. It appears that all research, whether motivated by sound scientific curiosity or directed to achieve relevant practical aims, can give intellectual satisfaction to the people involved in such work. The following case stories may illustrate this point.

The Carbonate Catalysis of the Formation of Complexes of Chromium (III)

Chromium(III) is a characteristically inert ion, that is, the formation of the complexes of chromium(III) from the hexaqua ion and the corresponding ligand is very slow. The study of the formation of thiocyanato complexes of chromium (III) by Niels Bjerrum (*1*) was the first investigation of the kinetics of successive complex formation. This system was reinvestigated by Bjerrum *et al.* (*2*) and by King and Postmus (*3*). It was a shocking experience for me as a young chemist to find widely divergent values published by famous authors. It was even more confusing that my very carefully determined rate constants were bigger than any of the previously found values. I tried to find different hypotheses to explain the controversial experience: the formation of isomeric species (thiocyanato and isothiocyanato complexes) with different rates; the relatively fast formation of mononuclear species followed by a slow dimerization process, *etc.* I was forced to reject each of these hypotheses. Then I meticulously compared the differences in the different studies. I prepared the chromium(III) perchlorate from chromium(III) chloride by precipitating chromium(III) carbonate followed by its dissolution in perchloric acid. This procedure was repeated until no chloride in the solution could be detected. Bjerrum and his coworkers prepared chromium(III) perchlorate by dissolving chromium(III) hydroxide in perchloric acid. Postmus and King prepared their chromium(III) perchlorate by reducing chromium acid in perchloric acid solution with formic acid.

The observed rate of the formation of complexes decreased in the given order of the procedures for the preparation of chromium(III) perchlorate solution. It followed from the nature of these procedures that the probability of the presence of carbonate or carbon dioxide in the solution was decreasing. Therefore a faint possibility of the effect of the carbon dioxide content in the reaction mixtures on the rate of the formation of the thiocyanato complexes of chromium(III) appeared. Experiments then proved beyond doubt the catalytic effect of carbon dioxide in this system. It could be concluded that a prerequisite of reproducible kinetic experiments in this system is the control of the amount of dissolved carbon dioxide. No explanation of the phenomenon could be given that time, and I am sure that when I gave a talk (*4*) about these findings at the 6th International Conference of Coordination Chemistry in London in 1959, very few if any of the chemists present believed me. Fortunately the catalytic effect of carbon dioxide on the formation of different chromium(III) complexes was found later in several laboratories (*5, 6*).

Further investigations permitted us to give the rationale for this effect: the carbon-dioxide and/or hydrogen carbonate ion reacts with the chromium(III) hexaqua ion extremely fast, because the formation of the hydrogencarbonato complex does not require the repture of the original chromium-oxygen bond.

Although the concentration of such a species is very small in acidic medium and its presence is not revealed by any spectral changes, the rate of further substitution reactions is enhanced. In the case of the reactions of $Cr(NH_3)_5(H_2O)^{3+}$ with different ligands, Earley et al. (6) found that the formation of the complexes takes place exclusively through the carbon dioxide-catalyzed path. In case of the successive formation of the dimethyl sulphoxide complexes of chromium(III), the contribution of the catalytic and noncatalytic routes could be exactly evaluated (7).

The moral of this experience is twofold. First, an apparently small difference in experimental conditions can result in great deviations in chemical behavior. Secondly, the belief in the unreactivity of certain chemical species based on wide and manifold experience can be deceptive. The case discussed refers to an implicit assumption regarding traces of carbon dioxide obviously irrelevant in the given system. More important examples, unfortunately not from my own work, refer to the for long unsuspected reactivity of noble gases and dinitrogen.

Reduction of Chromium(VI) in the Presence of Complex Forming Agents

When chromium(VI) is reduced in the presence of complex forming agents, a most interesting phenomenon can be observed. After the reduction, the chromium(III) is present in the form of the corresponding chromium(III) complex, although the formation of this complex from the hexaquachromium(III) ion and the ligand is very slow. Another less spectacular, but equally important, experience is the increase in the rate of reduction by the complex-forming substance.

When I first observed the fast formation of the thiocyanato and EDTA (ethylenediamine tetracetate) complexes of chromium(III), measurements were limited to this phenomenon only. I was pleased to find a simple and attractive solution to the puzzle: in the reduction of chromium(VI) the first "naked" chromium(III) ions are formed which then react fast with the ligands present. My good feeling did not last long. I read a paper by Pribil, Simon, and Dolezal (8) who a few years earlier described the same experience and the same explanation. However, I had not even finished reading this paper when it became obvious to me that their explanation could not be correct. First, even the formation of "naked" chromium(III) ions cannot be taken for granted: the reduction of chromate does not necessarily mean the rupture of the chromium-oxygen bonds. Secondly, assuming that "naked" chromium(III) is formed, obviously it will react with every kind of ligand present. On the other hand, the concentration of water exceeded that of the added ligands by many orders of magnitude; consequently at the end of the reduction most of the chromium(III) should be present in the form of an aquocomplex independent of the ligand.

The real explanation was found after years of kinetic investigations (9, 10). The effect of the ligands on the rate of reduction requires an interaction of the ligands with chromium(VI). The properties of the intermediate oxidation states of chromium also play important roles. The elucidation of the kinetics of this type of reaction was very interesting for us and permitted us to dig deeply into the mechanism. The most important feature of this whole adventurous investigation was that I had to realize that there is a great tendency to be less critical with our own ideas than with those of others. This became manifest for me in the above experience and helped me to be at least as scrupulous with my ideas as with those of others.

Much later some of our results were questioned in a paper (11). Originally we found that the rate and the conversion of the formation of the chromium(III) EDTA complex when chromium(VI) was reduced by hydrazine in the presence on EDTA changes with the concentration of EDTA according to a *saturation* curve. In this later paper it was reported that after reaching a maximum, the rate decreases with further increases in the concentration of EDTA. The worst feeling of a researcher is when his/her *factual* results are questioned. Obviously this controversial finding forced us to reconsider our experiments and to check their ones. There was a factual difference indeed, but it was a consequence of our critics' experiments being not well designed (12). By increasing the concentration of EDTA, the pH of their reaction mixture also increased and the maximum curve found by them was in fact a result of the effect of the EDTA concentration *and* the pH of the solution. This experience illustrates how important is to change but one parameter in a chemical system. (It must be borne in mind that strictly speaking this is impossible: a change in any parameter necessarily results in changes in other parameters to a certain extent. It is the art of experimentation to keep these changes on a negligible level.)

Recurrence and the Multiplicity of Problems

It is interesting to observe how study of a problem triggers new ones and how an old experience later emerges to solve other problems. The study of cyanocomplexes led me to such an experience. Cyanide forms very stable complexes with transition metal ions. The study of certain cyanocomplexes revealed extreme differences in the stability constants published in the literature. (In case of $Ni(CN)_4^{2-}$ the published values of log β_4 are in the range of 12–31!) This and similar findings made clear the necessity of a critical evaluation of stability constants of complexes (13). In addition, the extreme stability of the cyanocomplexes suggested ideas to solve two entirely different problems.

In the determination of metal ions by different methods, the complex forming substances present usually interfere strongly. For example, in the determination of metal ions by atomic absorption spectrometry, the value of the signal obviously depends on the immediate environment of the metal ion in question, which is determined by the quality and quantity of the complex forming ligands in the solution. However, if a ligand forming an extremely stable complex with this metal ion is added to the sample, the immediate environment will be the same, independent of the original composition. As applied to cyanide, this principle was used to eliminate the interfering effects of practically all ligands in the determination of Ru (*14*), Co (*15*), Pd (*16*) and Fe (*16*), and this trick probably can be applied in many other cases.

An entirely different consequence of their high stability is that in the primeval hydrosphere transition metal ions occurred in the form of cyanocomplexes (*17*) and their reactions should be considered from the point of view of prebiotic syntheses (*18*). This conclusion is necessarily valid if the prebiotic atmosphere was reducing and it can be valid even if it did not contain hydrogen cyanide. In this latter case cyanocomplexes could be formed (*19*) in the reaction of formaldehyde and hydroxylamine (*20*). The photolysis of octacyano-molybdenum (IV) in the presence of formaldehyde results in the formation of different amino acids, mainly glycine and adenine, with surprisingly high yields (*18*). This observation may be relevant from the point of view of the early phase of chemical evolution.

REFERENCES

1. Bjerrum, N.: 1921. *Z. Anorg. Chem.* 118, 131–164.
2. Poulsen, K.G., Bjerrum, J., and Poulsen, I.: 1954. *Acta Chem. Scand.* 8, 921–931.
3. Postmus, C. and King, E.L.: 1955. *J. Phys. Chem.* 59, 1216–1221.
4. Beck, M.T.: 1960. *J. Inorg. Nucl. Chem.* 15, 250–254.
5. Rao, V.K., Sunder, D.S., and Sastri, M.N.: 1966. *Z. Anal. Chem.* 218, 93–96.
6. Earley, J.E., Surd, D.J., Crone, L., and Quane, D.: 1970. *Chem. Commun.* 1401–1402.
7. Rábai, Gy., Bazsa, Gy., and Beck, M.T.: 1979. *Acta Chim. Hung.* 102, 223–234.
8. Pribil, R., Simon, V., and Dolezal, J.: 1951. *Collection* 16, 573–580.
9. Beck, M.T. and Bárdi, I.: 1961. *Acta Chim. Hung.* 29, 283–289.
10. Beck, M.T. and Durham, D.: 1970. *J. Inorg. Nucl. Chem.* 32, 1971–1977.
11. Ramanujam, V.M.S., Sundaram, S., and Venkatasubramanian, N.: 1975. *Inorg. Chim. Acta*, 13, 133–139.
12. Beck, M.T., Durham, D., and Rábai, Gy.: 1976. *Inorg. Chim. Acta*, 18, L17.
13. Beck, M.T.: 1977. *Pure Appl. Chem.* 49, 127–135.
14. El-Defrawy, M.M.M., Posta, J., and Beck, M.T.: 1978. *Anal. Chim. Acta* 102, 185–188.
15. El-Defrawy, M.M.M., Posta, J., and Beck, M.T.: 1980. *Anal. Chim. Acta* 115, 155–161.
16. El-Defrawy, M.M.M.: 1980. Dissertation, Debrecen.
17. Beck, M.T. and Ling, J.: 1977. *Naturwissenschaften* 64, 91.
18. Beck, M.T.: 1978. *Metal Ions in Biological Systems*, Vol. 7 (Sigel, H., ed.) pp. 1–28, Marcel Dekker, New York and Basel.
19. Sarkar, S. and Subramanian, P.: 1980. *Chim. Acta* 46, L67–L68.
20. Hatanaka, H. and Egami, F.: 1977. *J. Biochem.* 82, 499–502.

The Role of Clay Minerals in Prebiotic Protein Synthesis

CYRIL PONNAMPERUMA, ELAINE FRIEBELE,
and AKIRA SHIMOYAMA

C. PONNAMPERUMA is a native of SriLanka, and was first educated in SriLanka and India, obtained a B.A. degree from University of Madras. Then he proceeded to London where he received a B. Sc.degree in chemistry at Birkbeck College, University of London in 1959. In the same year, he moved to the University of California, Berkeley, where he was a graduate student in chemistry under the direction of Melvin Calvin. In 1962, after received a Ph.D. degree in chemistry, he joined NASA's Ames Research Center, Moffett Field, California as a research associate. Later he became a laboratory chief of the Chemical Evolution Branch. In 1971, he was appointed to the Professor of Chemistry and Director of the Laboratory of Chemical Evolution, the University of Maryland, College Park, Maryland. He was selected as a principal investigator in NASA's Appollo program, Viking program, and other space programs. He was also associated with many universities including Stanford University, the University of Nijmengen, and the Sorbonne University. He is also involved in the UNESCO programs concerning the development of science and culture in India and SriLanka. He was the first winner of Oparin Medal from the International Society for the Study of the Origin of Life in 1980. He is editor-in-chief of the International Journal "Origins of Life."

The importance of clay minerals in chemical evolution was suggested by Bernal as early as 1951 (1). He proposed that clays near the hydrosphere-lithosphere interface might have adsorbed organic micromolecules, thereby providing high local concentrations of reactants needed to form certain biologically important macromolecules and protection from destructive high energy radiation. Furthermore, he argued that clays could act as catalysts, giving rise to polymers of biological interest. Bernal's hypothesis of the possible role of clay minerals may be extended throughout the stages of chemical evolution on the primitive Earth in the following sequence:

1.   Clay minerals catalyze the reactions of biomonomer synthesis from gaseous constituents of the primordial atmosphere.

2.   Clay minerals adsorb biomonomers on their surfaces, providing a highly concentrated system in which those monomers have a specific orientation.

Clay minerals facilitate condensation reactions between adsorbed monomers forming biopolymers which are the building blocks of life. In addition, the surfaces of clay minerals might serve as templates for the specific adsorption and replication of organic molecules. This last role of clays as information-carrying crystals has been envisioned by Cairns-Smith (2). He proposed that after the evolution of a crystalline primitive gene, a "genetic metamorphosis" would occur, in which organic molecules would take over the control of information transfer and replication.

According to Bernal, clays can concentrate the organic compounds in the "primordial soup" by adsorption. The problem of clay mineral adsorption of relevant organic molecules (amino acids, peptides, proteins, nucleic acid bases, nucleosides, nucleotides, polynucleotides, and sugars) has been studied extensively by many investigators. Mortland (3) and Theng (4) have reviewed in detail the adsorption of organic compounds by clay minerals.

Adsorption of Amino Acids

Opinion differs concerning the possible upper limit of amino acids which can be adsorbed on the clay minerals. McLaren et al. (5) reported that montmorillonite can adsorb alanine in excess of its exchange capacity, while Sieskind (6) found that the exchange capacity of the clay mineral could not be saturated under similar conditions. The discrepancy in these results appears to be due to two different methods used for determining the quantity of amino acid adsorbed. McLaren et al., obtained their data indirectly, measuring the difference in the initial and the recovered concentrations of amino acids in solution at equilibrium. Sieskind estimated the quantity adsorbed directly, measuring the amount retained in the clay portion after washing with water. Cloos et al. (7) performed both types of measurements, analyzing quantities of amino acid in the equilibrium solution (indirect) and amino acids on washed clay residues (direct). They found that the indirect method gave a higher estimate of amino acids adsorbed. The difference between these two methods can be attributed in part to weakly adsorbed molecules lost during the wash of the clay in the method used by Cloos et al. It is also likely that the difference is partially due to experimental error involved in the indirect method. Generally, in an adsorption experiment the amount of amino acids remaining in solution is at least one order of magnitude greater than that adsorbed on the clay surface. Therefore, any error in estimating the amount adsorbed is

TABLE 1.   Adsorption of Equimolar Mixture of α- and β-Alanine on Na-montmorillonite[a]

	Amino acids	Initial (μm)	Amounts found in (μm)			Amounts adsorbed in (meq/100 g)	
			Supernt.	Clay	Total	Indirect	Direct
pH	α-Ala	200	165	19.3	184(92.0%)	8.8	4.8
	β-Ala	200	138	46.2	184(92.0%)	15.5	11.6
pH 7.0	α-Ala	200	178	3.67	182(91.0%)	5.5	0.9
	β-Ala	200	194	4.33	198(99.0%)	1.5	1.1
pH 10.0	α-Ala	200	181	1.94	183(91.5%)	4.8	0.5
	β-Ala	200	192	2.64	195(97.5%)	2.0	0.7

[a]Na-montmorillonite is from the Source Clay Repository (Univ. of Missouri, U.S.A.). 400 mg of the clay (less than 2 μm particle size) was suspended in 40 ml of water and was used for adsorption experiments at each pH.

TABLE 2.   Adsorption of Protein and Non-protein Amino Acids on Na-montmorillonite (10)

	pH	Amino acid	Adsorbed (%)			Non-adsorbed (%)	Total (%)
			Strongly	Weakly	Total		
C_2	3.0	Glycine	11.0	11.0	22.0	77.9	99.9
		Sarcosine[a]	11.0	11.3	22.3	72.7	95.0
	7.0	Glycine	1.1	6.4	7.5	87.7	95.2
		Sarciosne	1.3	4.1	5.4	83.8	89.2
	10.0	Glycine	1.7	4.9	6.6	92.2	98.8
		Sarcosine	0.7	4.5	5.2	94.5	99.7
C_3	3.0	DL-α-Alanine	11.6	10.8	22.4	78.6	101.0
		β-Alanine	36.3	9.6	45.9	57.6	103.5
	7.0	DL-α-Alanine	1.0	5.0	6.0	88.3	94.3
		β-Alanine	2.3	7.2	9.5	86.6	96.1
	10.0	DL-α-Alanine	0.8	5.1	5.9	96.4	102.3
		β-Alanine	0.8	4.9	5.7	91.0	96.7
C_4	3.0	DL-α-Aminobutyric acid	9.0	10.5	19.5	84.7	104.2
		γ-Aminobutyric acid	39.0	8.0	47.0	52.5	99.5
	7.0	DL-α-ABA	1.8	7.9	9.7	92.1	101.8
		γ-ABA	2.8	9.4	12.2	89.1	101.3
	10.0	DL-α-ABA	0.5	5.1	5.6	98.7	104.3
		γ-ABA	0.6	5.6	6.2	94.6	100.8
C_5	3.0	DL-Valine	11.8	10.4	22.2	74.4	96.6
		DL-Norvaline	11.4	10.3	21.7	77.7	99.4
	7.0	DL-Valine	1.1	6.1	7.2	87.1	94.3
		DL-Norvaline	1.0	5.8	6.8	87.3	94.1
	10.0	DL-Valine	0.7	4.5	5.2	93.1	98.3
		DL-Norvaline	0.7	4.5	5.2	93.5	98.7
C_6	3.0	L-Isoleucine	13.7	11.2	24.9	75.5	100.4
		D-Alloisoleucine	12.8	11.3	24.1	75.0	99.1
	7.0	L-Isoleucine	1.2	7.0	8.2	88.7	96.9
		D-Alloisoleucine	1.4	7.1	8.5	89.5	98.0
	10.0	L-Isoleucine	0.7	4.7	5.4	94.9	100.3
		D-Alloisoleucine	0.8	4.6	5.4	95.1	100.5

[a]Although sarcosine contains three carbon atoms, it is placed in the C_2 group because it is structurally related to glycine.

amplified by the indirect method. Furthermore, incomplete recovery of the molecules may occur, causing an overestimation of the quantity adsorbed if the indirect method of measurement is used.

An experiment in our laboratory illustrates the difference in results obtained by the two methods. Table 1 shows the data from an experiment in which equimolar quantities of α- and β-alanine were mixed together with Na-montmorillonite at three hydrogen ion concentrations. After adsorption had occurred, two fractions were recovered: (1) the "indirect" fraction—the supernatant after centrifugation, combined with a pH adjusted wash of the clay fraction, and (2) the "direct" fraction—the HCl extraction of the amino acids from the clay. The quantity of the two amino acids recovered in both fractions was estimated by an amino acid analyzer (Durrum D-500). The total recovery ranged from between 91 to 99% of the initial amount, and the amount of amino acids adsorbed estimated by direct and indirect methods differ significantly.

The adsorption of amino acids on the clay surface takes place mainly through cationic exchange, ion-dipole and coordination interactions, hydrogen bonding, and physical forces. The extent to which these various mechanisms occurs depends upon the isoelectric point, dipole moment, and molecular size and shape of the molecules in solution at a given pH.

The observation by many workers (6, 9) that adsorption of amino acids increases with decreasing pH indicates that the major mechanism of adsorption of low pH is cationic exchange, because under these conditions the amino acids are in cationic form. Friebele et al. (10) observed an additional mechanism of adsorption at low pH on Na-montmorillonite. The quantity of an amino acid which is strongly adsorbed (tightly held against washing with water) on the clay at pH 3 increases with increasing isoelectric point of the amino acid (Table 2). Thus, amino acids in the strongly adsorbed fraction are adsorbed by cationic exchange. The quantity of amino acids weakly adsorbed (washed from the clay with pH-adjusted water) is independent of the isoelectric point of the amino acid. Hydrogen bonding between the amino group and the oxygen of interlayer water molecules is a possible mechanism of adsorption for the weakly adsorbed fraction.

Under the conditions of the primitive oceans, which were probably slightly alkaline, due to dissolved ammonium ions (11), the majority of the amino acid molecules are in the zwitterion form; thus adsorption by cationic exchange can occur only to a limited degree. Greenland et al. (12) noted that no cations are liberated when amino acids and peptides are adsorbed on sodium and calcium clays at neutral pH. In Table 2, the major portion of adsorbed amino acids at pH 7 and pH 10 are weakly adsorbed, probably by hydrogen bonding. Thus, under the neutral conditions of the primitive oceans, mechanisms such as zwitterion association, hydrogen bonding, ion-dipole interactions, and van der Waals forces must have been responsible for most of the amino acid adsorption on clays. Other mechanisms or interactions which contribute to adsorption at neutral pH will be discussed in the section on peptides.

An important aspect of amino acid adsorption on clays in chemical evolution is the orientation of adsorbed molecules in the interlayer space of the clay minerals. Amino acids generally form single layer complexes as they are incorporated into the interlayer space (7, 8, 12, 13). Contraction of the interlayer space to a distance slightly smaller than the thickness of the adsorbed molecules has been observed by X-ray diffraction studies and explained by the above authors by a "keying" of the amino acid molecules into hexagonal holes of the clay surface.

Hsu (13) performed X-ray diffraction studies of various amino acids adsorbed on montmorillonite and found that increasing the number of carbon atoms between functional groups of the amino acid does not increase the interlamellar thickness of the clay. On the other hand, increasing the size of the side chain of α-amino acids does not increase the interlamellar spacing. He concluded that the adsorbed molecules are oriented so that the main chain of the amino acid (from one functional group to the other) lies parallel to the basal surface of the clay. Thus, the alkyl group substituted at the α-carbon projects into the interlamellar space at a large angle. His results were supported by the additional finding that the interlamellar spacing of the clay with the peptides, di-alanine and tri-alanine intercalated is the same as that when α-amino acid monomers are adsorbed.

Fripiat et al. (14) detected zwitterions of glycine and β-alanine adsorbed on montmorillonite with infrared studies of clay films, and they suggested a specific orientation of these molecules on the clay surface. Upon heating the film (140°–244°C), they found the formation of a secondary amide linkage. This experiment simulated somewhat the condition of a drying ocean beach and may indicate the importance of zwitterion association and orientation on the clay surface in chemical evolution.

Preferential Adsorption of Amino Acids

One of the most interesting questions concerning the role of clay minerals in chemical evolution is whether clays can preferentially adsorb certain types of organic molecules. Specifically, do clays adsorb (and thus concentrate) biological compounds more easily than non-biological ones? Friebele et al. (10) have recently investigated the question of whether clays selectively adsorb protein amino acids over non-protein amino acids. They mixed Na-montmorillonite with five different pairs of protein and non-protein amino acids and analyzed the quantities of amino acids in three fractions: non-adsorbed (indirect), weakly adsorbed (wash), and strongly adsorbed (remaining on clay, extracted with 1 N HCl). By estimating quantities of amino acids in all fractions ann obtaining complete recovery (100± 5%), they were assured that no error due to loss or degradation occurred. Their results, shown in Table 2, show that there is little difference in the adsorption of most protein and non-protein amino acids by Na-montmorillonite at differing hydrogen ion concentrations. The exception occurs at low pH, and to a slight degree at neutral pH, for the pairs containing an amino acid having more than one carbon atom separating the amino and carboxyl groups. The β- and γ-amino acids are preferentially adsorbed over their α-amino acid counterparts. Sieskind and Wey (9) obtained similar results with mixtures of single amino acids and montmorillonite and noted a linear relationship between adsorption of amino acids at low pH and the number of carbon atoms separating functional groups.

The only basis for differential adsorption of protein and non-protein amino acids appears to be differences in isoelectric points, which result in different degrees of cationic exchange occurring at low pH (β-alanine and α-aminobutyric acid have higher pI's than their α-amino acid counterparts). In this case, the difference in dipole moments of the molecules does not appear to influence their adsorption on Na-montmorillonite since there are no constant differences in the quantities of α- and non-α-amino acids in the adsorbed fractions over all hydrogen ion concentrations.

From these results, we can speculate that preferential adsorption by clays would result in removal of non-α-amino acids from the primitive ocean by adsorption, leaving α-amino acids in solution to react. The larger the distance between functional groups, the larger number of molecules removed (9) from the primordial soup. This differentiation would occur to a large extent only if the primitive ocean were acidic; it would occur to a smaller degree under neutral conditions, with any significant net change in the distribution of these amino acids requiring a large amount of time.

The Selection of Enantiomers of Amino Acids

The fact that proteins contain only L-optical isomers of amino acids and no D-isomers is an intriguing question in chemical evolution, since amino acids outside of biological systems tend to exist in the racemic state. Clay minerals might provide a mechanism for the concentration of homogeneous optical isomers if they exhibited selective adsorption of L-amino acids over D-amino acids. This ability to differentiate between the configurations of optical isomers would depend upon the presence of an asymmetric center on the clay mineral surface, which has not been confirmed. The possibility of chirality existing throughout kaolinite layers resulting from interlayer shifts and vacant octahedral sites in successive layers has been discussed by Bailey (15).

Jackson (16) reported that the quantity of L-phenylalanine adsorbed on Kaolinite was significantly greater than the quantity of D-phenylalanine adsorbed at pH 5.8. However, as demonstrated before, his indirect method of measurement, which was to take optical density measurements of the equilibrium solutions, could be subject to considerable error. This seems evident in the fact that the difference between the estimated quantities of D- and L-phenylalanine adsorbed was only slightly larger than the standard error of the measurements. Thus, the validity of this reported selective adsorption is questionable.

More recently, Bondy and Harrington (17) studied the adsorption of leucine, aspartic acid, and glucose enantiomers on bentonite using tritiated molecules. In their experiments, the addition of L-leucine inhibited the adsorption of L-(^{3}H) leucine, while the addition of D-leucine did not. However, in a similar experiment with labeled D-leucine, neither the addition of L-leucine nor D-leucine inhibited adsorption of D-(^{3}H) leucine. Similar results were obtained with L-aspartic acid and with D-glucose. This inconsistency in the results requires explanation, for if the L-isomer is preferentially adsorbed by a factor of ten, as these results imply, the adsorption of the L-isomer should inhibit adsorption of both of the tritiated D- and L-isomers. Perhaps the major problems with this work are that (1) adsorption of L- and D-isomers was measured in separate experiments; this reported differential adsorption was not tested within a single adsorption experiment with recemic amino acid mixtures; (2) an indirect method of measurement was used, i.e., the inhibitory effect of an isomer on the adsorption of marked isomers.

Direct measurements of amino acid enantiomers adsorbed on Na-montmorillonite in our laboratory have indicated that there is no large preferential adsorption of either enantiomer by Na-montmorillonite (18). There appears to be no doubt that the distribution of the amino acid enantiomers between the clay and the solution is due to abiotic processes, since complete recovery of the amino acids was obtained. Enantiomer ratios which were obtained by gas chromatography and a newer method employing chiral eluants in high performance liquid chromatography (19) are given for the adsorbed and non-adsorbed fractions in Tables 3 and 4.

The existence of a statistically significant difference between the standard and sample D : L ratios is denoted by an asterisk. In the strongly and weakly adsorbed fractions showing a significant difference, the D : L ratios are less than one, indicating a 0.5–2.0% greater quantity of the L-enantiomer adsorbed than D-enantiomer. The standard deviation of the mean % D- or L-amino acid obtained by gas chromatographic measurements ranges from 0.01 to 0.8% (a single exception was a standard deviation of 1.5% for α-aminobutyric acid at pH 10); thus, the difference in the quantities of L- and D-enantiomers in the adsorbed fractions is larger than analytical variation.

In order to be sure that these differences are not due to contamination, we calculated mass balances of the enantiomers adsorbed and nonadsorbed, obtaining a total quantity of D- and L-amino acids for each experiment. Since the experiments were begun with racemic mixtures, the total quantity of each enantiomer shouln be equal, within experimental error. In all experiments except for Experiment B at pH 3, the difference in the total measured quantities of D- and L-valine

TABLE 3. Gas Chromatographic Analysis of Amino Acid Enantiomers Adsorbed and Non-adsorbed on Na-montmorillonite

Sample	Valine				Norvaline				α-Aminobutyric acid			
	D:L	σ	n		D:L	σ	n		D:L	σ	n	
Standard	1.008	0.022	10	(A)					1.010	0.024	5	(A)
	1.011	0.014	10	(B)	1.006	0.031	10	(B)				
pH 3												
Adsorbed S	0.977*	0.013	5	(A)	0.986*	0.020	4	(B)	1.005	0.002	3	(A)
W	0.973*	0.026	3	(A)	1.017	0.011	3	(B)	0.985	0.012	4	(A)
Non-adsorbed	1.020*	0.005	3	(A)	1.006	0.027	6	(B)	1.014	0.015	4	(A)
pH 7												
Adsorbed S	0.996	0.013	6	(A)	0.979*	0.016	5	(B)	0.996	0.005	4	(A)
W	1.031*	0.021	3	(A)	0.995	0.013	4	(B)	0.993	0.032	3	(A)
Non-adsorbed	1.005	0.011	3	(A)	0.997	0.023	5	(B)	1.021	0.038	3	(A)
pH 10												
Adsorbed S	1.015	0.021	4	(B)	0.986	0.035	4	(B)	0.925*	0.046	7	(A)
W	1.008	0.021	3	(B)	1.011	0.025	3	(B)	0.988*	0.004	3	(A)
Non-adsorbed	1.010	0.011	3	(B)	1.000	0.012	3	(B)	1.012	0.021	4	(A)

S: strongly adsorbed; W: weakly adsorbed; *: statistically significant difference between standard and sample D:L ratios; n: number of GC analyses; (A) sterile conditions not used; (B) sterile conditions used.

TABLE 4. D:L Ratios of α-Alanine Adsorbed and Non-adsorbed on Na-montmorillonite

Sample		Analysis by gas chromatography			Analysis by liquid chromatography		
		D:L	σ	n	D:L	σ	n
Standard	(A)	1.028	0.029	10			
	(B)	1.005	0.014	4	0.997	0.016	15
pH 3							
Asdorbed S	(A)	1.022	0.009	5			
W	(A)	1.018	0.011	5			
Non-adsorbed	(A)	1.024	0.006	4			
pH 7							
Adsorbed S	(B)	0.976*	0.028	5	1.001	0.017	10
W	(B)	1.013	0.028	4	0.999	0.007	7
Non-adsorbed	(B)	1.013	0.020	6	1.003	0.018	4
pH 10							
Adsorbed S	(B)	0.953*	0.012	4	0.983*	0.008	10
W	(B)	0.979*	0.021	4	0.982	0.042	5
Non-adsorbed	(B)	1.011	0.006	3	1.008	0.015	8

is less than the variation in μmoles expected from the analytical method. Thus, we are assured that no loss or degradation of either enantiomer took place during the course of the experiments.

Although replicates of the experiments were performed, the reproducibility of the results was not thoroughly tested because some of the data sets were rejected as invalid when total measured D- and L-isomer quantities did not balance. An example of two replicate experiments of the adsorption of (D, L) valine at pH 3 is shown in Table 5. For comparison, the D : L ratios in experiments A and B are

TABLE 5.   Quantities of Valine Enantiomers Measued by Gas Chromatography in Three Fractions in Adsorption Experiments

	Fraction	μM D	μM L	Δ μM	σ μM
	pH 3.0				
	S	5.81	5.99	−0.18**	0.040
	W	5.11	5.29	−0.18**	0.070
Exp. A	Total adsorbed	10.92	11.28	−0.36	0.110
	Non-adsorbed	37.42	36.98	+0.44**	0.090
	Total	48.34	48.26	+0.08	0.200
	pH 3.0				
	S	5.66	5.84	−0.18*	0.024
	W	5.40	5.50	−0.10	0.034
Exp. B	Total adsorbed	11.06	11.34	−0.28*	0.058
	Non-adsorbed	35.23	35.66	−0.43*	0.25
	Total	46.29	47.0	−0.71	0.31
	pH 7.0				
	S	0.547	0.553	−0.006	0.004
	W	3.09	3.01	+0.08**	0.031
Exp. A	Total adsorbed	3.64	3.56	+0.086	0.035
	Non-adsorbed	43.46	43.64	−0.18	0.240
	Total	47.10	47.20	−0.10	0.275
	pH 10.0				
	S	0.601	0.599	+0.002	0.006
	W	2.30	2.31	−0.010	0.024
Exp. B	Total adsorbed	2.90	2.90	−0.008	0.030
	Non-adsorbed	43.05	43.05	0	0.24
	Total	45.95	45.95	−0.008	0.27

as follows: strongly adsorbed, 0.977 and 0.979, weakly adsorbed, 0.973 and 0.992, and non-adsorbed. 1.020 and 1.001, respectively. The adsorbed fractions both contain slightly larger quantities of the L-enantiomer; the actual quantities of the isomers in each fraction of the replicate experiments are very close. However, the fact that the D : L ratio for the non-adsorbed fraction is lower in Experiment B makes the total quantities of D- and L-enantiomers slightly unbalanced, and the data cannot be accepted for that reason.

Even with the knowledge that full recovery of amino acids was obtained and total recovered quantities of D- and L-enantiomers were equal within experimental error, it is difficult to be fully confident that these very small differences in the adsorbed fractions, which suggest preferential adsorption, are real. To determine whether the small differences in D : L ratios could be due to some artifact in the gas chromatographic analyses, enantiomer analysis of some of the samples was performed by liquid chromatography using a reversed phase column and an eluting buffer containing a copper-L-proline complex. Comparative results from GC and LC analyses of samples from the adsorption experiments are shown in Table 4. In the experiment at pH 7, the D : L ratios of all fractions are very close to 1.0, according to the LC analyses, and the preferential adsorption of L-α-alanine suggested by the GC analyses is not confirmed. In the pH 10 experiment, however, both GC and LC analyses support the adsorption of a larger quantity of the L-enantiomer by the clay. Thus, the LC data do not consistently confirm or deny the validity of the small preferential adsorption of L-amino acids by clays found with GC. Although small differences in the quantities of D- and L-amino acids adsorbed on Na-montmorillonite, which are slightly larger than analytical variation, have been detected, the evidence for a consistent, reproducible preferential adsorption of amino acid enantiomers by clays is inconclusive.

Polymerization of Amino Acids

I t is possible that clays might play several roles in promoting the reactions which produce macromolecules. The first role is that of catalyst; *i.e.*, the surface of the clay somehow initiates or promotes condensation reactions. A second function is to immobilize and organize adsorbed molecules and polymerization products on the clay surface, and in so doing, to protect the reaction products from decomposition. Finally, it is thought that clays might cause preferential polymerization, due to selective interactions with adsorbed monomers and steric restrictions in the interlayer space.

We will begin with the simplest aid to the condensation reaction: heat. The zwitterion form of amino acids adsorbed on the clay surface may undergo peptide formation when the clay-amino acid complex is heated to temperatures of 140°–244°C (*14*). Degens and Matheja (*20*) reported the formation of polymers containing aspartic acid and glutamic acid from mixtures of amino acids and kaolinite heated at 80°C for 7 days. By their own admission, it was surprising to find peptide bond formation at such a low temperature. In other experiments conducted at 140°C for short (63 hr) and long (80 days) time period, polymeric products with molecular weights ranging from 1,000 to 2,000 and containing principally aspartic and glutamic acid were formed in a mixture of 10 amino acids and kaolinite. Kaolinite was found to be a much more effective "catalyst" than montmorillonite. Other researchers (*21, 22*) have been unable to obtain polymerization of amino acids with kaolinite using heat as a source of energy. Degens and Matheja theorized that the apparent peptide bond formation is brought about by carboxyl activation, *i.e.*, the positively charged Al octahedral surface attracts the carboxyl group of the aminoacid, which becomes activated and undergoes subsequent attack by an amino group to produce the peptide bond. This can occur only to a small extent if montmorillonite is the mineral surface since the Al octahedra are "sandwiched" between SiO_4 layers.

Simulating a simple fluctuating primitive geologic system, Lahav *et al.* (*23*) subjected mixtures of glycine and Na-keolinite or Na-bentonite to wetting-drying and temperature fluctuation (25°–94°C) for a number of cycles and observed the formation of oligopeptides up to five glycine residues in length. Only trace amounts of diglycine formed in heated mixtures without clays. Wetting and drying cycles, when added to temperature fluctuations, enhanced the quantity and the chain length of peptides formed. They suggested that the monomers and peptides were redistributed on the clay during the wetting cycle, thus allowing greater contact between monomers and peptides for further polymerization. Lawless and Levi (24) studied the effect of different cationic forms of bentonite in this reaction during wetting-drying/heating. The effectiveness of the cations as "catalysts" is in the order $Cu^{2+} > Ni^{2+} > Zn^{2+} > Na^+$. However, the fact that these metal forms of clays are very rare in nature diminishes the similarity between conditions in these experiments and those in a primitive tidal lagoon or small lake.

Flegmann and Scholefield (25) analyzed the capacity of clays in polymerization reactions using theoretical and experimental approaches. They heated amino acids with and without kaolinite at 90°C for 30 to 60 days and observed no polymerization taking place. They then evaluated the thermodynamic feasibility of condensation reactions on clay surfaces by calculating the free energy of condensation in solution and comparing that to the free energy of the ion exchange reaction that would replace an amino acid monomer on the clay with a dipeptide. Their conclusion was that the thermodynamic barrier to "surface condensation" at 90°C is not lower than that for solution condensation. However, they did not evaluate the thermodynamics of condensation actually taking place on the clay, probably because no specific mechanism for surface condensation on the clay is known.

Aminonitriles, which have been known as one of the products of abiotic synthesis from a simulated primordial atmosphere, can produce amino acids upon hydrolysis (26) and can also form peptides without taking a pathway to amino acids. Hanafusa and Akabori (27) demonstrated the formation of di- and tri-glycines by heating aminoacetonitrile at 120°–140°C for several hours in the presence of kaolinite. (The reaction did not proceed in the absence of clays). In accordance with that work, Losse and Anders (28) synthesized an alanine polymer from α-aminopropionitrile by heating it in the presence of acidic clay (the equivalent of H-montmorillonite). Hanafusa and Akabori stated that the solid surfaces of the clay seems to play a role in this reaction which involves activation of the methylene group of polyglycine.

The question of preferential polymerization of free amino acid enantiomers in the presence of kaolinite and thermal energy has been raised. Degens et al. (20) reported that 25% of L-aspartic acid was polymerized, while only 3% of the D-isomer and 14% of the racemic mixture were polymerized in aqueous solutions heated to 90°C for 32 days in the presence of kaolinite. The quantity of polymerized amino acids (determined by biuret and total amino acids—free amino acids) fluctuated in each group over time. For instance, the quantity of material poly-merized from L-, D- and (D, L) aspartic acid after 21 days was only about 1% for all three. To support their results, Degens *et al.* (*20*) and Jackson (*16, 29*) presented results indicating the preferential adsorption of L-phenylalanine and also the cor-responding D-isomers on kaolinite. This preferential polymerization of aspartic acid in the presence of kaolinite was re-examined by Flores and Bonner (*21*), who de-termined enantiomeric, composition of extracted clay fractions, and by McCullough and Lemmon (*22*) who used L-Asp-2-^{14}C to search for polymer formation and also measured optical rotation of the supernatant. These investigators could observe neither the preferential adsorption of L-aspartic acid nor any peptide formation under the same experimental conditions used by the former investigators. The failure of other researchers to repeat the preferential polymerization under identical conditions emphasizes the importance of obtaining results by a direct confirmation of the compounds of interest isolated from the reaction mixture. Although claims of a large preferential adsorption and polymerization of amino acid enantiomers on clays have been made, these experimental results are difficult to explain in terms of clay structure and clay-amino acid interactions.

Although heat is the most "natural" condensing agent, chemical agents such as cyanate derivatives or polyphosphate can also initiate the condensation reaction. One experiment incorporated both types of condensing agents, with the naturally occurring mineral, apatite ($CaPO_4OH$) as the source of phosphate. Flores and Leckie (*30*) obtained dipeptides by heating cyanate, apatite, and glycine in the solid state at 95°C for a period of days. Probably condensation was brought about by formation of an amino acyl phosphate intermediate. Similar activated molecules are the amino acyl adenylates, which contain a phosphoanhydride bond between the carboxyl group of an amino acid and the phosphate group of the adenosine monophosphate. These compounds, found in contemporary living system, are precursors of polypeptides in the protein synthesis system.

Taking a clue from the biosynthetic pathway of proteins, Lewinsohn *et al.* (*31*) and Paecht-Horowitz and Katchalsky (*32*) proposed a role for aminoacyl adenyl-ates in the prebiotic formation of polypeptides under neutral to alkaline conditions at room temperature. Banda and Ponnamperuma (*33*) obtained peptides from the condensation of amino acid adenylates in alkaline solution. These studies were ex-tended with the use of montmorillonite (*34–37*). In the absence of the clay mineral, polycondensation took place, producing peptides up to 12 monomeric units long. The presence of montmorillonite resulted in not only a higher yield of the polymers, but also longer peptide skeletons. In the copolymerization of alanyl adenylate and seryl adenylate, peptides up to approximately 100 units long were reported. The polymerization on montmorillonite was further distinguished by the presence of a discrete series of peptides, which was not found in the absence of the clay mineral.

Paecht-Horowitz *et al.*, proposed difference mechanisms for polycondensation in the presence and absence of montmorillonite, based on the size distribution of the peptides synthesized. They postulated that in the presence of the clay mineral, the aminoacyl adenylates react to produce polymers without being hydrolyzed, whereas in the absence of the clay mineral the hydrolysis of the adenylates is a necessary step to initiate the reaction. Furthermore, they found that the aminoacyl adenylate is adsorbed on the face of the clay close to the edge, with the adenylic acid group protruding from the clay interlayer space. As polymerization proceeds, the lengthening peptide chain "creeps" over the face of the clay and remains protected from hydrolysis. Studies in which the face of the clay was completely covered with histidine, and the edge covered with Na-hexametaphosphate, showed that the amino acid adenylate must be attached to the clay at two points in order for maximum polymerization to occur: by the NH_3^+ group at the face of the clay, and by the phosphate group at the edges of the clay (37).

After studying the copolymerization of a variety of amino acid adenylates on montmorillonite, Paecht-Horowitz (36) observed a selective interaction; i.e., that peptide bonds between certain amino acids occurred at a higher frequency than bonds between other amino acids. It was suggested that montmorillonite could have been a prebiotic template for polypeptide synthesis which "selected" peptide sequences. However, Steinman and Cole (38) also observed different frequences of amino acid sequences in dipeptides formed in solution in the absence of clay with the condensing agent dicyanamide. Thus, whether or not montmorillonite plays a role in determining peptide sequence during polymerization is unclear.

Paecht-Horowitz and Katchalsky (39) have reported further that the formation and polymerization of alanyl adenylate is possible from a mixture of free alanine and ATP in the presence of montmorillonite and zeolite (Decalso F) at neutral pH at 37°C. They concluded that Decalso F aided the synthesis of alanyl adenylate, and that montmorillonite catalyzed subsequent peptide formation. However, Warden et al. (40) have presented negative results for adenylate synthesis and peptide formation under the same conditions as those in experiments performed by Paecht-Horowitz and Katchalsky (39). The only difference in the experiment by Warden et al. was the use of Decalso instead of Decalso F. Apparently, aminoacyl adenylates are necessary for the formation of peptides in this system, rather than a mixture of amino acids and ATP.

Most available studies show that clays alone do not act as "catalysts" or condensing agents in the polymerization of free amino acids under aqueous conditions. Clay minerals can provide protons, acting as a Bronsted acid, and they can accept electrons as a Lewis acid in the catalysis of organic chemical reactions (41, 42). These properties do not aid the dehydration-condensation reaction necessary for the formation of peptides or polynucleotides. The energy barrier of the reaction must be lowered, either by using activated monomers or condensing agents, or adding energy inputs. If polymerization is initiated by the presence of a condensing agent or energy source, the effect of clay minerals may be to promote the reaction by protecting the reaction products and allowing an area of greater concentration and interaction between reactants. Thus, polymers having longer chain lengths are formed when clays are present than when the reaction takes place in the absence of clays.

In summary, the role of clays in the origin of life, according to the experimental results to date, can be described as follows. Clay minerals facilitate, aid, or speed synthesis reactions in which biological monomers are formed from gaseous constituents of the primordial atmosphere. The presence of clays does not change the direction of the reaction path but it seems to promote the rate of synthesis. Clay minerals do absorb and concentrate by several orders of magnitude, these monomers from aqueous solution. The adsorption of amino acids occurs to a maximum degree by cation exchange at low pH, but adsorption is appreciable at neutral to slightly alkaline pH, where other adsorption mechanisms such as hydrogen bonding ann zwitterion association predominate. The latter condition is more similar to conditions which are thought to exist in the primitive oceans. Selective adsorption of biological monomers occurs only if there are large differences in the pK's of the monomers. Thus, more basic molecules are adsorbed more than neutral or acidic ones, with no distinction being made by the adsorbing clay between biological and non-biological molecules. Although several researchers have claimed to observe preferential adsorption of amino acid and sugar isomers, attempts to repeat their work have been unsuccessful. In thorough, careful work on this problem, there has been no conclusive evidence of asymmetric adsorption of isomers by clays. Finally, clays probably promote polymerization of amino acids by concentrating them, providing containment and a surface for immobilization of the reacting molecules, and protection for reaction products in the interlayer space from destructive energy sources and chemicals. There is no evidence, and further, no theoretical reason, for the initiation of condensation reactions between biological monomers by clays.

The role of clays in chemical evolution envisioned by Bernal as selective adsorber, concentrator, catalyst for polymerization, and organizer of molecules for their replication has been only partially borne out by experimental research. What laboratory experiments have shown us is that clays may have acted more as an inert surface on which organic molecules congregated, just as larger stable surfaces in the oceans are colonized by marine organisms. The reactions of chemical evolution producing first monomers and then polymers might have been promoted simply by the greater proximity of the concentrated molecules adsorbed on clays.

REFERENCES

1. Bernal, J.D.: 1951. *The Physical Basis of Life*, p. 34, Routledge and Kegan Paul, London.
2. Cairns-Smith, A.G.: 1965. *Theor. Biol.* 10, 53–88.

3. Mortland, M.M.: 1970. *Adv. Agron.* 22, 75–117.
4. Theng, B.K.G.: 1979. *The Chemistry of Clay Organic Reaction*, John Wiley and Sons, New York.
5. McLaren, A.D., Peterson, G.H., and Barshad, I.: 1958. *Soil Sci. Soc. Am. Proc.* 22, 239–244.
6. Sieskind, O.: 1960. *C. R.* 250, 2392–2393.
7. Cloos, P., Calcius, B., Fripiat, J.J., and McKay, K.: 1966. *Proc. Int. Clay Conf. Jerusalem* 1, 223–232.
8. Talibudeen, O.: 1955. *Trans. Faraday Soc.* 51, 582–590.
9. Sieskind, O. and Wey, R.: 1959. *C. R.* 248, 1652–1655.
10. Friebele, E., Shimoyama, A., and Ponnamperuma, C.: 1980a. *J. Mol. Evol.* 16, 269–278.
11. Bada, J.L. and Miller, S.L.: 1968. *Science* 159, 423–425.
12. Greenland, D.J., Laby, R.H., and Quirk, J.P.: 1962. *Trans. Faraday Soc.* 58, 829–841.
13. Hsu, S.C.: 1977. Ph.D. Diss., Polytechnic Inst. of New York, Brooklyn, New York.
14. Fripiat, J.J., Cloos, P., Calcius, B., and McKay, K.: 1966. *Proc. Int. Clay Conf. Jerusalem* 1, 233–246.
15. Bailey, S.W.: 1963. *Am. Mineral.* 48, 1196–1209.
16. Jackson, T.A.: 1971b. *Chem. Geol.* 7, 295–306.
17. Bondy, S. and Harrington, M.: 1979. *Science* 203, 1243–1244.
18. Friebele, E., Shimoyama, A., and Ponnamperuma, C.: 1980b. *Proc. 6th Int. Conf. Origins of Life*, Jerusalem, Israel.
19. Hare, P.E. and Gil-Av, E.: 1979. *Science* 204, 1226–1228.
20. Degens, E.T., Matheja, J., and Jackson, T.A.: 1970. *Nature* 227, 492–493.
21. Flores, J.J. and Bonner, W.A.: 1974. *J. Mol. Evol.* 3, 49–56.
22. McCullough, J.J. and Lemmon, R.M.: 1974. *J. Mol. Evol.* 3, 57–61.
23. Lahav, N., White, P., and Chang, S.: 1978. *Science* 201, 67–69.
24. Lawless, J. and Levi, N.: 1979. *J. Mol. Evol.* 13, 281–286.
25. Flegmann, A.W. and Scholefield, D.: 1978. *J. Mol. Evol.* 12, 101–112.
26. Ponnamperuma, C. and Woeller, F.H.: 1967. *Curr. Mod. Biol.* 1, 156–158.
27. Hanabusa, H. and Akabori, S.: 1959. *Bull. Chem. Soc. Japan.* 32, 626–630.
28. Losse, G. and Anders, K.: 1961. *Z. Physiol. Chem.* 323, 111–115.
29. Jackson, T.A.: 1971a. *Experientia* 27, 242–243.
30. Flores, W. and Leckie, J.O.: 1973. *Nature* 244, 435–437.
31. Lewinsohn, R., Paecht-Horowitz, M., and Katchalsky, A.: 1967. *Biochim. Biophys. Acta* 140, 24–36.
32. Paecht-Horowitz, M. and Katchalsky, A.: 1967. *Biochim. Biophys. Acta* 140, 14–23.
33. Banda, P.W. and Ponnamperuma, C.: 1971. *Space Life Sci.* 3, 54–62.
34. Paecht-Horowitz, M., Berger, J., and Katchalsky, A.: 1970. *Nature* 228, 636–639.
35. Paecht-Horowitz, M.: 1971. *Chemical Evolution and the Origin of Life* (Buret, R. and Ponnamperuma, C., eds.) pp. 245–251, North-Holland Publ. Co., Amsterdam.
36. Paecht-Horowitz, M.: 1973. *Angew. Chem. Int.* 12, 349–356.
37. Paecht-Horowitz, M.: 1974. *Origins of Life* 5, 173–187.
38. Steinmann, G.D. and Cole, M.N.: 1967. *Proc. Natl. Acad. Sci. USA* 58, 735–742.

39. Paecht-Horowitz, M. and Katchalsky, A.: 1973. *J. Mol. Evol.* 2, 91–98.
40. Warden, J.J., McCullough, J.J., Levinson, R.M., and Calvin, M.: 1974. *J. Mol. Evol.* 4, 189–194.
41. Solomon, D.H.: 1968. *Clays Clay Min.* 16, 31–39.
42. Hawthorne, D.G. and Solomon, D.H.: 1972. *Clays Clay Min.* 20, 75–78.

HCN as a Possible Precursor of the Amino Acids in Lunar Samples

S. YUASA and J. ORÓ

J. ORÓ received his licenciate in chemistry from the University of Barcelona, Spain in 1947, and his Ph.D. degree in biochemistry from the Baylor College of Medicine in 1956. He became associated with the University of Houston in 1955 where he is now a Full Professor in the Departments of Biophysical Sciences and Chemistry. His research has been primarily in the analysis of meteorites and lunar samples and on the synthesis of organic compounds under possible primitive Earth conditions in order to get a better understanding of the problem of the origin of life. He was a member of the Molecular Analysis Team of the NASA Viking Project. He has published some 180 publications and 10 books in these and related areas of research.

It is well known that several amino acids have been found in the acid hydrolyzed hot water extracts of lunar samples (1) in quantities totalling 2–70 ng/g. These include glycine, alanine, glutamic acid, aspartic acid, serine, and threonine. Investigations carried out by several groups indicate that the amino acids detected are not the result of common biological contamination, either on the lunar surface by the astronauts and spacecraft or during handling of the samples after return to Earth.

The question of whether the amino acids are present as such or are generated or synthesized during the analytical procedure has not yet been completely resolved, although consideration of the survival time of moderately complex organics under lunar conditions (2) and the experimental evidence suggest that they are derived from other sources associated with the lunar samples which have been designated as amino acid precursors. The nature of such precursor(s) and the mechanism of amino acid synthesis are not known. This still remains one of the most important unsolved problems in the studies of the organic chemistry of lunar samples.

A long time ago we reported (*3*) the formation of alanine, glycine, and aspartic acid from an aqueous alkaline solution 2.2 M in cyanide ions upon heating at 70°C for 25 days, and discussed the possible role in HCN polymers in their formation. Abelson (*4*) subsequently noted that dilute solutions of HCN at a pH 8–9 polymerize to give the tetramer (diaminomaleonitrile) as a major product, that UV radiation greatly increased the polymerization, and that 6N HCl hydrolysis of the irradiated solution yielded glycine, serine, alanine, and aspartic acid. Biemann (*5*) discussed the possibility that the amino acids found in lunar samples were produced from indigenous ammonia, cyanides, and aldehydes. During the extensive analysis of the lunar samples (*6*) several investigators reported finding HCN or DCN and other similar organic compounds by mass spectrometric analysis upon heating or acid dissolution with HCl or DCl of the lunar samples. On the basis of these results, we felt that HCN may have played an important role as one of several possible precursors in the origin of the amino acids observed in the lunar samples. We have, therefore, investigated the chemical behavior of HCN under conditions very similar to those used to extract the lunar samples for amino acids (*7*). Our work is described in this paper.

Experimental Procedures

H CN was generated by the reaction of H_2SO_4 on NaCN and condensed in a volume-calibrated cold trap. The liquid HCN was then added to a known volume of water to produce a solution of 0.2 molarity. Five-ml aliquots of the 0.2 M HCN (pH 4.7) were poured into chromic acid-cleaned Pyrex reflux tubes. The solution was frozen with liquid nitrogen and the tube volume was flushed with nitrogen during flame sealing of the tubes. The tubes were then maintained at temperatures of 0°, 50°, 80°, or 100°C for varying amounts of time. The appearance of the mixture was noted and the pH was measured following the heating period. The product mixture was then hydrolyzed, paper chromatographed, and derivatized for analysis by gas chromatography or gas chromatography-mass spectrometry (GC-MS) as described below. Blank runs omitting the HCN were made to verify the absence of any contamination or other artifacts.

The hydrolyses were carried out in 6 N HCl for 20 to 24 hr at 110°C (or for several days) in an air-evacuated, flame-sealed glass tube with the hydrolysis solution frozen during the evacuation and sealing. The hydrolyzed sample was brought to dryness by vacuum evaporation (this and all subsequent evaporations were carried out at room temperature) and then taken up in 2 ml of distilled water (all water was quadruply-distilled in an all-glass distillation apparatus). The sample was then charged onto a Dowex-50 (50–100 mesh) column (0.5 × 15 cm) for desalting and eventual elution of the amino acids with 30 ml of 2 N NH_4OH. The ammoniacal desalted eluant was concentrated to 1 ml by vacuum evaporation and analyzed by paper chromatography and GC-MS after appropriate derivatization as previously described (*8*).

Results and Discussion

The pH changes of the HCN solutions after being allowed to stand or being heated for 20 hr at different temperatures were as follows: 0°C (pH 4.8), 50°C (pH 5.4), 80°C (pH 8), and 100°C (pH 8.5). It can be seen that the 80° and 100°C samples had marked increases in pH. This is especially significant in view of the known basic pH requirement for the initiation of HCN polymerization. It was also noted that the 100°C sample contained black solids upon completion of the heating period. The 50° and 80°C samples appeared yellow and brown, respectively, while the 0°C sample did not change color.

No ninhydrin-positive compounds were obtained for the 0° and 50°C samples,

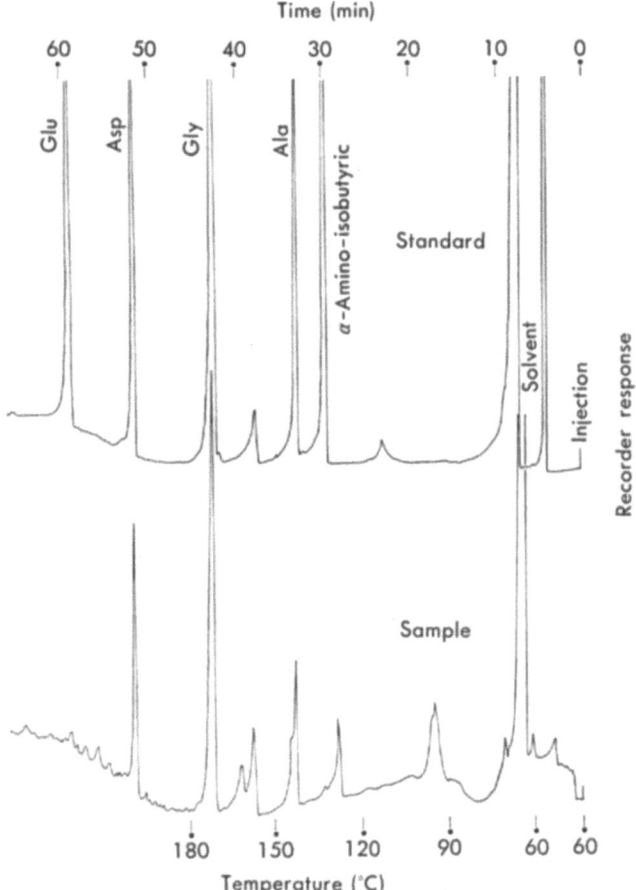

FIG. 1. GC-MS chromatogram of the hydrolyzed products from a 0.2 M HCN aqueous solution maintained at 100°C for 24 hr.

TABLE 1.   Amino Acids from Hydrolysis of a 0.2M HCN Aqueous Solution Refluxed at 100°C for 20 hr

Amino acids	μmol/mmol HCN	% Yield[a]
α-Aminobutyric	1	0.4
Ala	9	2.7
Gly	43	8.6
Asp	14	5.6
Glu	0.8	0.4

[a] Based upon input carbon of HCN

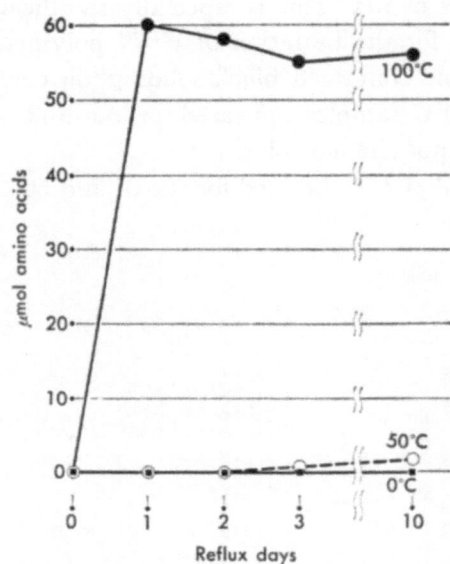

FIG. 2.   Total amino acids produced by hydrolysis of the products from a 0.2 M HCN aqueous solution maintained at 100°C.

but several ninhydrin-positive spots were detected in the 100°C sample. Therefore, this sample was analyzed in more detail by GC and GC-MS. Figure 1 is the gas chromatogram for the derivatized 100°C sample. The N-trifluoroacetyl-O-isopropyl ester derivatives were analyzed on an LKB-9000 gas chromatogarph-mass spectrometer. Retention times and mass spectra confirmed the presence of α-amino-isobutyric acid (m/e 43, 59, 154, 155), alanine (m/e 43, 140, 141, 168), glycine (m/e 43, 126, 127, 154, 197), aspartic acid (m/e 43, 139, 167, 184, 185, 212, 226, 227), and glutamic acid (m/e 43, 85, 153, 180, 181, 198, 226). Other unidentified peaks were also present.

Table 1 gives the amounts of the individual amino acids formed and their yields. The amounts were determined by integrating the GC peak areas and multiplying by calibration factors obtained by injection of known amounts of authentic standards. Figure 2 is a plot of the total amino acids formed at the three different temperatures as a function of time. No amino acids were detected by GC-MS in the 0°C samples after 10 days. Small amounts appeared in the 50°C samples after

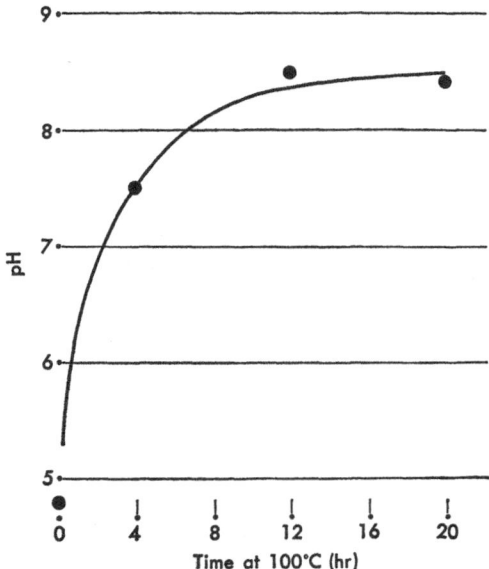

FIG. 3.   Change in pH of 0.2 M HCN solution as a function of time at 100°C.

7–8 days of heating. The 100°C samples attained a value of 55–60 μmol after one day and remained at that level (within the precision of our measurements) for the remaining nine days.

The observed formation of glycine, alanine, aspartic acid, glutamic acid, α-amino-isobutyric acid, and several other ninhydrin-positive compounds (diamino-propionic acid, *etc.*) by hydrolysis of the product mixture of an HCN solution which has simply been heated in a closed tube is quite remarkable, and is particularly significant for an interpretation of the genesis of the amino acids obtained from treatment of the lunar samples. It can be assumed that the hydrolysis of HCN during the heating period results in the formation of some basic compounds (NH_3, *etc.*) and thereby raises the pH (Fig. 3) to the point where polymerization of HCN can occur. As suggested by previous work (*3*, *9*), the products which hydrolyzed to amino acids are thought to be HCN oligomers and other HCN condensation products formed primarily *via* aminomalonodinitrile and diaminomaleodinitrile (HCN trimer and tetramer). By utilizing a few simple reactions, *i.e.*, non-oxidative deamination, HCN addition, hydrolysis of cyano groups, and decarboxylation, all of the amino acids found experimentally can be accounted for by HCN condensation through aminomalonodinitrile and diaminomaleodinitrile (*10*, *11*).

Hydrogen cyanide or its polymers could then be precursors of the amino acids observed in the lunar samples because the conditions of heating and hydrolysis used by various investigators (*1*, *7*) to extract the lunar samples provide conditions that are very similar to those we used in the above studies. Although HCN probably generates the majority of the amino acids detected in the lunar samples, this is not meant to imply that HCN is considered to be the only amino acid precursor.

It is possible that in addition to HCN and its oligomers, other compounds, such as formaldehyde, *etc.*, may also be precursors of the amino acids obtained from the lunar samples. Indeed, the presence of aldehydes may be necessary to generate serine and threonine as shown by our work (*12*) and that of Egami and coworkers (*13*).

The source of the HCN in the lunar samples is another problem which deserves additional attention. It is unfortunate that the Apollo lunar landing module produces significant HCN exhaust contamination, thereby making it more difficult to determine the presence of indigenous HCN. Many of the lunar samples reported to contain HCN, however, were collected at sampling sites located several hundred meters from the exhaust impingement area. Therefore, it is highly likely that the observed HCN is indigenous or has been incorporated in the lunar samples from other solar system sources. It is not the purpose of this paper to discuss fully the source of HCN and related amino acid precursors, but there are several candidates, such as meteorites, comets, and solar wind, which have probably contributed significantly to their formation (*1, 14, 15*).

In summary, the experimental results indicate that amino acids are formed by hydrolysis of the obtained by heating HCN at 100°C in a water solution. We consider these results significant for an interpretation of the genesis of the amino acids obtained from the analysis of lunar fines, since the nature of these amino acids and their relative amounts are very similar to those found in the lunar samples. These results may also have additional implications for abiotic syntheses in other extraterrestrial bodies as well as on the primitive Earth.

Acknowledgments

We wish to thank Drs. D. Flory, D. Nooner, and S.L. Miller for their suggestions. This research was supported in part by a grant from the National Aeronautics and Space Administration, NGR 44-005-002.

REFERENCES

1. Fox, S.W., Harada, K., and Hare, P.E.: 1972. *Geochim. Cosmochim. Acta* (Suppl. 3) 2, 2109–2118.
2. Sagan, C.: 1972. *Space Life Sci.* 3, 484–489.
3. Oró, J. and Kamat, S.S.: 1961. *Nature* 190, 442–443.
4. Abelson, P.H.: 1965–1966. *Carnegie Institution Year Book 65*, pp. 358–360, Carnegie Institution, Washington, D.C.
5. Bieman, K.: 1972. *Space Life Sci.* 3, 469–473.
6. Holland, P.T., Simoneit, B.R., Wszolek, P.D., and Burlingame, A.L.: 1972. *Geochim. Cosmochim. Acta* (Suppl. 3) 2, 2131–2147.
7. Kvenvolden, K.A.: 1972. *Space Life Sci.* 3, 330–341.
8. Oró, J., Gibert, J., Lichtenstein, H., Wikstrom, S., and Flory, D.A.: 1971. *Nature* 230, 105–106.

9. Yuasa, S. and Oró, J.: 1974. *Cosmochemical Evolution and the Origins of Life* (Oró, J., Miller, S.L., Ponnamperuma, C., and Young, R.S., eds.) pp. 295–299, Reidel Publishing Co., Dordrecht, Holland.

10. Oró, J.: 1963. *Ann. N.Y. Acad. Sci.* 108, 464–481.

11. Sanchez, R.A., Ferris, J.P., and Orgel, L.E.: 1967. *J. Mol. Biol.* 30, 223–253.

12. Oró, J., Kimball, A.P., Fritz, R., and Master, F.: 1959. *Arch. Biochem. Biophys.* 85, 115–130.

13. Yanagawa, H., Kobayashi, Y., and Egami, F.: 1980. *J. Biochem.* 87, 359–362.

14. Oró, J., Holzer, G., and Lazcano-Araujo, A.: 1980. *COSPAR Life Sciences and Space Research*, Vol. 8 (Holmquist, R., ed.) pp. 67–82, Pergamon Press, Oxford and New York.

15. Harada, K. and Hare, P.E.: 1980. *Biogeochemistry of Amino Acids* (Hare, P.E., Hoering, T., and King, Jr., K., eds.) pp. 169–181, John Wiley and Sons, New York.

A Model for Protocellular Coordination of Nucleic Acid and Protein Syntheses*

SIDNEY W. FOX

S. W. Fox was born March 24, 1912 in Los Angeles, California, and received his B.A. in chemistry from the University of California in 1933 and Ph.D. degree in biochemistry within the Division of Biology from the California Institute of Technology in 1940. He is Director and Research Professor at the Institute for Molecular and Cellular Evolution, which he founded at the University of Miami in 1964. During 1955–1964 he was Professor of Chemistry and Director of two institutes in turn at Florida State University. His research has included especially pioneering of studies in amino acid sequence in proteins, in evolution of protein molecules, and in the laboratory production of a protoorganism independent of nucleic acids. Theories of protobiogenesis and initial steps of the origin of the genetic code have emerged from the last-mentioned study.

One of the principal questions that must be answered in explaining the evolution of inanimate matter to modern living cells is that of the coordination of synthesis of proteins with coding by nucleic acids within the evolving cell. The proteinoid model has begun to provide an answer to this question *ab initio* from a geological matrix (Fig. 1). An essential part of the answer is the catalysis by lysine-rich proteinoid of (a) a synthesis of peptide bonds from amino acids and ATP (*1*) and (b) synthesis of internucleotide bonds from ATP (*2*). Microspheres containing lysine-rich proteinoid have been shown to carry out each kind of reaction.

* Presented at the Meeting of the International Society for the Study of the Origin of Life (*29*), and under the title of Informational Proteinoid as Matrix for the Genetic Mechanism in a Symposium on Biochemical Evolution of the Genetic Apparatus, in Potsdam, N.Y. on 2 July 1980.

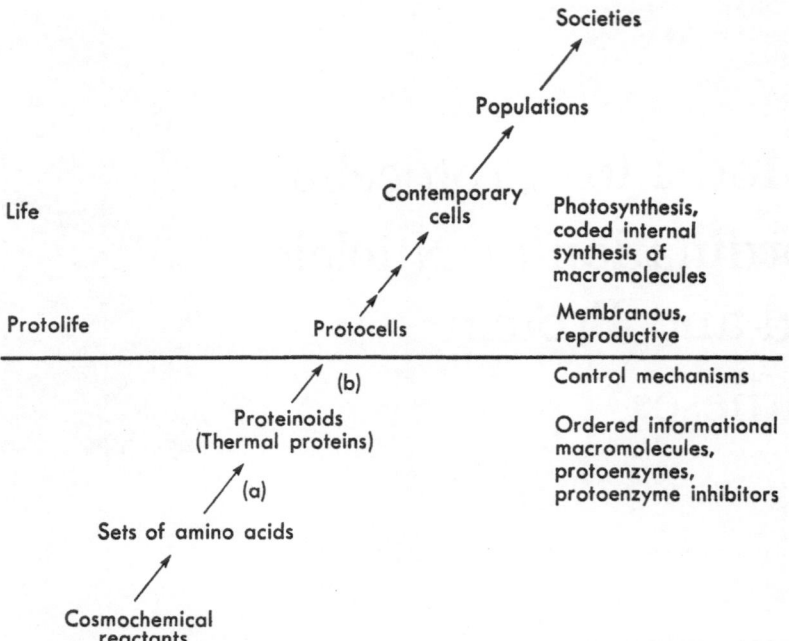

FIG. 1.   The flowsheet of cosmic evolution from a molecular perspective. Links (a) and (b) are connective; they replace the "missing link" between inanimate matter and cellular proto-life.

Although Fig. 1 extrapolates from experimentally demonstrated oligomers to the larger polymers of today's organisms, some investigators hypothesize that oligomers were sufficiently large initially (3). Even for today's organisms, trinucleotides are sufficiently large to be coding units (4).

Self-ordering of Amino Acids

The catalytic activity of lysine-rich proteinoid, or that of any other proteinoid that serves as a catalyst, would benefit from the existence of multiple copies of macromolecules of one kind, in contrast to the oft-assumed random array from polycondensation (5). Randomness is the antithesis of multiple copies. The nonrandom reality has been explained as due to the self-ordering of amino acids (6). The evidence for the self-ordering of amino acids is now from many laboratories and is otherwise extensive (7).

When the thermal polycondensation of glutamic acid, glycine, and tyrosine (8) was carried out, the most abundant fraction was found to be the tripeptide portion. This was found to comprise only two tripeptides: pyroglutamyltyrosylglycine and pyroglutamylglycyltyrosine, in an equimolar complex. This result is from the most thorough study on sequences from thermal polyamino acids yet reported (9–14) for sequences or terminal amino acid analyses. In their repetition of this study, Hartmann et al. (15) found these two tripeptides and established the absence of any not containing tyrosine. They also proposed an enlarged mechanism. The result is thus confirmed as highly nonrandom.

The second example is found in the proteinoids prepared by Snyder in our laboratory (16). When Snyder heated seawater salts with a mixture of common proteinous amino acids, there resulted simultaneously three acidic proteinoids and two basic proteinoids, all of which were cleanly separated by isoelectric focussing. Microspheres prepared from seawater proteinoid are stable at pHs above 9.0. Most remarkable is that acidic and basic proteinoids each ordered themselves during the polymerization.

Other evidences for self-ordering in nonrandom polymerization are many (16, 17).

Proteinoid Protocells First

One feature of all the processes in the formation of various microspheres stands out. This is the great tendency for such microparticles to form (18). Such prebiotic modelling is provided by sulphobes (19), marigranules (20), proteinoid microspheres (21), nucleoproteinoid microparticles (22), seawater proteinoid microparticles (16), lysine-rich proteinoid microparticles (23), coacervated proteinoid microspheres (24), and lumispheres (25), but not by coacervate droplets, which are made from modern polymers instead of amino acids or amino acid precursors (18). The polymerization of amino acids and the aggregation of the poly-amino acids is so facile as to represent the maximal conceivable ease of process (18, 26).

The existence of discrete microparticles containing hydrophobic zones and able e.g. to communicate with the environment through a membrane provides special features for evolution (18). In that light, in view of the ease of spontaneous formation of a protoorganism, no other conceivable pathway of evolution would have been able to compete. Moreover, no such pathway, such as DNA-first, has been demonstrated in the laboratory (27).

Synthesis of Peptide and Internucleotide Bonds

Proteinoids in general have been shown to catalyze a considerable number of modern biochemical reactions (17). These activities tend to be incorporated into microspheres when the proteinoids are aggregated (7). Most relevant to the present paper is the finding that lysine-rich proteinoids containing above a minimal content of lysine (percent not yet known) catalyze the formation of internucleotide bonds (2). In solution, adenine dinucleotide forms from ATP. In suspensions of microparticles composed of both acidic and lysine-rich proteinoids, adenine trinucleotide also forms from ATP.

The formation of peptide bonds from amino acids and ATP, as catalyzed by lysine-rich proteinoid, has been intensively studied in the last two years (*1, 28*). The reaction is observed in aqueous solution, in acidic: lysine-rich proteinoid particle suspension, and in suspensions of complexes of lysine-rich proteinoid and homopolyribonucleotides. All types of amino acid tested form peptides. In the nucleoproteinoid microparticle suspensions, all polynucleotides tested serve, and the results are selective between those constituents and the individual amino acids (*29*). This selectivity extends to sequences in oligopeptides synthesized. For example, glycylphenylalanine is mainly synthesized from glycine, phenylalanine, and ATP by poly A-lysine-rich proteinoid particles while phenylalanylglycine is principally synthesized by poly U-lysine-rich proteinoid particles.

Protocellular Orchestration of Biomacromolecular Synthesis

L ysine-rich proteinoid serves as catalyst for synthesis both of peptides from amino acids with ATP and the synthesis of oligonucleotides from the ATP without amino acids. In the production of peptides, ATP serves as an energy source. In the production of polynucleotides, ATP serves as a source of both energy and incorporated material. The plausible origin of ATP has been explained (*30*). The origins of most of the amino acids have been modelled, with varying degrees of plausibility (*17, 31*).

All of the plausible protocell models arise from amino acids (*9*) or amino acid precursors, which have been shown to exist in cosmic locales (*32, 33*). In all, or nearly all, of these syntheses, trifunctional amino acid plays a role, as it did not do in such protein-like polymers as nylon (*17*). Sets of amino acids from cosmic sources have been shown by Rohlfing to yield copolyamino acids that, in turn, yield microspheres (*33, 34*).

Essential to the new picture is (a) the experimentally based view that the primary protocell was already highly competent biochemically and cytophysically (*7, 28*). This state is explained by (b) the remarkable degree of self-ordering demonstrated for amino acids (*6*).

The concept of a protocell preceding coded genetics was seen already in the analysis of Ehrensvärd (*35*). Lacey and Mullins (*36*) have explained the benefit of isolated microsystems to the operation of the genetic code. The fact that lysine-rich proteinoid can serve two catalytic functions may have further aided in the economical use of microspace. Consistent with this emphasis are the indications that stereochemical interactions may well be at the base of genetic coding (*37, 38*).

Also related is the fact that lysine-rich proteinoid forms mixed particles readily with polyanions, such as polyribonucleotides or acidic proteinoids (*28*). It is these mixed particles that catalyze the synthesis of both internucleotide and peptide bonds (*28, 29*). The evolution is thus seen as a two-stage sequence in which geological synthesis of peptides evolved to protocellular synthesis of peptides and oligonucleotides (Fig. 2).

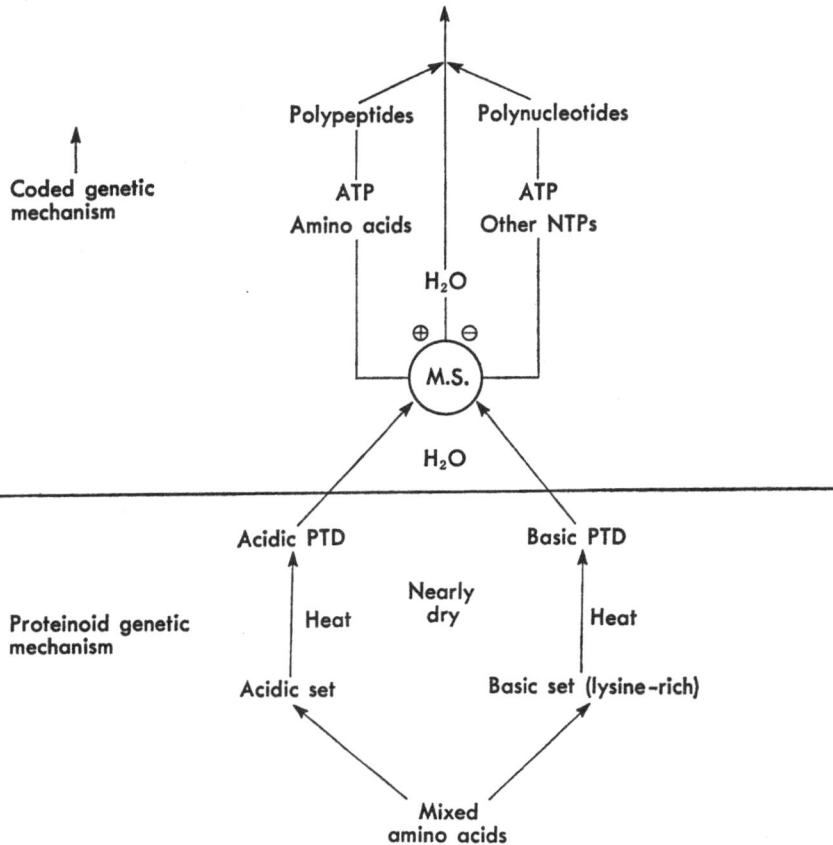

FIG. 2.   Evolution of proteinoid genetic mechanism to coded genetic mechanism. Thermal peptide bond synthesis (*17*) evolved to hydrous peptide and internucleotide bond synthesis using ATP.

The new picture (Fig. 2) coupled with the principle of the self-ordering of amino acids has already raised the question of the contribution of nucleic acids to evolution. Two of the answers that are proposed are: (a) the protection from the environment afforded to central cellular processes yielding polyamino acids by coding them and (b) the introduction of randomizing influences by nucleic acids during encoding and decoding; the greater variety available to selection processes would have broadened the variety available in evolution.

The possibility that the essential processes of peptide and internucleotide bond synthesis could have developed outside a cell is perhaps not rigorously ruled out; the protocell conceivably might have arisen later. The prior appearance of a protocell however seems most plausible, especially since the protocell could have provided hydrophobic zones in which both kinds of polymerization, amino acid and nucleotide, would have been favored. The basic predisposing factor is that protocells would have formed as easily as the experiments indicate.

The most popular conceptual alternative to the flowsheet of Fig. 2 is Fig. 3.

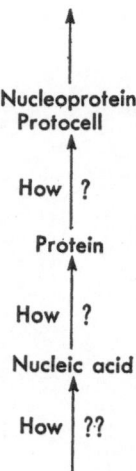

FIG. 3.   Conceptually alternative flowsheet to Fig. 2. While this sequence is concordant with the flow of information in the Central Dogma, the mechanisms for these steps in a primordial context have not been modelled.

The flow of information in Fig. 3 is, in accord with the Central Dogma, from nucleic acid to protein as in the modern cell. This is a popular view despite Crick's statement that the Central Dogma was not intended to apply to the origin of life nor to the origin of the genetic code (*39*). But no one has ever shown how nucleic acids would (a) arise without prior informational protein or (b) yield protein without the machinery of protein and ATP. In view of the ability of proteinoid to make internucleotide bonds (*2*) plus other experiments reported, there can be visualized an evolution of self-ordering amino acids→proteinoids→a nontemplated protein synthesis (*40, 41*)→a templated protein synthesis (Fig. 2).

Acknowledgment

The research yielding these results and interpretations has been supported by Grant No. NGR 10-007-008 of the National Aeronautics and Space Administration. Contribiuton No. 340 of the Institute for Molecular and Cellular Evolution.

REFERENCES

1.   Nakashima, T. and Fox, S.W.: 1980. *J. Mol. Evol.* 15, 161–168.
2.   Jungck, J.R. and Fox, S.W.: 1973. *Naturwissenschaften* 60, 425–427.
3.   White, D.H.: 1980. *J. Mol. Evol.* 16, 121–147.
4.   Nirenberg, M. and Leder, P.: 1964. *Science* 145, 1399–1407.
5.   Miller, S. and Orgel, L.: 1974. *The Origins of Life on the Earth*, p. 152, Prentice-Hall, Inc., Englewood Cliffs, NJ.
6.   Fox, S.W.: 1978. *The Nature of Life* (Heidcamp, W.H., ed.) pp. 23–92, University Park Press, Baltimore.
7.   Fox, S.W.: 1980. *Naturwissenschaften* 67, 378–383, 576–581.

8. Nakashima, T., Jungck, J.R., Fox, S.W., Lederer, E., and Das, B.C.: 1977. *Int. J. Quantum Chem.* QBS4, 65–72.

9. Fox, S.W. and Harada, K.: 1960. *J. Am. Chem. Soc.* 82, 3745–3751.

10. Harada, K. and Fox, S.W.: 1965. *Arch. Biochem. Biophys.* 109, 49–56.

11. Phillips, R.D. and Melius, P.: 1974. *Int. J. Peptide Protein Res.* 6, 309–319.

12. Melius, P. and Sheng, J. Y-P.: 1975. *Bioorg. Chem.* 4, 385–391.

13. Harada, K. and Fox, S.W.: 1975. *BioSystems* 7, 213–221.

14. Melius, P.: 1979. *BioSystems* 11, 125–132.

15. Hartmann, J., Brand, M.C., and Dose, K.: 1981. *BioSystems* 13, 141–147.

16. Snyder, W.D. and Fox, S.W.: 1975. *BioSystems* 7, 222–229.

17. Fox, S.W. and Dose, K.: 1977. *Molecular Evolution and the Origin of Life*, rev. ed., Marcel Dekker, New York.

18. Fox, S.W.: 1976. *Origins of Life* 7, 49–68.

19. Herrera, A.L.: 1942. *Science* 96, 14.

20. Yanagawa, H. and Egami, F.: 1978. *Origin of Life* (Noda, H., ed.) pp. 385–390, Japan Sci. Soc. Press, Tokyo.

21. Fox, S.W., Harada, K., and Kendrick, J.: 1959. *Science* 129, 1221–1223.

22. Waehneldt, T.V. and Fox, S.W.: 1968. *Biochim. Biophys. Acta* 160, 239–245.

23. Rohlfing, D.L.: 1975. *Origins of Life* 6, 203–209.

24. Smith, A.E. and Bellware, F.T.: 1966. *Science* 152, 362–363.

25. Heinz, B., Ried, W., and Pflug, H.D.: 1980. *Naturwissenschaften* 67, 178–181.

26. Young, R.S.: 1965. *The Origins of Prebiological Systems and of Their Molecular Matrices* (Fox, S.W., ed.) pp. 347–356, Academic Press, New York.

27. Florkin, M.: 1975. *Comprehensive Biochemistry*, Vol. 29B, pp. 231–260, Elsevier, Amsterdam.

28. Fox, S.W. and Nakashima, T.: 1980. *BioSystems* 12, 155–166.

29. Fox, S.W. and Nakashima, T.: 1981. *Origin of Life* (Wolman, Y., ed.) pp. 271–276, D. Reidel Publishing Co., Dordrecht.

30. Ryan, J. and Fox, S.W.: 1973. *BioSystems* 5, 115–118.

31. Hayatsu, R., Studier, M.H., and Anders, E.: 1971. *Geochim. Cosmochim. Acta* 35, 939–951.

32. Fox, S.W.: 1972. *Ann. N.Y. Acad. Sci.* 194, 71–85.

33. Saunders, M.A. and Rohlfing, D.L.: 1972. *Science* 176, 172–173.

34. Rohlfing, D.L.: 1975. *Bull. S. C. Acad. Sci.* 37, 186.

35. Ehrensvärd, G.: 1962. *Life: Origin and Development*, University of Chicago Press, Chicago.

36. Lacey, J.C., Jr. and Mullins, D.W., Jr.: 1974. *The Origin of Life and Evolutionary Biochemistry* (Dose, K., Fox, S.W., Deborin, G.A., and Pavlovskaya, T.E., eds.) pp. 311–319, Plenum Press, New York.

37. Fox, S.W.: 1974. *Mol. Cell. Biochem.* 3, 129–140.

38. Lacey, J.C. Jr., Stephens, D.P., and Fox, S.W.: 1979. *BioSystems* 11, 9–17.

39. Crick, F.H.C.: 1970. *Nature* 227, 561–563.

40. Dillon, L.S.: 1978. *The Genetic Mechanism and the Origin of Life*, p. 209, Plenum Press, New York.

41. Eigen, M. and Schuster, P.: 1978. *Naturwissenschaften* 65, 364.

Evolution of Uphill Electron Transfer in Photosynthesis

A. A. KRASNOVSKY

A. A. KRASNOVSKY was born on August 26, 1913 in Odessa (USSR). He graduated from Mendeleiev Chemical-Technological Institute in Moscow (1937). Degree candidate in chemistry (1940), doctor of biological sciences (1948). He was elected a corresponding member of the USSR Academy of Sciences (1962); full member (academician) (1976). He was elected a member of the American Society of Plant Physiology (1973), member of American Society of Photobiology (1976), Vice-President of the International Society for the Study of the Origin of Life (1977), member of the Executive Committee of the International Photobiological Society (1973–1976, 1980), member of the German Academy "Leopoldina." From 1944 up to now—in A. N. Back Institute of Biochemistry of the USSR Academy of Sciences Head of the Laboratory of Photobiochemistry—he is lecturing photobiology as Professor of the Moscow State University (1954). He was appointed as an editor-in-chief of the Journal "Biophysica" (USSR), member of the editorial board of the International Journals "Photosynthetica" and "Photobiochemistry and Photobiophysics." His main research interests: primary reactions of photosynthesis, photochemistry of chlorophylls, biosynthesis of pigments and their state of organisms, photobiology, evolution.

S table long-lived organic compounds, as well as molecular oxygen produced by plant photosynthesis, are used in metabolism of all organisms living on our planet. So, practical problems of photosynthesis include not only quantum energy conversion into potential chemical energy, but the longterm storage of that energy, which would provide its further utilization by various organisms.

During the evolution an efficient mechanism has been worked out for the conversion of the solar radiation in the chain of uphill electron transfer reactions. In contemporary photosynthetic cell the energy of light quanta absorbed by the bulk of the chlorophyll migrates to reaction centers directly participating in the electron transfer. The study of reaction centers revealed that the primary conversion of quantum energy results in the charge separation. In the chain of fast electron transfer processes occurring in the picosecond range, electrons and holes are stabilized at the ultimate molecular points of reaction center.

The evolution of uphill electron transfer in photosynthesis seems to follow the way of the most effective utilization of separated charges (electrons and electron vacances, the holes) produced under the action of light quanta of the photoreceptors. This required the development of membrane structures, which separate spatially the metabolic pathways of electrons and holes. We present here a brief outline of this concept. The pathways of the evolution of photosynthesis were considered by several authors (see, for example Refs. *1–10*).

Sources of Information

The question arises: does the origin of life depends on photosynthetic conversion of solar energy? A.I. Oparin (*6*) presented convincing arguments that photosynthesis was developed after the primary heterotrophic organisms originated their metabolism based on the use of abiogenic organic matter. But experimental approach is needed to solve the problem. Let us consider the sources of information available.

Investigation of space and planets, geology and paleontology, comparative evolutionary biology, and model experiments are the main approaches to experimental exploration of the problem.

The study of interstellar space reveals each year new organic molecules, and even possible ways of their polymerization following the tunnel mechanism. The investigation of planets by means of sample return missions has not yet revealed any types of life, but the future Mars explorations will be continued. The studies of the chemical composition and the thin structure of Precambrian rocks has shown a great many organic compounds formed abiogenically, or as a result of decomposition and penetration of biogenic organic matter. Extremely important was the discovery of microfossils resembling contemporary blue-green algae in the oldest Precambrian rocks dated 2–3 billions of years. These organisms are the most ancient species ever found by geologists.

Evolutionary and comparative biochemistry has been studying the possible ways of biological evolution and its manyfold branches. But among the organisms now existing on our planet are none filling the gap between chemical and biological evolution. The most primitive cells existing now are complex, having a reproductive apparatus based on the principle of the DNA double helix, and a great number of proteins and enzymes. There is a possibility to observe ancient metabolic patterns in the metabolism of contemporary cells. As far back as in 1939 Gaffron discovered that unicellular green algae are capable of bacterial type photosynthesis as a result of anaerobic adaptation, using the hydrogen as an electron donor, or they can evolve hydrogen under the action of light (*7*).

The most impressive experiments are the models created to simulate the photo and radiochemical reactions in the anaerobic atmosphere of the primitive

Earth. Acted upon by different kinds of radiation or electric discharges, on the mixtures of simple inorganic compounds (ammonia, water, carbon dioxide, methane) a vast number of organic compounds were discovered (see reviews *6, 8, 9*). Apart from monomeric molecules it was possible to find out polymeric molecules formed, probably, by the catalytic actions of the inorganic constituents of the Earth's crust.

Using polymeric substances, the macroscopic heterogenous structures were constructed in the laboratory and considered as models of primitive cells, *e.g.* Oparin's coacervates, Fox's microsheres, Egami's marigranules. Unfortunately, in the model experiments, it has not yet been possible to construct the simplest self-reproducing systems which may be considered a working model of the primary cell. We know rather definitely that for self-reproduction the principle of the DNA double helix won the competition, but probably along the way there were blind alleys of evolution that may be based on the use of inorganic matrices.

Is it possible to obtain knowledge of the jump from prebiological chemical development to the primary appearance of the self-reproducing cell? Perhaps the high risk science of planetary exploration will reveal the missing link, but the avenues of Precambrian geology, evolutionary biology, and creation of models in laboratory still look hopeful.

Prebiological Photochemistry

L et's consider the variety of photochemical processes may proceed on the primitive Earth.

Most investigators are of the opinion that atmo- and hydrosphere of the primitive Earth contained water, carbon dioxide, methane, ammonia, and hydrogen, apart from inorganic components of the Earth's crust. The primary atmosphere has been transparent for ultraviolet solar radiation, unlike the contemporary atmosphere which contains ozone absorbing shortwave ultraviolet (UV) up to 300 nm. At that ancient time the Earth had no oxygen atmosphere created in the course of the plant photosynthesis. Even though the oxygen could have been produced from the low-efficient water photolysis under the action of shortwave UV radiation, it was immediately bound by reduced products from the atmosphere and the primary ocean (*10*).

In that case the primary process of quantum energy conversion meant the cleavage of chemical bonds of simple molecules, and a subsequent formation of more complex organic substances storing a part of quantum energy. Under the action of radiation the covalent bonds in H_2O, NH_3, and CH_4 were broken, and free radicals and ions were formed with a subsequent formation of highly active compounds producing stable organic molecules. In Terenin's experiments (*11*) the aminoacids were produced by shortwave UV radiation of the vapours of water, ammonia, and methane. It is interesting to note that under the action of shortwave UV these organic compounds could be accumulated without their decomposition.

No doubt that inorganic constituents of the Earth's crust could play an important role. At the surface of Earth's crust minerals there could proceed an efficient inter-action of initially formed free radicals producing organic molecules, which being adsorbed had been extracted from the reaction sphere. It was found that the photo-dissociation of water and ammonia absorbed on alumina may proceed under the action of more longwave UV (up to 350 nm), as compared to shortwave UV which decomposes these molecules (12). The photochemical decomposition of ammonia forming the molecular nitrogen, was probably more efficient on inorganic matrices.

We proposed that the Earth's crust inorganic components could act as photo-sensitizers. In this case the inorganic matrix would be the photoreceptor absorbing the light quanta to use their energy in photochemical reactions to proceed. Most effective are the semiconductor photocatalysts. In this case an electron is trans-ferred by light quantum to the conductivity band followed by the charge separation. Electron and the hole can either be trapped, or can migrate by reaching the phase boundary to produce at the interface the electron-donor and electron-acceptor locuses, where the primary redox reactions may take place. Using titanium dioxide, zinc oxide, and tungsten trioxide we succeded in modelling some processes occurr-ing in plant chloroplasts (13, 14). They were models of Hill reaction—photochemical evolution of oxygen coupled to a reduction of such electron acceptors as ferric compounds, quinone, and oxygen. We have also observed photoreduction of methyl viologen having a redox potential of the hydrogen electrode (15). In the presence of bacterial hydrogenase the molecular hydrogen has been formed (16, 17). Such model reactions in some cases proceed with an uphill electron transfer (Fig. 1).

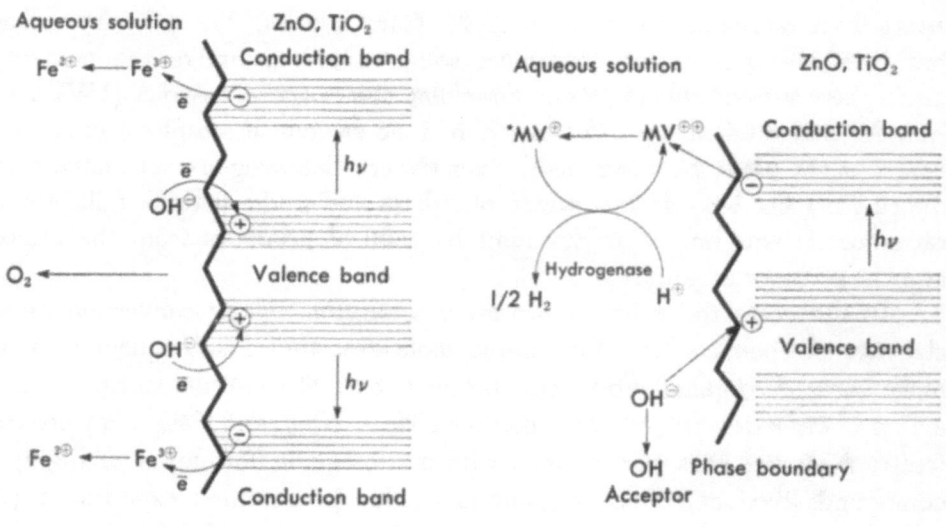

FIG. 1.   Photoinduced charge separation in the case of inorganic photocatalysts (titanium dioxide and zink oxide). Hypothetic scheme of oxygen (left) and hydrogen (right) evolution.

In the contemporary photosynthesis, probably, no inorganic photocatalysis was used, being perhaps a blind alley of the evolution. Synthesis of cycles containing a conjugated system of double bonds has been a prerequisite of the construction of extremely efficient photochemical systems, where $\pi \to \pi$ transitions might occur with the formation of excited singlet and triplet states.

Stable organic molecules produced abiogenically turned out to be more suitable as photoreceptors and photosensitizers. Among them such molecules as the porphyrins won the competition, being chemically stable and capable to complexing with metals.

In the gas mixture imitating the primary atmosphere, Hodgson and Ponnamperuma (18) observed the porphyrin formation. In our laboratory we observed the formation of porphyrins, chlorines and bacteriochlorines during pyrrol-formaldehyde interaction. The catalytic action of inorganic components of the Earth's crust was observed (19).

The ability of excited chlorophylls and porphyrins to undergo reversible photooxidation and photoreduction was studied in our laboratory for many years. As a result of electron transfer a pair of ion-radicals is primary formed and electron or a hole are probably delocalised in the π-electron system of conjugated double bonds (Fig. 2).

So, there is no doubt that during the prebiological period the solar energy

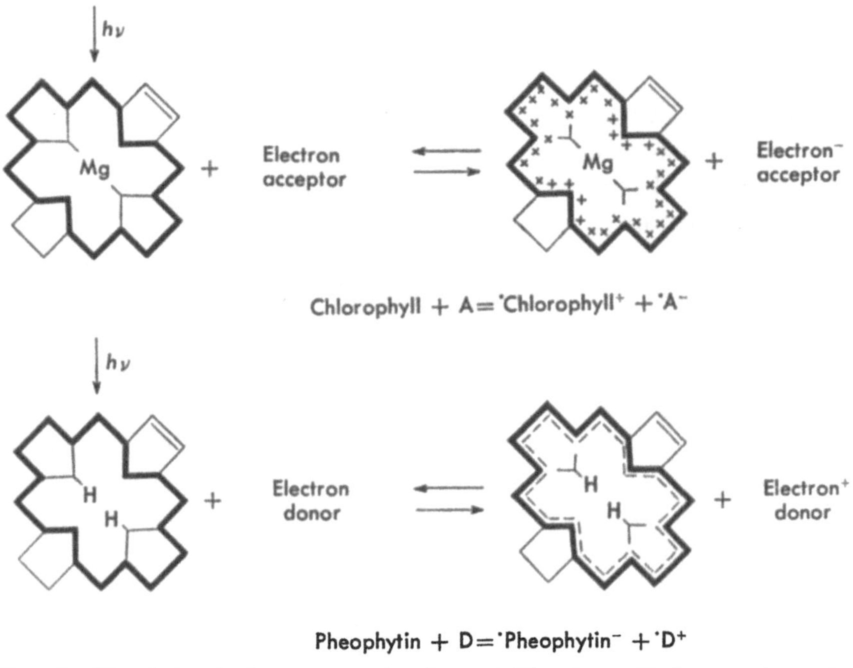

Fig. 2. Photoinduced charge separation in reversible photooxidation or photoreduction of chlorophyll and analogs.

could be converted into potential chemical energy in variety of reactions. The question is how this energy could be utilized by primary cells (probionts).

Photochemistry of Probionts

We can propose that among the primary cells there were heterotrophic and photoautotrophic variations. We accept the Baltscheffsky's hypothesis (*20*) that primary biological electron transfer might take place at the oxidoreduction level of hydrogen electrode and the further evolution followed the way to increase the redox potential of electron transferring metabolic systems up to oxygen electrode. We would like to emphasize that during this long way of evolution the use of excited photoreceptive molecules drastically increased their oxidoreductive capability. Excitation of photoreceptive molecule increased both their reductive and oxidative ability, *i.e.* to be an efficient donor or acceptor of an electron. Realization of the process depends on the redox properties of partner nonexcited molecule.

We have already suggested that activation by light of some biocatalysts in heterotrophic cells could be a kind of transition to photoautotrophy (*2*). So we have studied the light activation of NADH and NADPH aquiring the avility to reduce methyl viologen, ferredoxin and to evolve hydrogen (*21*). Recently in our laboratory it was revealed that light affects the hydrogen evolution of *Clostridium butyricum* considered as a possible model of primary heterotrophic cell. Probably flavines are the photoreceptors in this case (*22*).

Very peculiar is the ability to produce ATP by purple membranes of heterotrophic halobacteria. In this organelles the light driven proton translocation takes place but no primary photochemical charge separation has been ever seen. May be nevertheless there is an intramolecular charge separation (the shift of electron density) inside the bacteriorhodopsin molecule leading to longlived proton translocation? The question is on the situation of this type of ATP formation in the pattern of evolution. We may consider that case as one of the possible variation of the mode of activation by light of heterotrophic organisms, probably being a blind alley branch of the tree of evolution.

The incorporation of porphyrins into the primary cells increased their oxidoreductive capability being a prerequisite of efficient excitation by light and subsequent charge separation, and a crucial event in the evolution of photoautotrophy.

The concept is plausible that entire evolution of photosynthesis proceeded to find out the most effective mode of charge separation and the utilization of the stored energy by the metabolic pathways.

The study of isolated reaction centers from photosynthetic bacteria revealed that as a result of light quanta action the hole is localized at bacteriochlorophyll dimer (*23*) and electron passes through pheophytin to ubiquinone. The separated charges are longlived (up to 0.1 sec) in reaction center (Fig. 3). The electron and

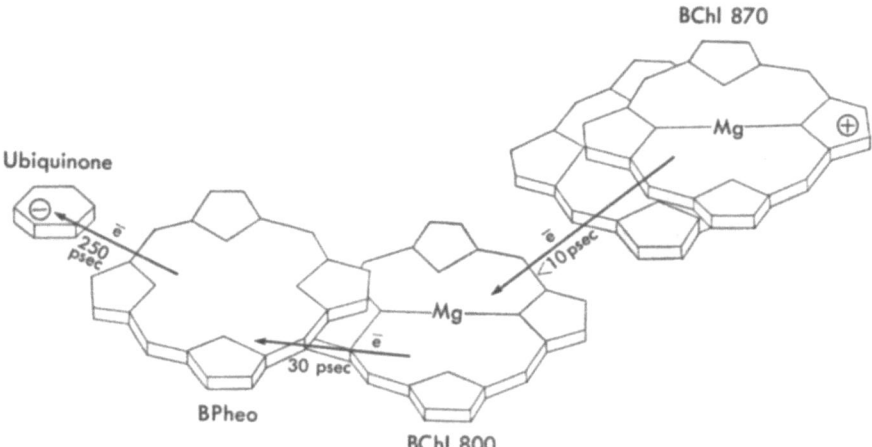

FIG. 3. Photoinduced charge separation in reaction centers of photosynthetic bacteria (see Ref. *24*).

the hole should be channelled by separate metabolic pathways. When the photoreceptor is incorporated into membrane the positive and negative charges may be localized on the different sides of the membrane. This phenomena may lead to a coupling of electron transfer with the proton translocation determined subsequent ATP formation. In this case the charge separation energy is used for the cyclic process when electron passing some intermediate carriers comes back to the hole performing synthesis of ATP.

Another possibility is noncyclic uphill electron transfer leading to ferredoxin or NADP reduction; the reduced NADH or NADPH are used in carbon dioxide reductive cycle. Both types of electron transfer are probably efficient in photosynthetic bacteria.

Noncyclic branch of electron transfer when water molecule was used as an electron donor and quinones as an electron acceptor was a great evolutionary achievement as widespread water was used as a source of electrons.

Most advanced was the coupling of two charge separation processes when the energy of two quanta was used to extract an electron from water molecule to reduce ferredoxin. Here two stages of charge separation proceeded occurring at different redox potential values. The ATP was synthesized on the way of electron through these two photosystems. As a result of molecular oxygen evolution, there happened an explosion of aerobic life on our planet.

The appearance of primitive cells inevitably required the coupling of energy conversion mechanism to the reproduction mechanism which used stored solar energy probably in the form of ATP.

So, the evolution of bioenergetic metabolism was closely connected with the evolution of nucleic acid reproduction cycles. The suggestion is plausible that in primary photoautotrophic cells the primary charge separation resulted in the synthesis of ATP used for cell reproduction. Probably this situation may be preserved in the contemporary photosynthetic bacteria or blue-green algae considered as the most ancient photosynthetic organisms.

The appearance of an efficient self-reproduction mechanism in combination with effective mechanisms of electron phototransfer seems to be the main condition which determined the origination of life based on the conversion of solar energy. If the construction of photochemical systems transferring an electron uphill can be realized *in vitro*, the creation of model self-reproducing stysems is a problem awaiting its solution still.

REFERENCES

1. Broda, E.: 1978. *The Evolution of the Bioenergetic Processes*, Pergamon Press, New York.
2. Krasnovsky, A.A.: 1959. *The Origin of Life on the Earth* (Oparin, A.I. *et al.*, eds.) pp. 606–618, Pergamon Press, New York.
3. Krasnovsky, A.A.: 1974. *The Origin of Life and Evolutionary Biochemistry* (Dose, K. *et al.*, eds) pp. 233–244, Plenum Press, New York.
4. Krasnovsky, A.A.: 1976. *Origins of Life* 7, 133–143.
5. Olson, J.M.: 1978. *Evolutionary Biology*, Vol. II (Hecht, M.K., ed.) pp. 45–68, Plenum Press, New York.
6. Oparin, A.I.: 1957. *The Origin of Life on the Earth*, USSR Academy of Sciences, Moscow (in Russian).
7. Gaffron, H.: 1972. *Horizons of Bioenergetics*, Academic Press, New York.
8. Pasynsky, A.G. and Pavlovskaya, T.E.: 1964. *Uspekhii Khimii* 33, 1198–1215 (in Russian).
9. Ponnamperuma, C.: 1972. *The Origin of Life*, Dutton Publ., New York.
10. Rutten, M.G.: 1971. *The Origin of Life*, Elsevier, Amsterdam.
11. Terenin, A.N.: 1959. *The Origin of Life on the Earth* (Oparin, A.I. *et al.*, eds.) pp. 136–139, Pergamon Press, New York.
12. Vilesov, F.I., Kotelnikov, V.A., and Lisachenko, A.A.: 1970. *Molecular Photonics*, pp. 318–334, Nauka, Leningrad (in Russian).
13. Krasnovsky, A.A. and Brin, G.P.: 1962. *Dokl. AN SSSR* 147, 656–659 (in Russian).
14. Krasnovsky, A.A. and Brin, G.P.: 1978. *Photosynthetic Oxygen Evolution* (Metzner, H., ed.) pp. 405–410, Academic Press, New York.
15. Krasnovsky, A.A. and Brin, G.P.: 1973. *Dokl. AN SSSR* 213, 1431–1434 (in Russian).
16. Krasnovsky, A.A., Brin, G.P., and Nikandrov, V.V.: 1976. *Dokl. AN SSSR* 229, 290–293 (in Russian).
17. Krasnovsky, A.A.: 1979. *Topics in Photosynthesis*, Vol. III (Barber, J., ed.) pp. 281–298, Elsevier, Amsterdam.
18. Hodgson, G.W. and Ponnamperuma, C.: 1968. *Proc. Natl. Acad. Sci. USA* 59, 22–28.
19. Krasnovsky, A.A. and Umrikhina, A.V.: 1972. *Molecular Evolution: Prebiological and Biological* (Rohlfing, D.L. and Oparin, A.I., eds.) pp. 141–150, Plenum Press, New York.
20. Baltscheffsky, H.: 1974. *Origins of Life* 5, 387–395.

21. Nikandrov, V.V., Brin, G.P., and Krasnovsky, A.A.: 1978. *Biokhimiya* 43, 636–645 (in Russian).

22. Zhukova, L.V., Nikandrov, V.V., and Krasnovsky, A.A.: 1980. *Biofizika* 25, 1095–1096 (in Russian).

23. Dutton, P.L., Kaufmann, K.J., Chance, B., and Rentzepis, P.M.: 1975. *FEBS Lett.* 60, 275–280.

24. Shuvalov, V.A., Krasnovsky, A.A.: 1981. *Biofizika* 26, 544–556 (in Russian).

Antiquity of Photoauto-trophy: A Frontier Revisited

MANFRED SCHIDLOWSKI

M. SCHIDLOWSKI was born in 1933, and received his Ph.D. degree in geology at Free University, West-Berlin, in 1961. He was a postdoctoral fellow at the Geology Department, University of Pretoria in 1961, a geologist in the South African mining industry from 1962 to 1963, and a research assistant and later "Privatdozent" at the Earth Science Department, Göttingen and Heidelberg, from 1963 to 1969. Since 1969, he is in charge of the Paleoenvironmental Research Group, Department of Air Chemistry, Max-Planck-Institut für Chemie, Mainz. From 1972 to 1973, he was a Honorary Research Fellow, Department of Geological Sciences, Harvard University; from 1979 to 1980, a member of Precambrian Paleobiology Research Group, University of California, Los Angeles. Since 1976, he is also adjunct professor at the Earth Science Department, University of Heidelberg.

Conditions on the Earth's surface are largely controlled by life processes, primarily as a result of a marked accumulation of negentropy by living systems imposing a thermodynamic gradient on all terrestrial near-surface environments. This gradient is the driving force for a number of important geochemical transformations in the exogenic cycle, a conspicuous example being the reactions triggered by free oxygen as by-product of water-splitting photosynthesis. Since the manifestations of biologically mediated geochemical activity are preserved in sediments, the respective features are propagated into the rock section of geochemical cycle; notably, the cycles of carbon, sulfur, and oxygen appear to be largely governed by the terrestrial biosphere. This feedback to the sedimentary record allows the biogeochemical evidence to be traced back in time, specifically into those remote sections of the record where unequivocal morphological vestiges of life are either missing or deplorably scanty. In the following, a brief review will be given of the history of the terrestrial carbon cycle, and of the potential of carbon isotope work in ancient sediments in efforts to temporally constrain the beginnings of photosynthetic carbon fixation, *i.e.*, the biochemical process ultimately responsible for the impact of life on the terrestrial environment.

The Earliest Record of Life

Prior to the appearance of the oldest metazoan faunas about 0.7×10^9 years ago, the paleontological record rests on two categories of evidence:

1) *Cellular structures* of microscopic dimensions, primarily of microfloral affinities (bacteria, algae). Excellently preserved microfloras have been reported from a fair number of Precambrian (notably Late Proterozoic) sediments, mostly from cherty facies. Until recently, the oldest demonstrably biogenic microstructures were those from the $\sim 2 \times 10^9$-year old Guntflint Iron Formation and the roughly coeval Pokegama Quartzite in North America which terminated the *generally accepted* cellular record of early life. Reports of Archaean microbiota, if not retracted (*e.g.*, the bacterial contaminants in earlier electron microscopic work on Swaziland sediment (*1*)), were usually viewed with scepticism. For details the reader is referred to the reviews by Schopf (*2*) and Cloud (*3*).

2) *Stromatolites*. These are not fossils in the proper sense, but lithified bio-sedimentary structures formed by mat-building communities of procaryotic microorganisms, mostly cyanophytes ("cyanobacteria"). The structures consist of sequences of bun-shaped interfering laminae representing consecutive stages in the growth of such communities. Although cellular relics of the original microscopic mat-builders have not been detected in all instances, the stromatolitic structures form their macroscopic "footprints". An encounter of such structures in the record would allow the minimum statement that benthic procaryotes were extant at the time when these sediments were formed. There are numerous classical occurrences of stromatolites in the Precambrian, the most ancient reported until lately being those from the 3.1×10^9-year old Pongola Supergroup of South Africa (*3*).

Biogeochemical Consequences of the Evolution of Life

It follows from our introductory considerations that, apart from the morphological record, there is also a geochemical record of life the most striking manifestation of which is the *impact of biosphere as a whole on the terrestiral carbon cycle*. Carbon is stored in the Earth's sedimentary shell as two different carbon species, namely,

1)   as *oxidized* or *carbonate carbon* (C_{carb}) as in limestones and dolomites, and
2)   as *reduced* or *organic* carbon (C_{org}), mostly in the form of *kerogen* which is defined as the acid-insoluble, polycondensed end-product of diagenetic alteration of sedimentary organic matter derived from living organisms and the products of their metabolism.

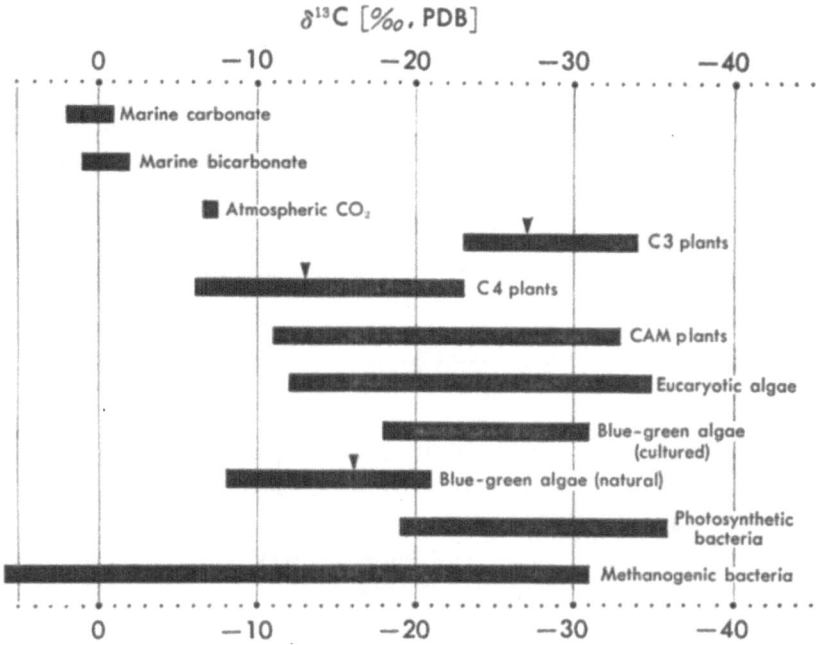

$\delta^{13}C$ [‰, PDB]

Fig. 1. Isotopic composition of higher plants, algae, and autotrophic bacteria compared to the environmental reservoirs of marine bicarbonate (HCO_3^-) and atmospheric carbon dioxide (CO_2) (4). $^{13}C/^{12}C$ ratios are expressed as $\delta^{13}C$ values, indicating differences in ^{13}C content of a sample (in permil) relative to the PDB standard (see footnote). As is indicated by the negative readings yielded by plant matter, organic substances are, on average, depleted in heavy carbon (^{13}C) by about 20–30 ‰ as compared to oceanic bicarbonate which is the most abundant inorganic carbon species in our environment (the reservoir of atmospheric CO_2 being smaller by almost two orders of magnitude). Note that the large spread of organic values contrasts markedly with the limited range for bicarbonate which is subsequently preserved in $\delta^{13}C$ values of sedimentary carbonates with a minor shift of about +1 ‰.

When looking at the precursors of these two carbon species in our contemporary environment, we will note a characteristic isotopic fractionation* between the principal forms of oxidized and reduced carbon (Fig. 1). Although displaying relatively large variations in their isotopic compositions, extant primary producers (higher plants, algae, autotrophic bacteria) are consistently depleted in heavy carbon (^{13}C) as compared to both oceanic bicarbonate (HCO_3^-) and atmospheric carbon dioxide (CO_2) (HCO_3^- is the prevalent form of inorganic carbon in the environment while the CO_2 of the atmosphere is just equivalent to about 1/60 of the bicarbonate reservoir). Accordingly, formation of reduced carbon compounds from a primary HCO_3^-–CO_2 pool must entail a sizable isotope fractionation with preferential accumulation of light carbon (^{12}C) in the organic substances.

---

* Differences in the relative abundances of the two stable carbon isotopes ^{12}C and ^{13}C are expressed in the conventional δ-notation, with

$$\delta^{13}C = \left[\frac{(^{13}C/^{12}C)_{sa}}{(^{13}C/^{12}C)_{st}} - 1 \right] \times 1,000 \quad [\text{‰, PDB}],$$

where sa=sample and st=standard. The standard is usually Peedee belemnite (PDB).

Low-temperature reduction of oxidized carbon as proceeding in terrestrial near-surface environments is, in essence, photosynthetic fixation of carbon dioxide. The principal mechanism by which this is achieved is known as the reductive pentose phosphate cycle ("Calvin cycle"), and the initial reaction in this cycle is the enzymatic fixation of CO_2 as a 3-carbon carboxylic acid (phosphoglycerate).

Two essential steps in biological carbon assimilation are (1) the (reversible) diffusion of external CO_2 into the plant tissue, and (2) the first carboxylation reaction by which "internal" CO_2 is being fed into the Calvin cycle, i.e.,

$$CO_{2(ext)} \underset{k_2}{\overset{k_1}{\rightleftharpoons}} CO_{2(int)} \overset{k_3}{\longrightarrow} R\text{--}COOH$$

with k_{1-3} standing for the rate constants of the respective reactions.

Inherent in both the enzymatic carboxylation reaction and in the preceding uptake of CO_2 and its diffusion to the carboxylation sites is a kinetic isotope effect which favors the reaction rates, and thereby preferential metabolization, of the light carbon isotope (^{12}C). Fractionation in the enzymatic step has been shown to be considerably larger (mostly between -17 and $-40‰$) than during CO_2 uptake and intracellular transport ($-4‰$) (for a detailed discussion see Ref. 4). Depending on which of the two reactions becomes rate-controlling in the specific instance, total fractionation may vary according to plant type and environment (Fig. 1). As a whole, the terrestrial biomass seems to be enriched by $25\pm5‰$ in light carbon as compared to the inorganic carbon pool of the atmosphere-ocean-crust system mainly composed of carbonate and bicarbonate with $\delta^{13}C \approx 0‰$.

Isotope Clues to the History of the Carbon Cycle

With both organic carbon and carbonate carbon preserved in sediments, we have a handle to trace the biological fractionation of the stable carbon isotopes back into the geologic past. The isotopic composition of marine bicarbonate has been always monitored by sedimentary carbonates with but minor changes ($\sim1‰$, cf. Fig. 1) while diagenetic alteration of organic matter may entail isotope shifts of several permil, notably as a result of the reconstitution of the organics along their maturation pathway in the sediment. Nevertheless, changes in the isotopic composition of the diagenetically stabilized kerogen fraction very rarely obscure the biological derivation of the precursor materials.

When tracing back in time the isotopic composition of both carbon species, we obtain a record as depicted in Figs. 2 and 3. It is apparent that the $\delta^{13}C$ values of marine carbonates have been always closely tethered to the zero permil line, with a mean approximating $+0.4\pm2.6‰$ (5, 6). The values for sedimentary organics mostly oscillate between -20 and $-35‰$, a plausible average being $-27\pm7‰$ (14–18). With the continuity of the record dating back to 3.5×10^9 years, the frac-

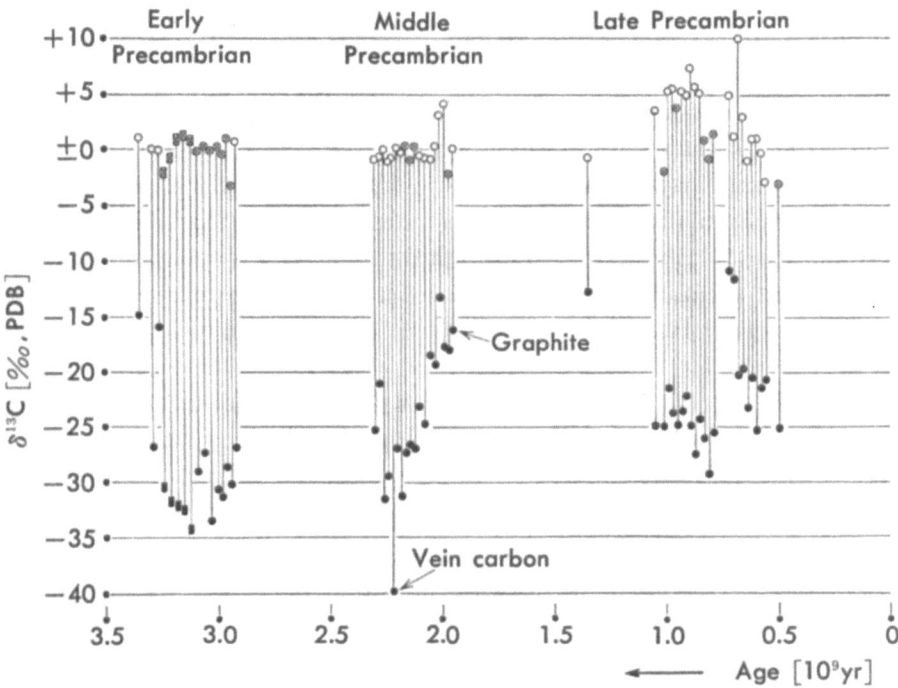

FIG. 2. Carbon isotope composition of coexisting pairs of sedimentary organic carbon (black dots) and carbonate carbon (white dots) over the last 3.5×10^9 years of the Earth's history (isotope values of this kind constitute the "jewels" of the available data base since C_{org} and C_{carb} occur in the same sample). Note that carbonate values lie close to zero permil ($\delta^{13}C_{carb} = +0.9 \pm 2.7$ ‰) while respective values for organic carbon display a considerable scatter around -25 ‰ ($\delta^{13}C_{org} = -24.7 \pm 6.0$ ‰), yielding an average fractionation $\Delta\delta = \delta^{13}C_{org} - \delta^{13}C_{carb} = -25.6$ ‰ (From Ref. *14*).

tionation between C_{org} and C_{carb} has, therefore, always been on average 25–27‰ over this time span, *i.e.*, of the same magnitude as observed in the contemporary environment (*cf.* Fig. 1).

As pointed out before, the prime responsibility for the characteristic enrichment of ^{12}C in ancient kerogens rests with the process which gave rise to the biological precursor materials and, in particular, with the step of enzymatic carboxylation during the Calvin cycle. Accordingly, the sedimentary $\delta^{13}C_{org}$ record would give a remarkably consistent isotopic signal of photoautotrophic carbon fixation on Earth as from 3.5×10^9 years ago. The uniformity of this signal would, furthermore, suggest an extreme conservatism of the basic enzymatic pathways of biological carbon fixation, notably those of the Calvin cycle (*4*).

For this concept to be invalidated, we have to postulate an inorganic geochemical process that had mimicked the carbon isotope fractionation in biological carbon fixation with almost incredible precision. There are, in principle, two processes which could be credited with an abiological synthesis of reduced carbon compounds, but both fail to meet completely the constraints of the available data.

Fischer-Tropsch processes have yielded fractionations considerably in excess of those of photosynthesis (between −50 and −100‰, *(7)*). Further, reactions of this type are high-temperature processes (∼400°K and more) which are unlikely to proceed in normal sedimentary environments. On the other hand, *Miller-Urey spark discharge syntheses* have consistently furnished values near the lower limit of biological isotope effects (between −4 and −12‰ for the bulk of the reaction products *(8)*), thus just marginally overlapping the range attributed to biological fractionations. We may safely assume that substantially reduced fractionations of this magnitude would lend themselves to a quick detection in unmetamorphosed sediments, their virtual absence precluding contributions from such sources to the kerogen content of ancient rocks.

Finally, it might be pointed out that sedimentary carbon isotopes have displayed a spectacular predictive power, having allowed first powerful conjectures about a possible involvement of the biosphere in the Archaean carbon cycle. When committing our notion to print that the apparent stabilization of the terrestrial carbon cycle as from $t \gtrsim 3.7 \times 10^9$ years ago was indicative of a corresponding continuity of life on Earth *(9, 10)*, the undisputed cellular record stopped at about Gunflint times (∼2 × 10⁹ years) and the oldest stromatolites known were some 3 × 10⁹ years old *(3)*. Meanwhile, a detailed investigation of the Warrawoona

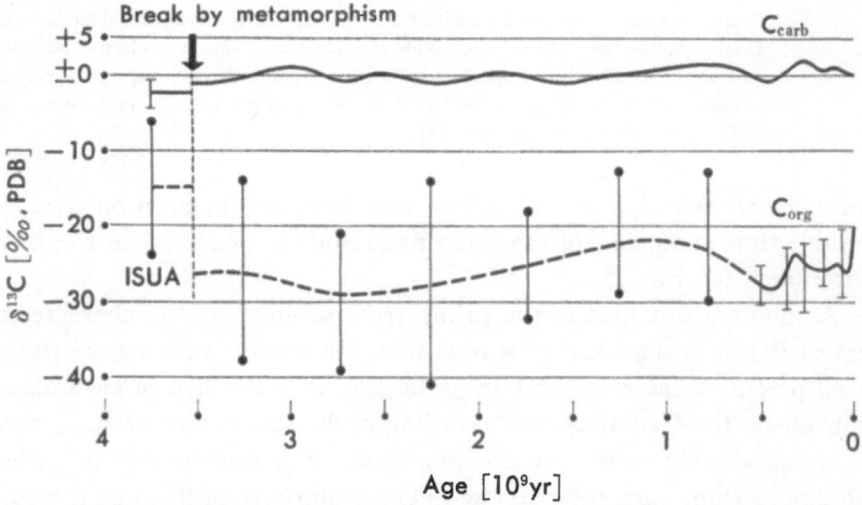

Fɪɢ. 3. Isotopic composition of carbonate carbon (C$_{carb}$) and organic carbon (C$_{org}$) over the presently known sedimentary record terminated by the Isua meta-sediments of West-Greenland (∼3.8 × 10⁹ years). Note continuity of both carbon records back to 3.5 × 10⁹ years, with carbonates close to zero permil with a small standard deviation (±2.6 ‰), and sedimentary organics oscillating around a mean between −25 and −27 ‰ with a relatively large scatter (s.d. ≈ ±7 ‰). Bars terminated by black dots denote total spread of values, others give standard deviations. The break at 3.5 × 10⁹ years B.P. is due to the amphibolite-grade metamorphism of the Isua sediments by which the original values of both carbon species have been reset as indicated. Data from Refs. *5, 6, 10, 14–18.*

Group of Western Australia (*11–13*) has pushed both records back to the benchmark of 3.5×10^9 years which hitherto terminates the occurrence of unmetamorphosed sediments, thus fully corroborating forecasts based on biogeochemical evidence alone.

The Isua Meta-Sedimentary Series: Rosetta Stone of Early Life?

The only conspicuous discontinuity in the carbon isotope record is the break between Isua and post-Isua sediments (Fig. 3). In contrast to the long-term averages for $\delta^{13}C_{carb}$ and $\delta^{13}C_{org}$ of about zero and $-26‰$, the corresponding means for the Isua sediments are -2.5 and $-15‰$, respectively.

Since the Isua suite has experienced amphibolite-grade metamorphism, there can be little doubt that the observed isotope shifts are the result of a post-depositional reconstitution of the rocks. There is ample published evidence to show that carbonates become isotopically lighter and sedimentary organics become heavier in response to progressive metamorphism (for an overview see Refs. *4, 10*). In the case of carbonates, this isotope shift is the result of metamorphic decarbonation reactions such as involved in the formation of calc-silicates like tremolite $[Ca_2Mg_5(OH)_2(Si_8O_{22})]$ which mineral is a widespread constituent of the Isua meta-sediments and is formed from dolomite $[CaMg(CO_3)_2]$ and quartz (SiO_2) by the reaction

$$5 \text{ dol} + 8 \text{ qtz} + H_2O \longrightarrow 1 \text{ tremolite} + 3 \text{ calcite} + 7 \text{ CO}_2.$$

Fractionation factors calculated by Bottinga (*19*) show that the volatile CO_2 phase released at metamorphic temperatures is enriched in ^{13}C, thus making the residual calcite $(CaCO_3)$ isotopically lighter. In the case of organic carbon, the enrichment of metamorphosed or "high-rank" kerogens in ^{13}C is mainly due to preferential cracking of $^{12}C-^{12}C$ bonds during increasing thermocatalytic stress, with subsequent mobilization of the light (usually hydrogenated) carbon fraction.

With both carbon species in the Isua suite having responded to metamorphism in a fairly predictible manner, *i.e.*, in keeping with presently available thermodynamical data and observations from other metamorphic terranes, a metamorphic interpretation of these isotope shifts is very hard to challenge. If this interpretation is correct, then both carbonate and reduced carbon in the Isua suite originally possessed δ-values close to those of their geologically younger counterparts. Hence, is should be justified to extrapolate the previous trends of $\delta^{13}C_{carb}$ and $\delta^{13}C_{org}$ right back to 3.8×10^9 years ago, with all attendant corollaries and consequences for the terrestrial carbon cycle. The principal implication would be a minimum age of 3.8×10^9 years for the beginnings of photoautotrophy as the quantitatively most important process of biological carbon fixation. Such conclusion would be consistent with results of (hitherto unpublished) C_{org} assays showing the reduced carbon content of Isua rocks ($\sim$0.5%) to lie within the range of geologically younger sediments. Microstructures of suspected biological affinities recently described by Pflug and Jaeschke-Boyer (20) from the Isua suite are objects of a current controversy (21).

Acknowledgments

Studies leading to the ideas expressed in this paper have been sponsored by the Deutsche Forschungsgemeinschaft (SFB-73). They have, furthermore, benefited from my association with the Precambrian Paleobiology Research Group, University of California, Los Angeles, supported by NASA Grant NSG 7489 and NSF Grant DEB 77-225/B (Waterman Award) to J.W. Schopf.

REFERENCES

1. Barghoorn, E.S. and Schopf, J.W.: 1966. *Science* 152, 758–763.
2. Schopf, J.W.: 1975. *Annu. Rev. Earth Planet. Sci.* 3, 213–249.
3. Cloud, P.E.: 1976. *Paleobiology* 2, 351–387.
4. Schidlowski, M., Hayes, J.M., and Kaplan, I.R.: 1981. *Origin and Evolution of Earth's Earliest Biosphere* (Schopf, J.W., ed.), Princeton Univ. Press, Princeton, N.J., in press.
5. Schidlowski, M., Eichmann, R., and Junge, C.E.: 1975. *Precambrian Res.* 2, 1–69.
6. Veizer, J., Holser, W.T., and Wilgus, C.K.: 1980. *Geochim. Cosmochim. Acta* 44, 579–587.
7. Lancet, M.S. and Anders, E.: 1970. *Science* 170, 980–982.
8. Chang, S., Des Marais, D., Mack, R., Miller, S.L., and Strathearn, G.: 1981. *Origin and Evolution of Earth's Earliest Biosphere* (Schopf, J.W., ed.), Princeton Univ. Press, Princeton, N.J., in press.
9. Junge, C.E., Schidlowski, M., Eichmann, R., and Pietrek, H.: 1975. *J. Geophys. Res.* 80, 4542–4552.
10. Schidlowski, M., Appel, P.W.U., Eichmann, R., and Junge, C.E.: 1979. *Geochim. Cosmochim. Acta* 43, 189–199.
11. Dunlop, J.S.R., Muir, M.D., Milne, V.A., and Groves, D.I.: 1978. *Nature* 274, 676–678.
12. Walter, M.R., Buick, R., and Dunlop, J.S.R.: 1980. *Nature* 284, 443–445.
13. Awramik, S.M., Schopf, J.W., Walter, M.R., and Buick, R.: 1981. *Science*, in press.
14. Eichmann, R. and Schidlowski, M.: 1975. *Geochim. Cosmochim. Acta* 39, 585–595.
15. Degens, E.T.: 1969. *Organic Geochemistry* (Eglinton, G. and Murphy, M.T.J., eds.) pp. 304–329, Springer-Verlag, Berlin.
16. Galimov, E.M.: 1980. *Kerogen* (Durand, B., ed.) pp. 271–299, Editions Technip, Paris.
17. Schidlowski, M., Eichmann, R., and Fiebiger, W.: 1976. *Neues Jahrb. Miner. Mh.* 1976, 344–353.
18. Welte, D.H., Kalkreuth, W., and Hoefs, J.: 1975. *Naturwissenschaften* 62, 482–483.
19. Bottinga, Y.: 1969. *Geochim. Cosmochim. Acta* 33, 49–64.
20. Pflug, H.D. and Jaeschke-Boyer, H.: 1979. *Nature* 280, 483–486.

21. Bridgwater, D., Allaart, J.H., Schopf, J.W., Klein, C., Walter, M.R., Barghoorn, E.S., Strother, P., Knoll, A.H., and Gorman, B.E.: 1981. *Nature* 289, 51–53.

Fermentations Dependent on Accessory Oxidants, and Their Significance in Biochemical Evolution

HOWARD GEST

H. GEST received the Ph.D. degree in microbiology from Washington University (St. Louis) in 1949, and then became a member of the faculty of the Department of Microbiology of Case Western Reserve University School of Medicine. In 1959, he returned to Washington University as Professor of Microbiology, and in 1966 joined the faculty of Indiana University (Bloomington). He was designated Distinguished Professor of Microbiology, Department of Biology, Indiana University (Bloomington) in 1978. He has investigated numerous aspects of microbial physiology and biochemistry, especially with photosynthetic bacteria (N_2 fixation and H_2 metabolism, regulation of amino acid biosynthesis, photophosphorylation and other bioenergetic mechanisms, *etc.*). During the tenure of a Guggenheim Fellowship in 1970, he did research at Imperial College (London) and the University of Stockholm: and was a Visiting Professor of the Department of Biophysics and Biochemistry, University of Tokyo and the Japan Society for the Promotion of Science. A second Guggenheim Fellowship during 1979–1980 was devoted, in part, to study of problems of biochemical evolution at the University of California, Los Angeles, where he was a member of the Precambrian Paleobiology Research Group.

The biochemical evolution of energy transducing systems has been one of Professor Egami's interests for many years and this article focuses on new developments related to his conception of "nitrate fermentation (1)." This term is meant to suggest a modification of the Embden-Meyerhof-Parnas (EMP) pathway in which nitrate acts as the terminal electron acceptor of reducing equivalents generated in the triose phosphate dehydrogenase reaction. Professor Egami has suggested (2) that "nitrate fermentation" may have been an important stage in the early evolution of anaerobic energy conversion systems, leading eventually to aerobic respiration.

The Earliest Fermentation Mechanisms

The special significance of sugars in biochemical evolution has been aptly summarized by Quayle and Ferenci (*3*) as follows: "The occurrence of sugars in
the macromolecules (nucleic acids, polysaccharides, membranes, cell walls, *etc.*)
and intermediary metabolites of all known cells attests to the importance of sugars
in the earliest development of self-replicating organisms." In addition, the relative
simplicity of the energy conversion mechanism associated with sugar fermentation,
"substrate level" phosphorylation, supports the view that a process of this kind
was the first energy conservation scheme successfully exploited by primeval cells
growing in the biosphere of the anaerobic Earth.

To most biochemists, "fermentation" suggests only the classical energy-yielding
mechanisms by which glucose is anaerobically converted to lactic acid, or ethanol
plus CO_2. This is probably due in large measure to the fact that elucidation of
these particular fermentations constituted one of the major thrusts of modern biochemistry. The processes noted, however, are relatively simple in comparison with
the fermentation patterns of typical saccharolytic clostridia and various other
anaerobes that produce a multitude of extracellular end-products from sugars.

What is the significance of the various "complex" fermentation patterns
observed in contemporary anaerobes? In my view, these represent the evolutionary
results of natural "experiments" aimed at solving the problem of achieving oxida-
tion-reduction (O/R) balance in the most propitious way. For each molecule of
sugar fermented *via* the EMP pathway, 2 $NADH_2$ must be reoxidized to regenerate
the NAD required for continued fermentation. In the "classic" fermentations this
occurs simply by reduction of organic fermentation intermediates, notably pyruvate
or an immediate derivative such as acetaldehyde or acetyl-Coenzyme A.

The disposal of H from $NADH_2$ can be seen as a crucial feature in the sugar
metabolism of growing fermentative anaerobes. If this is obligatorily coupled to
pyruvate reduction, this important intermediate (or its precursor phosphoenol-
pyruvate (PEP)) becomes unavailable for biosynthesis. I envisage that one of the
early tides of metabolic evolution consisted of "trials" to improve efficiency by
minimizing the use of pyruvate simply as a H (electron) "dump." The large variety
of bacterial fermentations now known suggests that the "trials" must have been
very extensive. It would appear that only the more successful of these survived in
the forms of the many anaerobes that show "specialized" fermentation schemes.

Accessory Oxidant-dependent Fermentations

The most familiar fermentations can be regarded as being "internally balanced"
from the standpoint of O/R, that is, the terminal electron acceptor for the
energy-yielding O/R reaction of the fermentation mechanism is generated from

the fermentation substrate. Do fermentations exist that are balanced "externally," *i.e.*, by reduction of an oxidant *not* derived from the organic fermentation substrate? Professor Egami and his colleagues were very prescient in suggesting, in 1957, that certain kinds of clostridia (as well as other systems) may use inorganic nitrate in this fashion (*1*). Thus, they proposed the term "nitrate fermentation" to describe sugar fermentation in which $NADH_2$ is recycled to NAD by H transfer to a soluble system that reduces nitrate to nitrite. This interesting idea of "inorganic fermentations," supported only by rather indirect experimental observations, was interpreted in subsequent papers (*2*) as an important stage in the early evolution of energy-transducing systems.

It is reasonable to assume that biological energy conversion mechanisms evolved from relatively inefficient anaerobic fermentations to more efficient processes, and that this occurred through a lengthy sequence of gradual and interdependent improvements in both bioenergetic and biosynthetic capacities. This consideration underlies the author's recent theory (*4*) on the evolution of fermentation mechanisms and the origin of anaerobic photosynthetic energy conversion. In essence, the theory suggests that the earliest fermentations were internally balanced in respect to O/R and became altered through the employment of "accessory" oxidants, thus sparing pyruvate (or PEP) for biosynthetic use. Although nitrate could, in principle, serve the function of an accessory oxidant, the theory is modeled on exploitation of CO_2 for this purpose. Thus, carboxylation of pyruvate (or PEP) supplies C_4 dicarboxylic acids that function as electron "sinks":

$$CO_2+\text{pyruvate (or PEP)} \longrightarrow \text{oxaloacetate} \xrightarrow{2H} \text{malate} \longrightarrow \text{fumarate} \xrightarrow{2H} \text{succinate.}$$

This sequence can fulfill the O/R requirements of glucose fermentation, using only one of the C_3 moieties derived from cleavage of the C_6 sugar chain. Thus, the other C_3 fragment is potentially available for synthesis of cell materials. The reductive conversion of oxaloacetate to succinate is construed as a reaction sequence that later was modified to operate in the reverse direction as part of the "oxidative" citric acid cycle. The wide-spread occurrence of the capacity to use fumarate as an electron sink is fermentative metabolism is striking, being found in numerous and diverse prokaryotes (*5*) as well as in many invertebrates that can survive long periods in essentially anaerobic habitats (*6*).

Dark Fermentation Mechanisms in Photosynthetic Bacteria

A number of photosynthetic bacteria can utilize sugars as sole carbon sources for anaerobic *photoheterotrophic* growth, but common experience over many years indicated that sugars could not support dark anaerobic multiplication. It was, consequently, somewhat surprising when Yen and Marrs (*7*) described experiments which demonstrated that *Rhodopseudomonas capsulata* and closely related species can grow anaerobically in darkness in synthetic media with glucose as the sole carbon and energy source provided that dimethyl sulfoxide (DMSO) is supplied. During growth, the DMSO is reduced to dimethyl sulfide, leading Yen and Marrs to suggest that under the conditions described, ATP is produced either in association with an "anaerobic respiration" in which DMSO functions as the terminal electron acceptor or by an unusual fermentation (substrate-level phosphorylation) process that requires an accessory oxidant. More detailed studies with *R. capsulata* using trimethylamine-N-oxide (TMAO) as the oxidant, in place of DMSO, have established that the energy-yielding process involved is indeed a fermentation dependent on an accessory oxidant (*8–10*). Thus: (i) only potentially fermentable substrates serve as energy sources for TMAO-dependent dark anaerobic growth, (ii) anaerobic dark growth of *R. capsulata* on fructose plus TMAO is characterized by the formation of classical fermentation end-products in substantial amounts, as in numerous fermentations catalyzed by heterotrophic anaerobes or facultative anaerobes, and (iii) the electron transport system responsible for the oxidation of $NADH_2$ by TMAO in *R. capsulata* is localized in the cytoplasmic (soluble) fraction of the cell.

Why does fermentative growth of *R. capsulata* on sugars require an accessory oxidant? This organism apparently contains virtually all of the "classical" fermentation machinery found in many heterotrophic anaerobes, but evidently cannot effectively manage O/R balance using only fermentation intermediates as terminal electron acceptors. For reasons still unknown, during anaerobic dark growth on hexoses the organism is unable to reoxidize $NADH_2$ by making highly reduced organic end-products (or H_2), and an exogenous "accessory" electron acceptor

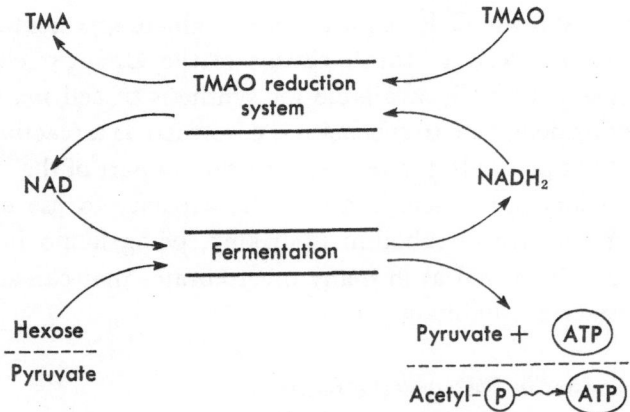

Fig. 1. Schematic representation of the dark accessory oxidant-dependent fermentative metabolism of *R. capsulata*. Energy-yielding catabolism of hexose to pyruvate and the conversion of pyruvate to acetyl phosphate (a phosphoryl donor for ATP synthesis) require NAD as a H (electron) acceptor. In order for fermentation to proceed, NAD must be continuously regenerated from $NADH_2$. This is accomplished by transfer of H to the accessory oxidant TMAO, which is thereby reduced to trimethylamine (TMA). (From Madigan *et al.* (*9*)).

(TMAO or DMSO) must therefore be provided. The essence of this interpretation is shown in Fig. 1. As indicated, NAD is continually supplied to the fermentation apparatus by transfer of H from NADH$_2$ to the TMAO-reduction system. The latter functions simply as a dumping-ground, or "sink," for electrons. According to this interpretation, the energy for growth is provided by substrate level phosphorylation associated with hexose catabolism to pyruvate, and from the sequence:

$$\text{pyruvate} \longrightarrow \text{acetyl-CoA} \longrightarrow \text{acetylphosphate} \xrightarrow{\text{ADP}} \text{acetate} + \text{ATP}.$$

Evolution of Energy-transducing Systems

Early accessory oxidant systems that could effect reoxidation of NADH$_2$, produced in the course of fermentation, presumably were relatively simple and soluble (cytoplasmic), and functioned only to ensure redox balance. The ubiquity of fumarate-reducing systems and the obviously close connection of the C$_4$ dicarboxylic acids with other aspects of metabolism suggest that fumarate was an important accessory oxidant prototype. Progression from a soluble fumarate-reducing mechanism to the kind now observed in many anaerobes would have required the gradual addition of intermediary electron carriers (including cytochrome *b*) and incorporation of most of the systen into the cytoplasmic membrane. Thus, the simple original electron "sink" could have been transformed eventually to a membrane-associated system with the vectorial properties requisite for coupling phosphorylation with electron flow. Numerous fumarate-reducing anaerobes, in fact, show the capacity to couple ATP synthesis with electron transfer from NADH$_2$ to fumarate (*ca.* 1 ATP/NADH$_2$; see Refs. *5, 11*).

If we define "electrophosphorylation" as any process in which phosphorylation of ADP (yielding ATP) is dependent upon electron flow *via* membrane-bound carriers (*4*), we can say that many heterotrophic anaerobes that obtain the bulk of their ATP by conventional fermentation (*i.e.*, substrate-level phosphorylation) also operate an electrophosphorylation "module" that simultaneously facilitates O/R balance and yields additional ATP.

Recognition of the fact that various heterotrophic anaerobes contain membrane-associated electron transport systems capable of electrophosphorylation suggests a solution to the puzzle of how anaerobic photosynthetic energy conversion (of the kind now observed in purple and green bacteria) could have arisen in an anaerobic world populated only by heterotrophic fermenters. I have proposed (*4*) that this important event in biochemical evolution may have occurred through "fusion" of an electrophosphorylation-competent sequence of carriers from an accessory oxidant-dependent fermentation with a magnesium porphyrin pigment complex. Thus, useless recombination of light-induced separation of charges in the pigment complex could be circumvented by an alternative route of step-wise electron transfer, geared to conservation of energy in forms utilizable for biosynthesis.

Concluding Remarks

Figure 1 is, in fact, based on a similar diagram published by Egami *et al.* (*1*) in 1957, to describe their concept of "nitrate fermentation" in systems lacking cytochromes. In the meantime, some experimental evidence indicating catalysis of "nitrate fermentation" by certain clostridia has been obtained by Hasan and Hall (*12*). More recently, Cole and Brown (*13*) have described experiments indicating that anaerobic reduction of nitrite to ammonia during sugar fermentation by several facultative anaerobes results from nonphosphorylative oxidation of NADH$_2$ with nitrite as the electron acceptor. Although it is usually assumed that "dissimilatory" sulfate reduction by bacteria represents energy conservation by electrophosphorylation, it now appears that growth of certain *Desulfotomaculum* species on lactate+sulfate is supported by substrate-level phosphorylation (*14*); sulfate apparently functions as an accessory oxidant for the non-phosphorylative conversion of lactate to pyruvate. Additional examples of accessory oxidant-dependent fermentations by heterotrophic anaerobes are noted in Ref. *4*.

Thus, it is fitting on Professor Egami's 70th birthday that recognition of the existence of accessory oxidant-dependent fermentations of the kind he envisaged in connection with his pioneering research on nitrate metabolism has finally arrived. Moreover, his idea that such processes were important in the evolution of energy transducing systems is obviously still alive and well.

Acknowledgment

The author is the recipient of a fellowship from the John Simon Guggenheim Memorial Foundation.

REFERENCES

1. Egami, F., Ohmachi, K., Iida, K., and Taniguchi, S.: 1957. *Biokhimiya* 22, 122–134.
2. Egami, F.: 1974. *Origins of Life* 5, 405–413.
3. Quayle, J.R. and Ferenci, T.: 1978. *Microbiol. Rev.* 42, 251–273.
4. Gest, H.: 1980. *FEMS Microbiol. Lett.* 7, 73–77.
5. Thauer, R.K., Jungermann, K., and Decker, K.: 1977. *Bacteriol. Rev.* 41, 100–180.
6. Hochachka, P.W.: 1976. *Biochemical Adaptation to Environmental Change* (Smellie, R.M.S. and Pennock, J.F., eds.) (Biochem. Soc. Symp. 41) pp. 3–31, The Biochemical Society, London.
7. Yen, H.-C. and Marrs, B.: 1977. *Arch. Biochem. Biophys.* 181, 411–418.
8. Madigan, M.T. and Gest, H.: 1978. *Arch. Microbiol.* 117, 119–122.
9. Madigan, M.T., Cox, J.C., and Gest, H.: 1980. *J. Bacteriol.* 142, 908–915.
10. Cox, J.C., Madigan, M.T., Favinger, J.L., and Gest, H.: 1980. *Arch. Biochem. Biophys.* 204, 10–17.

11. Kröger, A.: 1978. *Biochim. Biophys. Acta* 505, 129–145.
12. Hasan, M. and Hall, J.B.: 1977. *Z. Allg. Mikrobiol.* 17, 501–506.
13. Cole, J.A. and Brown, C.M.: 1980. *FEMS Microbiol. Lett.* 7, 65–72.
14. Liu, C.-L. and Peck, H.D., Jr.: 1981. *J. Bacteriol.* 145, 966–973.

Comparative Cytochrome *c* Biochemistry

MARTIN D. KAMEN and TERRANCE E. MEYER

M.D. KAMEN was born in Tronto in 1913. He had all his schooling in Chicago, earning a Ph.D. degree in physical chemistry at the University in 1936 with a thesis on neutron-proton interactions. This early start in nuclear processes and chemistry brought him to the Radiation Laboratory (Berkeley, California) from 1936–1945 during which his research interests gradually shifted to isotope tracer studies on photosynthesis and comparative microbial metabolism with S. Ruben who collaborated with him in the discovery of the long-lived radioactive carbon isotope, ^{14}C. For the next decade he worked in these areas while holding various academic and administrative appointments at Washington University (St. Louis) and Brandeis University (Waltham, Massachusetts). These studies developed into a general approach to comparative biochemistry of redox components in bacterial and algal energy transduction in particular the cytochromes and non-haem iron proteins. He has pursued these interests for over two decades at the University of California (San Diego) and the University of Southern California. He held a number of administrative and editorial assignments there and in France where he headed a research unit for the CNRS at Gif-sur-Yvette (1967–1969). Recently, he became Professor Emeritus at both UCSD and USC.

Over the past two decades, studies on prokaryote cytochrome systems have provided a basis for construction of a comparative biochemistry of cytochromes, particularly those of "*c*"-type. There is a remarkable homology, in particular between the mitochondrial cytochromes *c* of eukaryotes and a certain class of prokaryotic *c*-type cytochromes ("cytochromes c_2"). Some 16 primary structures of this class can be compared with a selection of the 80-odd sequences now known for mitochondrial cytochromes *c*, as shown in Table 1. The mitochondrial examples include the whole range of aerobic animals, plants, algae, and facultative bacteria.

TABLE 1.   Percentage Difference in Amino Acid Sequence for Cytochromes c_2 and Mitochondrial Cytochromes c

Species	Percentage differences
Horse (Equus caballus)	25
Brandling worm (Eisenia foetida)	46 · 41
Wheat (Triticum aestivum)	38 · 37 · 25
Enteromorpha intestinalis	43 · 49 · 54 · 51
Euglena gracilis	44 · 45 · 50 · 46 · 45
Crithidia oncopelti	45 · 39 · 41 · 43 · 56 · 53
Saccharomyces oviformis	50 · 42 · 44 · 40 · 54 · 50 · 24
Candida crusei	44 · 38 · 41 · 41 · 55 · 50 · 25 · 22
Debaromyces kloeckeri	29 · 35 · 41 · 38 · 47 · 43 · 33 · 35 · 27
Schizosaccharomyces pombe	43 · 37 · 42 · 41 · 48 · 48 · 34 · 32 · 33 · 25
Humicola lanuginosa	34 · 39 · 45 · 41 · 47 · 55 · 33 · 35 · 35 · 30 · 23
Neurospora crassa	44 · 37 · 43 · 42 · 49 · 46 · 32 · 28 · 32 · 35 · 28 · 30
Ustilago sphaerogena	52 · 55 · 62 · 60 · 51 · 62 · 62 · 63 · 60 · 54 · 56 · 57 · 61
Tetrahymena pyriformis	58 · 62 · 63 · 61 · 62 · 58 · 65 · 66 · 63 · 57 · 60 · 62 · 66 · 59
Rhodospirillum rubrum	54 · 54 · 50 · 50 · 55 · 56 · 51 · 60 · 57 · 53 · 55 · 57 · 60 · 62 · 42
Rhodospirillum photometricum	61 · 59 · 59 · 58 · 63 · 64 · 65 · 66 · 60 · 62 · 63 · 66 · 60 · 59 · 59 · 60
Rhodopseudomonas spheroides	67 · 74 · 73 · 69 · 71 · 65 · 73 · 72 · 71 · 69 · 71 · 67 · 66 · 59 · 59 · 66 · 53
Rhodopseudomonas capsulata TJ 12	58 · 66 · 63 · 63 · 61 · 58 · 65 · 63 · 58 · 64 · 61 · 68 · 60 · 58 · 61 · 43 · 29
Rhodopseudomonas capsulata SL	61 · 62 · 65 · 59 · 67 · 62 · 64 · 66 · 64 · 63 · 62 · 63 · 62 · 64 · 56 · 54 · 48 · 44
Paracoccus denitrificans	57 · 59 · 60 · 57 · 57 · 52 · 61 · 65 · 59 · 57 · 61 · 63 · 55 · 62 · 59 · 66 · 56 · 57 · 58
Rhodopseudomonas palustris 6	50 · 51 · 49 · 50 · 56 · 53 · 58 · 59 · 57 · 49 · 52 · 54 · 54 · 53 · 52 · 59 · 66 · 58 · 62 · 48
Rhodomicrobium vannielii	46 · 50 · 52 · 49 · 55 · 47 · 52 · 54 · 55 · 51 · 50 · 55 · 45 · 55 · 59 · 63 · 64 · 59 · 60 · 53 · 47
Rhodopseudomonas viridis	53 · 51 · 54 · 56 · 53 · 55 · 57 · 58 · 60 · 51 · 56 · 54 · 52 · 53 · 55 · 63 · 62 · 61 · 52 · 43 · 42
Rhodopseudomonas acidophila	56 · 55 · 54 · 56 · 49 · 46 · 58 · 62 · 57 · 53 · 57 · 61 · 59 · 62 · 63 · 62 · 68 · 66 · 67 · 59 · 53 · 53 · 42
Rhodopseudomonas globiformis	66 · 69 · 66 · 65 · 60 · 68 · 68 · 70 · 69 · 67 · 63 · 67 · 66 · 68 · 64 · 66 · 70 · 63 · 57 · 65 · 61 · 65 · 63 · 53
Ectothiorhodospira mobilis	60 · 58 · 58 · 57 · 56 · 59 · 61 · 55 · 58 · 59 · 56 · 60 · 58 · 64 · 67 · 65 · 70 · 63 · 65 · 69 · 58 · 58 · 54 · 58 · 65 · 63
Rhodospirillum molischianum iso‑1	59 · 61 · 58 · 57 · 56 · 56 · 61 · 55 · 50 · 54 · 59 · 57 · 62 · 66 · 63 · 57 · 65 · 63 · 63 · 55 · 56 · 55 · 58 · 54 · 63 · 65 · 11
Rhodospirillum fulvum iso‑1	57 · 58 · 57 · 55 · 55 · 58 · 52 · 57 · 55 · 51 · 57 · 59 · 57 · 65 · 69 · 59 · 65 · 57 · 59 · 57 · 59 · 43 · 51 · 56 · 65 · 63 · 34 · 31
Rhodospirillum molischianum iso‑2	54 · 54 · 53 · 51 · 56 · 55 · 58 · 61 · 57 · 49 · 53 · 57 · 54 · 55 · 60 · 60 · 61 · 57 · 42 · 51 · 50 · 55 · 42 · 43 · 51 · 59 · 65 · 34 · 30
Rhodospirillum fulvum iso‑2	54 · 54 · 53 · 51 · 56 · 55 · 58 · 61 · 57 · 49 · 53 · 57 · 54 · 55 · 60 · 60 · 61 · 57 · 42 · 51 · 50 · 55 · 42 · 43 · 51 · 50 · 58 · 64 · 30 · 34 · 12

TABLE 2.   Homologies between Cytochromes c and c_2

Source			I	S	Σ
c		c_2			
Tuna	vs.	R. rubrum	36	42	78
	vs.	Rps. capsulata	36	34	68
	vs.	P. denitrificans	43	34	77
	vs.	Rps. palustris	34	34	68
	vs.	Rps. spheroides	31	43	74
	vs.	Rm. vannielii	51	35	86
	vs.	R. molischianum	39	40	79
Horse	vs.	Rm. vannielii	51	35	86
	vs.	R. viridis	55	29	84
	vs.	R. rubrum	42	39	81
	vs.	Euglena gracilis	56	34	80

Numbers shown are identities (I), identities obtainable by one-base substitutions in codons (S), and summations of both (Σ).

The data on cytochromes c_2 are those provided by the studies of R.P. Ambler and his associates (1) working with samples isolated and purified in our laboratory. The percentage differences in amino acid sequences for these two classes of c-type cytochromes show overlaps, particularly between the bacterial proteins and those from protozoa and yeast. If one takes into account conservative substitutions, as well as identities, the homologies become even more marked (Table 2). As an example, the *Rhodopseudomonas viridis* cytochrome c_2 displays 54% identity when compared to the primary structure of horse heart cytochrome c—a match obtained by only a single deletion in the total chain of 104 residues (2).

Nevertheless, there are major functional differences. Thus, the cytochromes c_2 from the purple photosynthetic bacteria are usually an order of magnitude or more less reactive in the mitochondrial oxidase system and exhibit standard redox potentials as much as 100 mV more oxidizing than those of the mitochrondial type (3, 4). The functional role of cytochrome c_2, where established, is as an electron donor in the anaerobic photophosphorylation process during cyclic electron transport in bacterial photochemical energy storage and conversion. In at least one case a cytochrome c_2 also is apparently involved in an aerobic process (5–10). In a nonphotosynthetic denitrifier, *Paracoccus denitrificans*, the cytochrome c_2 (also called "c-550") appears to operate *in vivo* in a coupled oxidative phosphorylation, analogous to the function displayed in mitochondria by cytochrome c, but is still a poor donor to mitochondrial oxidase (11).

Other classes of prokaryotic cytochromes c so far established (12) retain the characteristic 3-D structure, as in *Chlorobium* cytochrome c-555 (13) and *Pseudomonas aeruginosa* cytochrome c-551 (14), but diverge in primary structure and exhibit no functional resemblance to mitochondrial cytochrome c. Finally, we may mention two major classes of bacterial cytochromes c which have entirely different structures—the high-spin cytochromes c' and the low-spin cytochromes c_3. Many others among the soluble forms remain to be classified. In addition, membrane-bound functional c-type prokaryotic forms, of which the *Chromatium* cytochrome "c-552, c-556" is an example (*15, 16*), the eukaryotic cytochrome c_1 and the chloroplast cytochrome f remain to be characterized.

The use of biological redox systems, obtained in varying degrees of purity and involving usually only one or at most a limited range of substrate cytochrome c concentrations, has demonstrated in a qualitative, or semi-quantitative manner the widely variable functional behavior of cytochrome c from both prokaryotic and eukaryotic sources (*17*). Dr. B. Errede, in our laboratory, using a standardized "Complex IV" oxidase from beef mitochondria (*18*), has established for the first time a general rate law covering an adequate range of cytochrome c concentrations, amplifying earlier studies by Minnaert (*19*) and Hollocher (*20*). The approach used has been to evaluate binding constants and relative first order activities based on a reasonable kinetic mechanism, assuming productive intermediate complexes formed between substrate and enzyme. Comparison of these parameters using a variety of prokaryotic and eukaryotic cytochromes c and c_2 have defined the site of active binding to involve certain lysines (those numbered in the horse heart sequence as 13, 25, 27, 72, 73, 86, and 87) clustered on the face of the protein from which the haem group projects slightly. These results agree well with those obtained recently by others, using chemical modifications of these surface lysines (*21–23*) as well as in earlier studies in Japan (*24*). One of the prokaryotic proteins which most resembles horse heart cytochrome c, that of *Rhodomicrobium vannielii*, is the least reactive of any cytochrome c_2 so far examined. This fact shows that surface perturbations are not determined solely by increase in size, owing to bulk insertions such as one finds in many cytochromes c_2.

A major finding, still to be explained, is the good reactivity of cytochromes c_2 with the mitochondrial reductase (*18*). It is claimed that the active binding sites for both oxidase and reductase coincide, insofar as the relevant lysines are concerned (*25, 26*). A particularly interesting case is that of the two cytochromes c-555 from the photosynthetic green sulfur bacteria, *Chlorobium thiosulfatophilum* and *Prosthecochloris aestuarii*. The former, for which the 3-D structure has been published (*13*), shows a strategic placement of a few surface lysines for oxidase interaction but still behaves atypically in not reacting with the reductase (*18*). The latter reacts with neither (*18*). This finding suggests that some structural feature, other than lysine placement, is absent from these cytochromes as contrasted with cytochromes c and c_2. It will be most interesting to examine the 3-D structure of the *P. aestuarii* protein, when it is known, for possible clarification of this anomaly.

Because of many features—their widespread occurrence, accessibility, relatively small size, central role in bioenergetics, and solubility—cytochromes c offer unique opportunities for studies in molecular evolution. Good correlations between phylogenetic sequences deduced from primary structures and those from paleontological analyses have been found for eukaryotic cytochromes c (*27*)—a not surprising result when it is remembered that mitochondrial cytochromes c are functionally restricted to a common role as bridge between terminal redox complexes in coupled oxidative phosphorylation. However, a considerable degree of optimism is needed if one wishes to extend such correlations to the more functionally and structurally diverse prokaryotic cytochromes c. As an example, there are suggestions for evolutionary systems based on differential rates of reaction with pseudomonad nitrite reductase as compared with mitochondrial cytochrome c oxidase (17) but these have limited application because they do not include studies on concentration dependence or other kinetic parameters.

Further observations raise the possibility of on-going evolution in bacteria. Thus, in the sulfur oxidizer *Thiobacillus novellus* two c-type cytochromes exist, one of which looks like a mitochondrial cytochrome c functionally, whereas the other does not (28). Again, there is the example of an a-type oxidase in the photosynthetic facultative bacterium, *Rhodopseudomonas spheroides*, together with a simultaneously ocurring "o"-type oxidase (a variant of the B cytochrome group) and a cytochrome c_2 which appears to function with its own cytochrome a-type oxidase (5–9). Recently, a report has appeared of a mutant cytochrome c_2 in which one oxidase remains functional, while the other does not (10).

A case which has received particular attention is the denitrifying facultative bacterium, *P. denitrificans* which has been proposed on the basis of many biochemical and physiological similarities to support the notion that an evolutionary transition to the inner mitochondrial membrane from the plasmid membrane of an ancient aerobic prokaryote occurred, and that this was an ancestral form of *P. denitrificans* (29). Unfortunately, the major argument—the presence of a cytochrome c in the present bacterium—is rendered implausible because the cytochrome involved is not a c, but a c_2. It is quite homologous in size and sequence to those of the photosynthetic non-sulfur purple bacteria, *Rps. spheroides* and *Rps. capsulatus*. Comparisons of 3-D structures for different types of cytochromes, related distantly or hardly at all—mitochondrial cytochrome c, two cytochromes c_2, a pseudomonad cytochrome c-551, and *Chlorobium* cytochrome c-555, with function ranging from full activity in the mitochondrial system to none at all—show, despite great differences in chain lengths and standard redox potentials, a common overall folding pattern that is still recognizable.

It has been proposed (31) that an ancestral form of non-sulfur photosynthetic bacterium gave rise by divergent evolution to prokaryotes, with the cyclic photophosphorylation function on the one hand, and to the respiratory aerobic coupled oxidative phosphorylation of eukaryotes on the other, the two processes having existed together in the ancestral form. This occurred by loss of the photosynthetic function in the ancestral prokaryote. There are many difficulties with so simplistic a proposal, not the least being that the gross changes in cell organization between prokaryotic and eukaryotic cells are not rationalized. Moreover, the mere existence of a cytochrome c in one case and a c_2 in the other does not guarantee that they are adequate representatives of the entire electron transport system in which they are found, nor is there certainty phylogeny has not been confused irretrievably by genetic transfer (*1*). However, there is evidence that present-day *Rhodospirillaceae* include at least one strain that can grow anaerobically with nitrate in the dark (*32*) and that this strain also contains a copper protein, which has nitrite reductase activity (*33*). Moreover, *P. denitrificans* has a "cytochrome *cd* type" nitrite reductase. This leaves open the question whether the cytochrome c_2 present is there because of gene transfer or whether this is a *bona fide* case of the kind of hypothetical evolutionary process suggested above; that is, whether this bacterium is a colorless mutant of a photosynthetic bacterium related to *Rps. spheroides*, for example, or has received its cytochrome c_2 by gene transfer.

The prospects for a great enrichment of present knowledge about structure-function relations exists, in view not only of the remarkably diverse characters of the cytochromes *c* isolated already in such large numbers and over so great a range of metabolic patterns, but because of the great expansion of physico-chemical methodologies based on X-ray analyses and all the forms of spectroscopies-ultra-violet, visible, infra-red, and microwave. One attempt to characterize a 3-D structure without X-ray crystal analysis has been described—that of a *Desulfovibris vulgaris* cytochrome c_3 (*34*). Unfortunately the structure so deduced from a combination of results from EPR, NMR, and other spectroscopies on the protein in solution has not been verified by the actual structure derived from X-ray analysis of crystals (*35*). Comparison studies using NMR to probe fine structure differences between two well-characterized structures, those of horse heart cytochrome *c* and *Rhodospirillum rubrum* cytochrome c_2 have been reported (*36*).

One major obstacle is the continuing inability to develop technologies which will solve structures of bound forms as they exist in functional form in membranes. Recently, the oxidase of the prokaryote, *P. denitrificans*, has been purified (*37*). This could lead for the first time to a comparison of a prokaryotic oxidase with the mitochondrial complex. Correlation of changes in primary and tertiary structure with evolutionary history of cytochromes *c* is a more distant objective, because it is essential to investigate in depth not only interactions of various cytochromes *c* with a particular oxido-reductase system, as in mitochondria, but to extend observations, with care for proper rigor in kinetic analyses, to oxido-reductase systems of all organisms from which the selected cytochromes *c* have been derived. A beginning in this area has been described (*17*).

Acknowledgment

Data cited from our researches come from projects supported by grants from the National Science Foundation (BMS 75-13608), the Department of Energy (DE-ATO3-76ER70293) and the National Institutes of Health (GMS-18528).

REFERENCES

1. Ambler, R.P.: 1977. *The Evolution of Metalloenzymes, Metalloproteins and Related Materials* (Leigh, G.J., ed.) pp. 100–118, Symposium Press, London.
2. Ambler, R.P., Meyer, T.E., and Kamen, M.D.: 1976. *Proc. Natl. Acad. Sci. USA* 73, 472–475.
3. Pettigrew, G.W., Meyer, T.E., Bartsch, R.G., and Kamen, M.D.: 1975. *Biochim. Biophys. Acta* 430, 197–208.
4. Pettigrew, G.W., Bartsch, R.G., Meyer, T.E., and Kamen, M.D.: 1978. *Biochim. Biophys. Acta* 503, 509–523.
5. Saunders, V.A. and Jones, O.T.G.: 1974. *Biochim. Biophys. Acta* 333, 439–445.
6. Scholes, P.B., McLain, G., and Smith, L.: 1971. *Biochemistry* 10, 2072–2076.
7. Dutton, P.L. and Jackson, J.B.: 1972. *Eur. J. Biochem.* 30, 495–510.
8. Prince, R.C. and Dutton, P.L.: 1975. *Biochim. Biophys. Acta* 387, 609–613.
9. Baccarini-Melandri, A., Jones, O.T.G., and Hauska, G.: 1978. *FEBS Lett.* 86, 151–154.
10. Zannoni, D., Prince, R.C., Dutton, P.L., and Marrs, B.L.: 1980. *FEBS Lett.* 113, 289–293.
11. Smith, L., Newton, N., and Scholes, P.: 1966. *Hemes and Hemoproteins* (Chance, B., Estabrook, R.N., and Yonetani, T., eds.) pp. 395–403, Academic Press, New York.
12. Kamen, M.D. and Horio, T.: 1970. *Annu. Rev. Biochem.* 39, 673–700.
13. Korszun, Z.R. and Salemme, F.R.: 1977. *Proc. Natl. Acad. Sci. USA* 74, 5244–5247.
14. Almassy, R.J. and Dickerson, R.E.: 1978. *Proc. Natl. Acad. Sci. USA* 75, 2674–2678.
15. Cusanovich, M.A., Bartsch, R.G., and Kamen, M.D.: 1968. *Biochim. Biophys. Acta* 153, 397–417.
16. Kennel, S.J. and Kamen, M.D.: 1971. *Biochim. Biophys. Acta* 234, 458–467.
17. Yamanaka, T.: 1973. *Space Life Sci.* 4, 490–504.
18. Errede, B. and Kamen, M.D.: 1978. *Biochemistry* 17, 1015–1027.
19. Minnaert, K.: 1961. *Biochim. Biophys. Acta* 50, 23–24.
20. Hollocher, T.C.: 1964. *Nature* 202, 1006–1007.
21. Ferguson-Miller, S., Brautigan, D.L., and Margoliash, E.: 1976. *J. Biol. Chem.* 251, 1104–1115.
22. Staudenmayer, N., Smith, M.B., Smith, H.T., Spees, F.K., Jr., and Millett, F.: 1976. *Biochemistry* 15, 3198–3205.
23. Staudemayer, N., Ng, S., Smith, M.B., and Millett, G.: 1977. *Biochemistry* 16, 600–604.
24. Wada, K. and Okunuki, K.: 1968. *J. Biochem.* 64, 667–681; 1969. 66, 249–262, 263–272.
25. Ferguson-Miller, S., Brautigan, D.L., and Margoliash, E.: 1978. *J. Biol. Chem.* 253, 149–159.
26. Konig, B.W., Osheroff, N., Wilms, J., Muijsers, A.O., Dekker, H.L., and Margoliash, E.: 1980. *FEBS Lett.* 111, 395–398.
27. Langley, C.H. and Fitch, W.M.: 1974. *J. Mol. Evol.* 3, 161–177.

28.   Yamanaka, T., Takenami, S., Akiyama, N., and Okunuki, K.: 1979. *J. Biochem.* 70, 349–358.
29.   John, P. and Whatley, F.R.: 1975. *Nature* 254, 495–498.
30.   Timkovich, R. and Dickerson, R.E.: 1973. *J. Mol. Biol.* 79, 39–56.
31.   Dickerson, R.E., Timkovich, R., and Almassy, R.J.: 1976. *J. Mol. Biol.* 100, 473–491.
32.   Satoh, T., Hoshino, Y., and Kitamura, H.: 1970. *J. Biochem.* 108, 265–269.
33.   Samada, E., Satoh, T., and Kitamura, H.: 1978. *Plant Cell Physiol.* 19, 1339–1351.
34.   Dobson, C.M., Hoyle, N.J., Geraldes, C.F., Wright, P.E., and Williams, R.J.P.: 1974. *Nature* 249, 425–429.
35.   Haser, R., Pierrot, M., Frey, M., Payan, F., Astier, J.P., Bruschi, M., and LeGall, J.: 1979. *Nature* 282, 806–810.
36.   Smith, G.M. and Kamen, M.D.: 1974. *Proc. Natl. Acad. Sci. USA* 72, 4303–4306.
37.   Ludwig, B. and Schatz, G.: 1980. *Proc. Natl. Acad. Sci. USA* 77, 196–200.

Biochemical Diversity and Evolution

SEYMOUR S. COHEN

S.S. COHEN received his Ph.D. degree at Columbia in 1941 as the student of Erwin Chargaff. He then worked with Wendell Stanley at the Rockefeller Institute. In 1943 studies on the typhus vaccine took him to the Department of Pediatrics of the University of Pennsylvania, where he did his work on bacteriophage and virus-induced enzymes. He became an American Cancer Society Professor of Biochemistry in 1957 and Chairman of the Department of Therapeutic Research from 1963 to 1971, and in these positions he initiated studies on D-arabinosyl nucleosides and on polyamines. In 1971 he moved to the University of Colorado Medical Center as Professor of Microbiology and until 1976 also served the American Cancer Society as Chairman of its Council of Analysis and Projection. In 1976, he became Distinguished Professor of Pharmacological Sciences at the State University of New York at Stony Brook. He is working on polyamine biosynthesis in a virus infection of chloroplasts and has recently described a general approach to the chemotherapy of infectious disease.

The biochemists who have worked in their discipline from the 1940's to the present time have witnessed an extraordinary growth and development in this field. Only some of this group have attempted to continue to grasp the vast proliferation of biochemical knowledge and to organize this breadth of knowledge both to satisfy their own intellectual needs and to fulfill their commitment to teaching and to research. And within this still curious and now quite sparse group, only a very few, scarcely a handful in any country, have discovered a solution to containing the explosion of new phenomena and data, a solution which depends on the very breadth and diversity of the data itself. Rejecting the early easy solution of the biochemists and primitive molecular biologists of the 40's to the 60's who stressed the similarities of *Escherichia coli* and the elephant, a very few biochemists suggested that the differences among organisms, which were demonstrated within the burgeoning forest of publications and data, could be understood best by attempting to understand where and how organisms and cells arose and evolved. These biochemists believed that an attempt to imprint an evolutionary flow chart, *i.e.* the parameters of time and direction, upon biochemical diversity might and indeed did simplify the understanding and structure of the growing body of biochemical data. Professor Egami is one of those in Japan who perceived this possible solution very early and who, in following this insight, has maintained his grasp of this discipline, has thus fulfilled his commitment to his colleagues and students, and, not least, has continued to enjoy the very evolution of the field itself.

In recent years, many workers, initially students of aminoacid sequences and protein conformation, followed by specialists in the dissection of the nucleic acids, *via* hybridization, recombinant techniques, and sequence analysis, have also noted the relevance and significance of their studies to the characterization of evolutionary relationships. Although they have made important contributions to biochemical evolution, it is of interest to the history of science that few of this younger group have any interest in the relevance of the large body of data provided in the study of intermediary metabolism. And, on the other hand, the phenomena and sense of vigorous metabolic motion and transformation developed from the 30's to the 60's are now pursued mainly as details of metabolic regulation by a new generation of specialized enzymologists. To these individuals the aims, language and training of the molecular geneticists are a world apart, and both of these major fragments of biochemistry, *i.e.* metabolism and molecular biology, now appear quite separate from the outlook of the classical biochemist. Nevertheless it must be asked if the fragmentation of our discipline is not self-defeating, if a knowledge of the growth of the trunk of the tree of biochemistry and the nutritional base of that growth can not strengthen the extended and separated branches in important ways. I believe that individuals, such as Professor Egami, who span the period of the growth, from the development of intermediary metabolism after the late 30's, to molecular biology and polymer structure in the 60's, to the present, can provide an important perspective. Their view of the continuity and interconnections of our discipline helps to restrain fragmentation and strengthens personal and professional ties within the field.

In the last few years most biologists and biochemists have begun to see that attempts to analyze biochemical evolution have not been merely a sterile exercise in self-preservation when confronted by an engulfing inchoate mass of apparently isolated individual papers. Quite the contrary, the medically oriented student of biochemical diversity recognized even in the 1940's that the differences among organisms are the very basis of the therapeutic efficacy of sulfa drugs and of antibiotics, both groups of which restrain or kill bacterial parasites without damaging the host. The student oriented to general biology soon saw that animals, including biochemists, were entirely dependent on photoautotrophic organisms, and many other apparently esoteric organisms which metabolized such unusual substances as nitrite and nitrogen gas. Biochemical diversity then was a subject which encompassed such practical pursuits as the control of infectious disease and the improvement of agriculture. The study of the nature and origin of such diversity should help in solving the problems of the human condition on a shrinking planet.

For my part I have been teaching comparative biochemistry for about thirty years, and have attempted to understand developments in molecular biology to which my own work in biochemical virology has made some contributions. The possibilities of bringing theoretical and laboratory research and discovery to practical application have always intrigued me. A Department of Pharmacological Sciences, in which I now work, is a site in which basic and applied science can and sometimes do intersect. It is in fact a major vantage point from which the innumerable difficulties preventing a unity of theory and practice in medicine can be viewed.

In the past few years I have realized that the solutions to certain of our planet-wide biomedical problems in fact require the elimination of the apparently major discontinuity of biochemistry, i.e. the separation of intermediary metabolism from molecular biology. Each of these branches of biochemistry, old and new, has provided a piece of information, essential at this time to the design of therapeutic drugs capable of selectively inhibiting the multiplication of any organism pathogenic for man or other mammals (1, 2). Comparative biochemistry, particularly that branch of which has focussed on the differences in metabolic systems among numerous organisms, has defined systems essential to energy development or essential steps of metabolism necessary to provide the building blocks and polymers of the organisms (2). It is possible to examine any pathogenic organism and to define crucial differences which may define potential targets for chemotherapeutic strategies. Such potentially exploitable systems include bacteria, viruses, protozoa, yeasts, or even some metazoan parasites. For example, the characteristic and unique cell walls or ribosomes of bacteria are well known as specific targets of certain antibiotics. The ferredoxin-based anaerobic metabolism of amoebae can be exploited by certain reducible nitro-containing compounds, chitin synthesis might be attacked in arresting the development of some yeasts or the production of schistosomal eggs in a mammalian host, and so forth. The few relatively specific agents which are known to be antagonistic to virus multiplication without marked damage to the host take advantage of unique virus-determined metabolic systems (2).

Until recently a knowledge of the existence of such systems was insufficient to permit a practical approach to their inhibition. The isolation and characterization of an enzyme was far more difficult than it has recently become. The design of an inhibitor for a key enzyme or inhibitor rested on relatively primitive theory. The observation that sulfanilamide, as an analogue of p-aminobenzoic acid, was chemotherapeutically effective in blocking synthesis of folic acid in bacteria did not extend to other coenzymes, although certain inhibitors of dihydrofolate reductase were useful in the specific treatment of malaria, or of some bacterial infection (3). For many analogous enzymes, i.e. enzymes of similar function, it has not proven easy to exploit differences at the active sites (4). Thus in inhibiting the essential enzyme, thymidylate synthetase, fluorodeoxyuridylate substitutes for a substrate in a complex active site catalysing a complex and subtle reaction sequence. The mechanism of this reaction has imposed similarities on the structure of this site in different thymidylate synthetases. The similarities minimize the possible chemical variations of a molecule which can occupy and inhibit this catalytic site. For this reason, it will take a sophisticated chemical and enzymological effort to design inhibitory site-reactive agents capable of discriminating between mammalian and parasitic thymidylate synthetases, and indeed between other pairs of analogous essential enzymes as well.

Nevertheless modern molecular biology has provided the assurances that the structures of these analogous enzymes, i.e. both their active sites and other domains, will be sufficiently different to warrant such research efforts. The absence of serological crossreactions between proteins of host and parasite has been confirmed and extended by the absence of cross hydridization among the nucleic acids and ultimately by the major differences in base compositions and sequences of the DNAs of the parasite and its host. Thus analogous functions of the enzymes and other proteins are fulfilled by proteins which are necessarily structurally different and only distantly related at most. In recent years, molecular biology has assured us that organisms as distant as parasites and their hosts contain different nucleic acids and therefore different proteins. The latter can provide the specific targets pinpointed as essential by the comparative metabolic studies of an older biochemistry.

Although we have known for twenty-five years that the proteins of parasites e.g. induced by viruses, were different from host proteins (5), it is only in the past decade that the technology and science of protein chemistry have achieved the competence to make a strategy based on inhibiting proteins appear practicable. The isolation of pure proteins and enzymes from very small amounts of complex mixtures is now well developed, as well as the definition of amino acid sequence, surface conformation and side chain reactivities. With such information the synthesis of complementary tight-binding or irreversible inhibitors has become feasible. For example three of the surface antigenic sites of lysozyme have been simulated in synthetic polypeptides and these have been used to design complementary sequences of other specifically binding polypeptides (6). Significant progress has been attained in improving the penetrability into cells of a peptide that prevents crystallization of sickle cell hemoglobin or another that irreversibly inhibits a polio virus-induced protease.

Our ability to devise inhibitors specific for the essential enzymes or proteins of defined organisms comes directly from the continuity of development of biological, biochemical, and chemical knowledge over the past forty years. Such a capability can be the foundation of a practical strategy for the development of a specific chemotherapy for any infectious disease. It is recognizable for the most part primarily by individuals prepared to look at the long term growth of biochemistry and other related disciplines and to overcome the discontinuities and fragmentation of their own discipline. The realization of such a capability, as in the development of antiviral agents in the developed world, or, by the discovery of drugs capable of effectively treating and/or preventing malaria, trypanosomiasis, leishmaniasis, or schistosomiasis in the developing world, turns on the evolution of policies of using the biochemical science which is now emerging. It seems reasonable to propose that the biochemists who have contributed to the maturation of their science should also participate in the formulation of policies for the application of this science.

REFERENCES

1. Cohen, S.S.: 1977. *Science* 197, 431–432.
2. Cohen, S.S.: 1979. *Science* 205, 964–971.
3. Albert, A.: 1973. *Selective Toxicity*, Chapman and Hall, London.
4. Mansour, T.E.: 1962. *J. Pharmacol. Exp. Ther.* 135, 94–101.
5. Flaks, J.G. and Cohen, S.S.: 1957. *Biochim. Biophys. Acta* 25, 667–668.
6. Atassi, M.Z. and Zablocki, W.: 1977. *J. Biol. Chem.* 252, 8784–8787.

Trace Elements and the Synthesis of Acetate by *Clostridium thermoaceticum*

LARS G. LJUNGDAHL

L. G. LJUNGDAHL was born in Stockholm in 1926. He studied chemical engineering at Stockholm Technical Institute, and received B.S. degree in 1945. He was a research technician at the Department of Biochemistry, Karolinska Institute, Stockholm, with Pehr Edman and Eric Jorpes from 1943 to 1946, a research chemist at Stockholm Brewery Company from 1947 to 1958. In 1958, he moved to Western Reserve University, Cleveland, Ohio. He received the Ph.D. degree from Western Reserve University in 1964 under Harland G. Wood with a thesis on "Fixation of carbon dioxide and formation of acetate by *Clostridium thermoaceticum*," and then became a member of the faculty of the Department of Biochemistry, Western Reserve University. He moved to University of Georgia, Athens, in 1967 where he is now Professor of Biochemistry and Microbiology. He received the Humboldt-Preis in 1974 from Alexander von Humboldt Stiftung, Bundesrepublik Deutschland, and was Visiting Professor of Universität Göttingen (Institut für Mikrobiologie), Göttingen, from 1974 to 1975.

The subject of this essay has been chosen to reflect Professor Egami's interest and contribution with regard to trace elements and the origin and evolution of life. I became aware of this interest of Dr. Egami's during a meeting in 1975 when he showed an exceptional curiosity in a finding by Jan Andreesen from Göttingen, Germany. Jan Andreesen had, in my laboratory, just discovered a possible involvement of tungsten in formate dehydrogenase in *Clostridium thermoaceticum*. It is correct to say that Dr. Egami greatly increased my interest in the origin of life, and it is an honor and a pleasure to be allowed to contribute to the celebration of Dr. Egami's 70th birthday.

C. thermoaceticum, a small rod-shaped anaerobic and thermophilic bacterium, was discovered by Fontaine *et al.* (*1*). It has the capacity to ferment one mol of glucose or fructose to three mol of acetate, one of which is apparently synthesized from carbon dioxide (2, 3). This bacterium was introduced to me by Harland G. Wood when I arrived to start graduate studies at Western Reserve University in Cleveland, Ohio. I was to elucidate the pathway of acetate synthesis from CO_2. As many problems do, this problem turned out to be much more interesting than what was anticipated. For instance, the pathway of acetate synthesis from CO_2 involves derivatives of tetrahydrofolate and vitamin B_{12} (4), a transcarboxylation involving the carboxyl group of pyruvate and presumably a Co-methylcorrinoid (5), a formate dehydrogenase containing non-heme iron, selenium and tungsten and/or molybdenum (6, 7), an anaerobic electron transport system (8), and as recently has been discovered, a nickel enzyme (9, 10).

In this essay I would like to discuss the importance of trace elements in the synthesis of acetate from CO_2 as it occurs in C. thermoaceticum.

Pathway of Acetate Synthesis from CO_2

The earlier tracer studies using $^{14}CO_2$ indicated that acetate was totally synthesized from CO_2 (2, 3). However, this is not correct. The methyl group of the acetate is formed from CO_2 in a series of reductive steps involving formate and one-carbon derivatives of tetrahydrofolate and a corrinoid, most likely 5-methoxy-benzimidazolylcobamide (11), according to the following sequence:

$$CO_2 \rightarrow formate \rightarrow formyl\text{-}H_4folate \rightarrow$$

$$methenyl\text{-}H_4folate \rightarrow methylene\text{-}H_4folate \rightarrow$$

$$methyl\text{-}H_4folate \rightarrow methylcorrinoid.$$

The carboxyl of acetate is formed from the carboxyl group of pyruvate (5). However, the carboxyl group of pyruvate is in isotope equilibrium with CO_2, a reaction which is catalyzed by pyruvate: ferredoxin oxidoreductase. Consequently, isotopically labeled CO_2 is incorporated into the carboxyl group of acetate with the result of an apparent direct fixation of CO_2.

The final steps of acetate synthesis involve methyltetrahydrofolate and pyruvate. The methyl group of the tetrahydrofolate derivative is first transferred to the corrinoid, which appears to be protein bound, to form the methylcorrinoid. This transmethylation reaction is then followed by a "transcarboxylation" reaction in which the methyl group of the methylcorrinoid and the carboxyl group of pyruvate are joined together to form acetate. These final steps seem to be very complex and only recently has progress been made in the elucidation of the mechanism and the enzymes involved.

Drake et al. (12) have found that the synthesis of acetate from pyruvate and methyltetrahydrofolate requires four protein components, F_1, F_2, F_3, and F_4, plus ferredoxin. Protein F_3 contains nickel and catalyzes the oxidation of CO to CO_2 as follows (9):

$$CO + H_2O \longrightarrow CO_2 + 2H^+ + 2e.$$

The most recent findings by Drake et al. (12) are that a synthesis of acetate from CO and methyl-H_4folate is catalyzed by the nickel enzyme (F_3) in combination with the protein fraction F_2. Pyruvate is not involved. This has led to the speculation that the carboxyl group of pyruvate and carbon monoxide may both yield a hypothetical "C_1" unit which is the ultimate precursor of the carboxyl group of acetate. It is clear that studies of the synthesis of acetate from CO_2 continues to yield unexpected and most interesting results. A review of the formation of acetate from CO_2 will soon be published (13).

Formate Dehydrogenase, a Non-Heme Iron-Selenium-Tungsten (Molybdenum) Enzyme from C. thermoaceticum

The first step in the synthesis of the methyl group of acetate from CO_2 is the reduction of CO_2 to formate. In *C. thermoaceticum* a NADP-dependent formate dehydrogenase (Formate:NADP$^+$ oxidoreductase E.C. 1.2.1.43) catalyzes the reaction:

$$CO_2 + NADPH \rightleftharpoons HCOO^- + NADP^+.$$

That formate could be an intermediate in the synthesis of the methyl group was strongly suggested by results obtained by Lentz and Wood (14). They found that cell suspensions of *C. thermoaceticum* catalyze an isotope exchange between CO_2 and formate and that formate is a better precursor to the methyl group of acetate than CO_2. Later studies using cell-free extracts revealed the NADP linked formate dehydrogenase (15) which catalyzes the reversible reduction of CO_2 with NADPH (16, 17). The enzyme was found to be very sensitive toward oxygen. However, it has a high thermostability.

During the spring of 1970 Jan Andreesen started work on the formate dehydrogenase in *C. thermoaceticum*. His studies eventually led to the discovery that selenite-tungstate or selenite-molybdate were needed in the growth medium for the cells to have high formate dehydrogenase activity (6, 17). Interestingly, with the combination selenite-tungstate the formate dehydrogenase activity was higher than with selenite-molybdate. However, the highest activity was obtained with selenite plus both tungstate and molybdate, which may indicate that there is not a simple replacement of tungstate with molybdate or *vice versa*.

The discovery of the positive selenite-tungstate effect came about as follows: normally boiled tomato juice from tin cans was added to the growth medium. Cells grown in the presence of the tomato juice were high on formate dehydrogenase activity, whereas cells grown without the juice had low activity. Pinsent (18) in an important, but as I believe, little recognized paper, had already in 1954 noticed that selenite and molybdate were needed for formation of formate dehydrogenase in *Escherichia coli*. These salts were added to the *C. thermoaceticum* medium and, sure enough, they replaced and outdid the tomato juice in promoting formation of formate dehydrogenase. The effect of selenite and molybdate became known in the laboratory as the "tin can effect."

Tungstate is known as an antagonist toward molybdate (*19*). Therefore, tungstate was added to the medium. The idea, of course, was to inhibit the formation of active formate dehydrogenase and thus obtain an indication that this enzyme was a molybdenum enzyme. Surprisingly, the cells grown with tungstate had a higher activity than those grown with molybdate in the medium. The result raised the possibility that *C. thermoaceticum* formate dehydrogenase is a selenium-tungstate enzyme. The enzyme is inhibited by cyanide and EDTA. This is consistent with a role of a metal in the activity of the enzyme. Other inhibitors of the enzyme are hypophosphite and azide.

The enzyme has been partially purified using an anaerobic buffer system containing a thioglycolate-ferrous complex (*17*). Such preparations of the enzyme from cells grown in media containing ^{75}Se-selenite or [185]W-tungstate were radioactive which indicated an incorporation of the two elements in the active enzyme (*6, 20*). Furthermore, the incorporation of tungsten from [185]W-tungstate into the enzyme was not affected by the presence of molybdate in the growth medium even when the concentration of molybdate was 100 times that of tungstate (*21*). These observations suggest that the *C. thermoaceticum* formate dehydrogenase is a selenium:tungstate enzyme.

Recently, Takashi Saiki, from the University of Tokyo, while visiting the University of Georgia, developed a new purification procedure for the formate dehydrogenase (*7*). Thioglycolate is no longer used in the buffer since it appears that the pure enzyme is destabilized by compounds containing sulfhydryl groups. Instead the enzyme is stabilized by glycerol, formate, sodium azide, and sodium dithionate. Sodium azide has earlier been used to stabilize the formate dehydrogenase from *Clostridium pasteurianum*. an enzyme which contains molybdenum, non-heme iron and inorganic sulfur, but not tungsten or selenium (*22*). That glycerol stabilizes the formate dehydrogenase from *C. thermoaceticum* has been noted earlier (*17*).

In the new procedure, preparations of the enzyme with a specific activity (μmol, min^{-1}, mg^{-1}) of 170 to 200 are routinely obtained. These preparations are not homogeneous. Two major protein bands, both with formate dehydrogenase activity, and two minor bands are obtained during disc gel electrophoresis. The two major bands constitute about 90% of the total protein, and consequently the enzyme may be 90% pure. The reason for the appearance of two proteins with formate dehydrogenase activity is not yet clear.

Chemical and atomic absorption analyses as well as studies using [185]W-tungsten incorporation showed that the purified enzyme contains per mol 0.62 g-at

Se, 0.2–0.3 g-at W, and 0.09–0.16 g-at Mo. In addition the enzyme has exceptionally high contents of non-heme iron and inorganic sulfur; about 70 g-at of iron and 40 g-at of inorganic sulfur were found per mol. The molecular weight of the enzyme was assumed to be 300,000 as was earlier determined using gel filtration (17). I feel it is safe now to conclude that the C. thermoaceticum formate dehydrogenase contains 1 g-at each of tungsten (perhaps partly substituted by molybdenum) and selenium per mol and that it is a non-heme iron sulfur protein.

In addition to the formate dehydrogenase, two additional fractions containing tungsten can be obtained from extracts of C. thermoaceticum (7, 20). One of these is a protein with a molecular weight of 60,000. The other fraction has a molecular weight of less than 10,000 and it is likely non-proteinaceous. At present there is no knowledge concerning the function of these tungsten fractions. One may speculate that they play a role in the processing of tungstate and in the incorporation of tungsten into the formate dehydrogenase in a way similar to what has been proposed for a molybdenum cofactor and a molybdenum carrier protein (23).

The formate dehydrogenase of C. thermoaceticum may not be the only tungsten containing enzyme. The formation of formate dehydrogenase in Clostridium acidiurici and Clostridium formicoaceticum also depends on the presence of tungstate and selenite in the growth medium. However, formate dehydrogenases from other anaerobic bacteria including several clostridia require molybdenum; tungsten is either inactive or even inhibitory (19).

Iron and Electron Transfer Proteins in C. thermoaceticum

In addition to the formate dehydrogenase, C. thermoaceticum contains other non-heme iron proteins such as ferredoxin and rubredoxin as well as a b-type cytochrome. These are likely involved in electron transfer reactions. When hexoses are fermented by C. thermoaceticum NADH and reduced ferredoxin are generated. The reduction of CO_2 in the synthesis of acetate is dependent on NADPH and most likely $FADH_2$. This necessitates a transfer of electrons from NADH and reduced ferredoxin to NADP and FAD. A most obvious enzyme in this transfer would have been a NADH:NADP transhydrogenase. However, this enzyme has not been found in C. thermoaceticum in spite of an extensive search for it. The electron transfer reactions remain to be elucidated.

Ferredoxin, first demonstrated in C. thermoaceticum by Poston and Stadtman (24), has been purified by Shiow-Shong Yang, a former graduate student of mine (25). It has a molecular weight and many other properties that are typical for ferredoxins found in other clostridia. However, it contains only one $[Fe_4S_4]$ cluster, whereas all other reported clostridial ferredoxins contain two $[Fe_4S_4]$ clusters. It is also unusually thermostable, which may not be surprising since it is a component of a thermophilic bacterium.

The physiological role of the ferredoxin in C. thermoaceticum is at least partly understood. It functions as the electron acceptor in the pyruvate: ferredoxin oxido-reductase reaction and it serves as a donor of electrons to NADP (*25*). It is required in the synthesis of acetate from either methyl-H_4folate or methyl-B_{12} and pyruvate (*12, 24*). The function of the ferredoxin in this reaction is not clear, however it may be related to the role of pyruvate. Interestingly, it functions also as an electron acceptor in the newly discovered carbon monoxide dehydrogenase reaction (*9*). As discussed earlier this enzyme contains nickel.

Two rubredoxin, each containing one iron atom, have been isolated from *C. thermoaceticum* by Yang *et al.* (*26*). The role of these in the metabolism of the bacterium is not known. An enzyme containing FMN (rubredoxin reductase) catalyzes the transfer of electrons from NADH and NADPH to the two rubredoxins as well as to dichlorophenolindophenol and mammalian cytochrome *c*, but not to feredoxin. The electron transfer from the nucleotides to the enzyme and to the acceptors appears irreversible, and consequently the rubredoxin reductase is not functioning as a NADH:NADP transhydrogenase.

It has been generally accepted that clostridia lack cytochromes. However, a few years ago in cooperation with Andreesen's group in Göttingen, and LeGall in Marseille, we were able to demonstrate membrane bound *b*-type cytochromes both in *C. thermoaceticum* and *C. formicoaceticum* (*27*). In *C. formicoaceticum* the cytochrome functions with fumarate reductase. A role for the cytochrome in *C. thermoaceticum* has not been established. However, it is of interest that it functions, in addition to ferredoxin, as an electron acceptor in the carbon monoxide dehydrogenase reaction (*9*).

Concluding Remarks

The work to elucidate the synthesis of acetate from CO_2 in *C. thermoaceticum* has been, and still is, exciting. A pathway of CO_2 fixation *via* formate has been established. This pathway may well be used for autotrophic CO_2 fixation as has been suggested for *Acetobacterium woodii* (*28*). This microorganism grows autotrophically by producing acetate from CO_2 and H_2 (*29*). Several new and perhaps unique enzymes have been discovered. Some of these enzymes depend on trace elements for activity. These are cobalt, iron, selenium, tungsten (molybdenum), and nickel. Cobalt is a constituent of the corrinoids which are involved in the final steps of acetate synthesis. Iron is a constituent of ferredoxin, rubredoxins, cytochrome *b*, and the formate dehydrogenase. The latter requires also selenium, tungsten, and perhaps molybdenum for activity. The last exiting finding is obviously that of the nickel containing carbon monoxide dehydrogenase (*9, 10*), an enzyme which also may be involved in the last steps of acetate synthesis (*12*).

Egami (*30–32*) has advanced theories that trace elements and especially transition metals have been involved in the genesis and evolution of life. It is interesting that clostridia, which are considered among the primitive organisms, depend on so many of these elements. This fact seems to support the theories of Egami. Furthermore, the synthesis of acetate from CO_2 has been considered to be one of the pathways that developed early (33). Finally, I believe that a case can be made for the theory that thermophilic anaerobic microorganisms were among the first to develop (34). These may have preceded mesophilic anaerobes from which aerobic and thermophilic aerobic microorganisms originate.

Acknowledgment

Work on the synthesis of acetate at the University of Georgia is supported by Public Health Service grant AM 27,323 from the National Institute of Arthritis, Metabolic and Digestive Diseases and grant PCM-7,726,054 from the National Science Foundation.

REFERENCES

1. Fontaine, F.E., Peterson, W.H., McCoy, E., Johnson, M.J., and Ritter, G.J.: 1942. *J. Bacteriol.* 43, 701–715.
2. Barker, H.A. and Kamen, M.D.: 1945. *Proc. Natl. Acad. Sci. USA* 31, 219–225.
3. Wood, H.G.: 1952. *J. Biol. Chem.* 194, 905–931.
4. Ljungdahl, L.G. and Wood, H.G.: 1969. *Annu. Rev. Microbiol.* 23, 515–538.
5. Schulman, M., Ghambeer, R.K., Ljungdahl, L.G., and Wood, H.G.: 1973. *J. Biol. Chem.* 248, 6255–6261.
6. Andreesen, J.R. and Ljungdahl, L.G.: 1973. *J. Bacteriol.* 116, 867–873.
7. Saiki, T., Shackleford, G., and Ljungdahl, L.G.: 1981. *Second Int. Symposium on Selenium in Biology and Medicine*, Lubbock, Texas, in press.
8. Andreesen, J.R., Schaupp, A., Neurauter, C., Brown, A., and Ljungdahl, L.G.: 1973. *J. Bacteriol.* 114, 743–751.
9. Drake, H.L., Hu, S.-I., and Wood, H.G.: 1980. *J. Biol. Chem.* 255, 7174–7180.
10. Diekert, G. and Thauer, R.K.: 1980. *FEMS Microbiol. Lett.* 7, 187–189.
11. Ljungdahl, L.G., Irion, E., and Wood, H.G.: 1965. *Biochemistry* 4, 2771–2780.
12. Drake, H.L., Hu, S.-I., and Wood, H.G.: 1979. XIth Int. Congr. Biochem., Toronto and personal communications.
13. Ljungdahl, L.G. and Wood, H.G.: 1981. *Vitamin B_{12}* (Dolphin, D., ed.) Wiley-Interscience, New York, in press.
14. Lentz, K. and Wood, H.G.: 1955. *J. Biol. Chem.* 215, 645–654.
15. Li, L.-F., Ljungdahl, L.G., and Wood, H.G.: 1966. *J. Bacteriol.* 92, 405–412.
16. Thauer, R.K.: 1972. *FEBS Lett.* 27, 111–115.
17. Andreesen, J.R. and Ljungdahl, L.G.: 1974. *J. Bacteriol.* 120, 6–14.
18. Pinsent, J.: 1954. *Biochem. J.* 57, 10–16.
19. Ljungdahl, L.G.: 1980. *Molybdenum and Molybdenum-Containing Enzymes* (Coughlan, M.P., ed.) pp. 463–486, Pergamon Press, Oxford.
20. Ljungdahl, L.G. and Andreesen, J.R.: 1975. *FEBS Lett.* 54, 279–282.
21. Ljungdahl, L.G. and Andreesen, J.R.: 1976. *Microbial Production and Utilization of Gases (H_2, CH_4, CO)* (Schlegel, H.G., Gottschalk, G., and Pfennig, N., eds.) pp. 163–172, E. Goltze KG, Göttingen.

22. Scherer, P.A. and Thauer, R.K.: 1978. *Eur. J. Biochem.* 85, 125–135.
23. Johnson, J.L.: 1980. *Molybdenum and Molybdenum-Containing Enzymes* (Coughlan, M.P., ed.) pp. 345–383, Pergamon Press, Oxford.
24. Poston, J.M. and Stadtman, E.R.: 1967. *Biochem. Biophys. Res. Commun.* 26, 550–555.
25. Yang, S.-S., Ljungdahl, L.G., and LeGall, J.: 1977. *J. Bacteriol.* 130, 1084–1090.
26. Yang, S.-S., Ljungdahl, L.G., DerVartanian, D.V., and Watt, G.D.: 1980. *Biochim. Biophys. Acta* 590, 24–33.
27. Gottwald, M., Andreesen, J.R., LeGall, J., and Ljungdahl, L.G.: 1975. *J. Bacteriol.* 122, 325–328.
28. Tanner, R.S., Wolfe, R.S., and Ljungdahl, L.G.: 1978. *J. Bacteriol.* 134, 668–670.
29. Balch, W.E., Schoberth, S., Tanner, R.S., and Wolfe, R.S.: 1977. *Int. J. Syst. Bacteriol.* 27, 355–361.
30. Egami, F.: 1974. *J. Mol. Evol.* 4, 113–120.
31. Egami, F.: 1975. *J. Biochem.* 77, 1165–1169.
32. Yanagawa, H., Kobayashi, Y., and Egami, F.: 1980. *J. Biochem.* 87, 359–362.
33. Hartman, H.: 1975. *J. Mol. Evol.* 4, 359–370.
34. Ljungdahl, L.G.: 1979. *Adv. Microbial Physiol.* 19, 149–243.

Enzyme Classification and Nomenclature: Origin, Development, and Future Problems

OTTO HOFFMANN-OSTENHOF

O. HOFFMANN-OSTENHOF was born in 1914 in Vienna. He studied chemistry and medicine and received his Ph.D. degree in chemistry at the University of Zürich in 1939. During World War II he was a soldier in the German army. Professor of Biochemistry at the University of Vienna in 1959, at present Head of the Institute of General Biochemistry at University of Vienna. Main scientific activities: research on quinone action, phosphate metabolism, automutagenic substances in higher plants, sporulation of yeast, enzymes of the biosynthesis and the degradation of *myo*-inositol, and the other inositols and cyclitols. Acted as Visiting Professor at Universities in Australia, Canada, China, India, Iran, Israel, Mexico, Nigeria, USA, and several European countries. Served as member of the council of the International Union for Biochemistry, 1955–1967; President of the Federation of European Biochemical Societies, 1964–1965; Chairman of the Commission on Biochemical Nomenclature, 1965–1976; president of the Österreichische Biochemische Gesellschaft, 1963–1965 and 1973–1975. Recipient of the Ernst-Späth-prize of the Austrian Academy of Science; fellow of the New York Academy of Science; Diplôme d'honneur of the Federation of European Biochemical Societies.

The Original Problem

Most of the enzymes known until about 1930 had one action in common; they hydrolyzed their substrates. There were very few exceptions like catalase or peroxidase. The only existing rule for the naming of enzymes had been proposed by Duclaux (*1*) in 1883; it postulates that an enzyme should be named after the substrate upon which it acts, and that the substrate name should be followed by the ending -*ase*, *e.g.* esterase, peptidase, and asparaginase. This rule was generally accepted for newly discovered enzymes, whereas some previously known proteolytic enzymes like pepsin, trypsin, and papain had to retain their old names with the ending *-in*, because the rule of Duclaux would not have made it possible to differentiate between them and could therefore not be applied.

The first attempts to classify the enzymes were made in the twenties of this century. A distinction was made between enzymes catalyzing the hydrolysis of their substrates, which form the class of "hydrolases," and all other enzymes, *i.e.* those that attack their substrates by any action other than hydrolysis. This second class of enzymes was usually called "desmolases" (2).

From around 1930 the number of enzymes discovered every year increased considerably and at the same time many newly found enzymes exhibited hitherto unknown types of action. It therefore became a necessity to come to a more differentiating classification of the enzymes. The first step in this direction was made by Ernest Baldwin (3) who distinguished between three main classes of enzymes namely: (1) *Splitting enzymes*, including hydrolases, phosphorylases, and a group of enzymes oddly called "adding enzymes," which comprised decarboxylases, dehydratases, and similar enzymes. (2) *Transferring enzymes*, including all enzymes catalyzing oxidoreductions, furthermore all group-transferring enzymes like aminotransferases or phosphotransferases. (3) *Isomerases and racemases*. Although Baldwin's classification of the enzymes was certainly a progress, it still did not achieve the wanted goal, namely, to find a place for each known enzyme within the classification.

With respect to enzyme nomenclature, the increasing number of newly discovered enzymes had led to a completely uncontrolled situation. Until 1955 the naming of the enzymes was exclusively left to the discoverer. In the older literature we therefore find enzyme names having neither a relation to the substrate name nor to the catalyzed reactions. A few examples will illustrate this point; the presently recommended name is given in parentheses: "Zwischenferment" (glucose-6-phosphate dehydrogenase), "old yellow enzyme" (NADPH dehydrogenase), "diaphorase" (dihydrolipoamide reductase), "condensing enzyme" (citrate (*si*)-synthase), "histozyme" (aminoacylase), "Lohmann's enzyme" (creatine kinase), "P-enzyme" (phosphorylase), "Q-enzyme" (1,4-α-glucan branching enzyme). In some cases two authors discovering the same enzyme gave it different names; in other cases two different enzymes got the same name.

It was the present author who from 1950 on tried to put the attention of the workers in the field of enzymology to the difficulties in the classification and nomenclature of enzymes (4, 5). These attempts were not always favourably met, as many authors as discoverers of enzymes considered it their privilege to name "their enzyme," or other authors just did not want to change the enzyme names, to which they were used. Eventually, however, he found an enlightened audience in the General Assembly of the then newly founded International Union of Biochemistry. At a meeting of this body in Brussels in 1955 it was decided to form a commission to investigate these problems and to make suggestions about possible improvements of the situation. The International Commission of Enzymes was established in 1956 by the President of the International Union, Professor Marcel Florkin, who had consulted the national committees of the Union and an *ad hoc* committee for suggestions concerning membership.

The following members were appointed by the Union: A.E. Braunstein (USSR), S.P. Colowick (USA), P.A.E. Desnuelle (France), M. Dixon (U.K.), W.A. Engel-hardt (USSR), E.F. Gale (U.K.), O. Hoffmann-Ostenhof (Austria), A.L. Lehninger (USA), K. Linderstrom-Lang (Denmark), F. Lynen (Germany). Furthermore F. Egami (Japan) and L.F. Leloir (Argentinia) were made corresponding members. Malcolm Dixon was elected chirman, O. Hoffmann-Ostenhof became the secretary of the commission.

The Work of the Enzyme Commission

This commission had a formidable task, namely to conceive a classification wherein all enzymes known until then had a place, further to coin a rational nomenclature for all enzymes, and also to review the names of coenzymes, cyto-chromes, the units of enzyme activity and the symbols of enzyme kinetics. From 1956 to 1961 the commission met every year for five days, but a good part of its work was also done by correspondence.

First of all the commission agreed on three general principles:

(a)   Only single enzymes, *i.e.* single catalytic entities, should be named by a name with -*ase* and should be classified. Systems containing more than one enzyme should be named by a term containing the word "system," *e.g.* succinate oxidase system, pyruvate dehydrogenase system.

(b)   Enzymes are to be classified and named according to the overall chemical reaction they catalyze. Although other possibilities were considered, it was found that at that time—and this is still true at present—the chemical reaction is the only generally applicable criterion for our purposes. It follows that an enzyme cannot be classified and be given a systematic name as long as the chemical reaction is not yet completely known.

A very important consequence of this principle is that a certain name does not designate a single enzyme protein but a group of proteins with the same cata-lytic property. Thus enzymes from different sources—bacteria, plants, animals— are classified as one entry and are to be given the same name.

Furthermore this rule implies the disapproval of the designation of a natural substance or even of a hypothetical active principle, responsible for a physiological or biophysical phenomenon that cannot be described in terms of a definite chemical reaction, by the name of the phenomenon in conjugation with the suffix -*ase*, which implies an individual enzyme. Thus the use of names like permease, translocase, reparase, joinase, and codase, frequently found in the biological literature, should be avoided and discouraged.

(c)   The subdivision of enzymes into groups on the basis of the type of reaction catalyzed together with the name of the substrate provides the basis for naming individual enzymes.

The commission devised a system which simultaneously provides a classification of enzymes and a basis to number them according to a decimal system.

For the classification it was originally suggested that five main classes should be sufficient to allocate all enzymes known at the time; during the years of the commission's work, however, it became clear that a sixth main class had to be added to the scheme. The six main classes for the classification of enzymes are: (1) *Oxidoreductases*. (2) *Transferases*—enzymes catalyzing the transfer of atoms or atom groups from a donor to an acceptor, excluding, however, oxidoreductases, and enzymes with water as acceptor, *i.e.* hydrolases. (3) *Hydrolases*. (4) *Lyases*— enzymes cleaving C–C, C–O, C–N, and other bonds by means other than hydrolysis or oxidation; they differ from other enzymes in that two substrates are involved in one reaction direction, but only one in the other direction. When acting on the single substrate, a molecule is eliminated leaving an unsaturated residue. Examples: Decarboxylases, dehydratases. (5) *Isomerases*—enzymes catalyzing changes within one molecule. (6) *Ligases (Synthetases)*—the main class to be added later; enzymes catalyzing the joining of two molecules with concomitant hydrolysis of the pyrophosphate bond in adenosine triphosphate or another nucleoside triphosphate.

The mentioned main classes were of different size; the smallest one comprised originally 56 enzymes, other ones contained several hundred. In all cases it was necessary to form subclasses, often also sub-subclasses. Although with respect to the subclasses some changes had to be made during the 20 years since the system was first published, in general the principles of classification are still considered satisfactory.

A very important innovation proposed by the first commission dealing with enzyme nomenclature was the numbering system for individual enzymes (EC = enzyme code). Each enzyme number contains four elements, separated by points, and arranged on the following principles:

(a)   The first figure shows to which of the six main classes the particular enzyme belongs.

(b)   The second figure indicates the subclass. For the oxidoreductases it shows the type of group in the donors that undergoes oxidation (1 denoting a –CHOH– group, 2 a carbonyl group, and so on); for the transferases it indicates the nature of the group transferred; for the hydrolases it shows the type of bond hydrolyzed; for the lyases it denotes the type of link broken between the group removed and the remainder; for the isomerases it indicates the type of isomerization catalyzed; for the ligases it shows the type of bond formed.

TABLE 1. Some Examples Illustrating the Numbering System of the Enzymes

Recommended name	Main class	Subclass	Sub-subclass	EC-number[a]
Lactate dehydrogenase	Oxidoreductases (digit 1)	Acting on the CHOH group of donors (digit 2)	With NAD+ or NADP+ as acceptor (digit 1)	1.1.1.27
Sucrose phosphorylase	Transferases (digit 2)	Glycosyltransferase (digit 4)	Hexosyltransferases (digit 1)	2.4.1.7
AMP nucleosidase	Hydrolases (digit 3)	Glycosidase (digit 2)	Hydrolysing N-glycosyl compounds (digit 2)	3.2.2.4
Homocysteine desulphhydrase	Lyases (digit 4)	Carbon-sulphur lyases (digit 4)	No sub-subclass in this subclass (digit 1)	4.4.1.2
Lactate racemase	Isomerases (digit 5)	Racemases and epimerases (digit 1)	Acting on hydroxyacids and derivatives (digit 2)	5.1.2.1
Glutamine synthase	Ligases (digit 6)	Forming carbon-nitrogen bonds (digit 3)	Acid-ammonia ligases (digit 1)	6.3.1.2

[a] The fourth digit is the serial number of the enzyme in its sub-subclass.

(c) The third figure indicates the sub-subclass. For the oxidoreductases it shows for each type of donor the type of acceptor involved (1 denoting a coenzyme—NAD+ or NADP+, 2 a cytochrome, 3 molecular O_2, and so on); analogously the third number indicates further details of the enzymes of the other main classes.

(d) The fourth figure is the serial number of the enzyme in its sub-subclass.

A few examples illustrating the working of the system are given in Table 1.

This system greatly helps to identify each enzyme described in the literature. It was intended that the numbers should remain permanently attached to the same enzymes as a definite means of identification. It must be admitted that in several cases, this idea could not be upheld. When enzymes originally were thought to belong to a certain group in the classification and later results showed that their allocation was wrong, a change of the number became necessary, but the old number was never to be used again.

Another important point was the question of a systematic, logical nomenclature for enzymes. A consistent nomenclature of this kind was not possible without making great changes in the existing nomenclature; furthermore such a really systematic nomenclature of enzymes must contain the complete names of their substrates. Some of them are complex substances with long chemical names; thus, the resulting enzyme names turn out to be too long for ordinary use.

After long discussions it was decided to have two kinds of names, one systematic, and one working or trivial. The systematic name is to be formed in accordance with definite rules, has to identify the enzyme precisely, and to show the action of the enzyme as exactly as possible. The trivial name is to be short, not necessarily very axact or systematic, but should be precise enough to avoid misunderstanginds. In many cases the name already in use could be retained.

The commission set up a set of rules for systematic and trivial nomenclature of enzymes, consisting of two parts, namely general rules, to be applied for all enzymes, and rules for particular classes of enzymes. Most of these rules have survived almost without change during the last 20 years.

One of the actions of the commission, which at its time was most controversial, was the renaming of the nicotinamide coenzymes. Various objections against the former names of these compounds—diphosphopyridine nucleotide (DPN) and triphosphopyridine nucleotide (TPN)—amongst them the most serious one that they are really the chemical names of other compounds, led to the renaming of the compounds to nicotinamide-adenine dinucleotide (NAD) and nicotinamide adenine dinucleotide phosphate (NADP). This action caused strong opposition amongst many workers in the field, especially in the United States, but in the course of time it seems that the new names and abbreviations have been adopted by the great majority of biochemists.

The work of the commission was always made in close contact with the Nomenclature Commission for Organic Chemistry of the International Union of Pure and Applied Chemistry (IUPAC); furthermore the advice of many experts in the field, sometimes assembled in subcommissions, was considered. The final outcome was a report in book form (6), which contained chapters on enzyme units, symbols of enzyme kinetics, nomenclature of coenzymes, classification and nomenclature of cytochromes, classification and nomenclature of enzymes, terminology of enzyme formation, general conclusions; and as appendices, recommended symbols for enzyme kinetics, list of cytochromes, key to numbering and classification of enzyme, and a list of enzymes. The report was presented to the General Assembly of the International Union of Biochemistry at its meeting in Moscow 1961 and was duly adopted.

It can safely be said that this first body working on these problems, the Enzyme Commission, really laid the fundaments for enzyme nomenclature and classification. Most of the concepts and rules elaborated by it have persisted until now, although the number of enzymes has almost trebled since the first report.

The Further Development

With the adoption of the report the Enzyme Commission was dissolved and a Standing Committee on Enzymes, consisting of S.P. Colowick, O. Hoffmann-Ostenhof, A.J. Lehninger, and S.P. Webb was set up with the task to consider the comments and criticisms received on the published report and eventually to prepare a revised version, which was to be considered a definitive document. It came out in 1964 (7); then the committee was also dissolved.

The rapidly growing number of enzymes (Table 2) soon made it clear that

TABLE 2. Number of Individual Enzymes Classified and Named in the Official Publications on Enzyme Nomenclature

Year	Number
1961	712
1964	875
1972	1,770
1975	1,974
1978	2,122

within a few years a new, revised, edition of the book on enzyme nomenclature would become necessary. In 1965, the two international unions involved, *i.e.* IUB and IUPAC, transformed the old IUPAC Commission on the Nomenclature of Biological Chemistry into a new joint venture, the Commission on Biochemical Nomenctalure (CBN); the present writer was elected chirman of this body.

In 1969 it was decided by CBN that a revised edition of the Recommendations on Enzyme Nomenclature should be prepared with the main purpose to include the great number of enzymes discovered in the time since the previous edition. An expert committee with E.C. Webb (Australia) as convener and A.E. Braunstein (USSR), J.S. Fruton (USA), O. Hoffmann-Ostenhof (Austria), B.L. Horecker (USA), W.P. Jakoby (USA), P. Karlson (Germany), B. Keil (France), E.C. Slater (Netherlands), and W.J. Whelan (USA) as members was formed. Once again a subcommittee of experts and suggestions coming from authors and editors helped the work of the expert group. The third, completely revised, edition was published 1973 under the auspices of both unions (*8*). A supplement to the enzyme list, mainly covering new entries, deletions, and corrections, was prepared during 1974 and 1975 and came out in Biochimica and Biophysica Acta (*9*).

In 1976 a reorganization of the nomenclature bodies dealing with biochemistry took place. Again enzymes became the responsibility of a new nomenclature committee of IUB. This union arranged a cooperation with the National Institutes of Health, Bethesda, Maryland, to enter the enzyme list on computer tape and prepare future versions of it as a computer print-out. In 1979 once again a revised edition of Enzyme Nomenclature (*10*) was published, the first one containing the enzyme list as computer print-out.

In the seventies a change of emphasis took place following the suggestions of many journal editors and workers in the field. Whereas previously the frequently cumbersome systematic name was given first place, the committee now shifted prominence to the recommended trivial names. This change is reflected in the newest edition of "Enzyme Nomenclature." However, the systematic name was retained because it serves as a basis for classification as well as for identification. The code number cannot completely replace it as it is only useful if an enzyme list is at hand, whereas the systematic name is self-explanatory. Editors of biochemical journals usually request their authors to identify enzymes that are the main subject of a paper by code number or alternatively by the reaction equation, systematic name, and source at the first mention in a paper; thereafter the recommended name may be used. Discoverers of new enzymes are not supposed to assign code numbers; this should be left to the committee.

Open Problems of Enzyme Nomenclature; Future Trends

There is no doubt that every attempt to classify natural phenomena like biocatalysis meets with considerable difficulties and is never absolutely perfect. The present system of enzyme classification based on the equations of the reactions catalyzed, wherefrom enzyme nomenclature is derived, is at the time being certainly better than any system that can be constructed on another basis. But this will not necessarily be so in the future. A system based on the nature of the active centre of the enzymes or rather on their intimate mechanisms of action can be imagined. In fact even at present use of these criteria is already made in the subclass of the peptidases where we have sub-subclasses 3.1.16 Serine carboxypeptidases, 3.1.17 Metallo-carboxypeptidases, 3.1.21 Serine proteinases, 3.4.22 Thiol proteinases, 2.4.22 Carboxyl (acid) peptidases, 3.4.23 Metalloproteinases, and 3.4.99 Proteinases of unknown catalytic mechanism. In all these terms reference is made to the nature of the active centre or to the catalytic mechanism. However, it would be difficult and in many cases even impossible to extend this system of classification to other classes of enzymes.

In this context a problem should be mentioned. There are enzymes from different sources, catalyzing the same reaction, but obviously acting by different mechanisms. They are listed in "Enzyme Nomenclature" under the same name and the same code number, e.g. Fructose-bisphosphate aldolase (EC 4.1.2.13). Here the animal enzyme acts by a Schiff base mechanism, whereas the enzymes from yeast or bacteria are zinc-proteins and no Schiff base is formed as intermediate. The present author is of the opinion that in such cases at least two different entries would be desirable.

Another problem that has been discussed by Peter Karlson (11), the present chairman of the nomenclature committee, is the nomenclature of the various single components of the so-called multienzyme complexes like e.g. the pyruvate dehydrogenase system. Here the difficulty is that enzymes—as all catalysts— are defined as components that accelerate a reaction without in the overall reaction undergoing any change. As some reactions catalyzed by the single components of such multienzyme complexes cannot be formulated in a way that the enzyme protein remains unchanged, i.e. that a complete catalytic reaction cycle takes place, so far no satisfactory way for the nomenclature of these components of multienzyme complexes has been found.

The present state of enzyme classification and nomenclature as it has developed during the last 25 years has proved to be adequate for most purposes in biochemistry. However, as there is a steady increase in the number of enzymes and there are still some open problems in nomenclature, it will be necessary to retain a standing body of experts to keep an eye on these questions and developments. Probably at a later period it will also be the task of such a committee to discuss the question whether the principles that form the basis of the system should not be reconsidered. Time is certainly not yet ripe to change to a system exclusively based on the nature of the active site and the reaction mechanisms, but with out increasing knowledge of the nature of enzymes it can be expected that in not too far a future such a system will become feasible and may offer advantages over the present one.

REFERENCES

1. Duclaux, E.: 1883. *Traité de Microbiologie*, Masson, Paris.
2. Ammon, R. and Dirscherl, W.: 1948. *Fermente, Vitamine und Hormone*, 2nd ed., Thieme, Leipzig.
3. Baldwin, E.: 1947. *Dynamic Aspects of Biochemistry*, Cambridge University Press, Cambridge.
4. Hoffmann-Ostenhof, O.: 1953. *Adv. Enzymol.* 14, 219–260.
5. Hoffmann-Ostenhof, O.: 1954. *Enzymologie*, Springer-Verlag, Vienna.
6. *Report of the Commission on Enzymes of the International Union of Biochemistry*, Pergamon Press, Oxford, 1961.
7. *Enzyme Nomenclature—Recommendations 1964 of the International Union of Biochemistry*, Elsevier, Amsterdam, 1965.
8. *Enzyme Nomenclature—Recommendations 1972 of the International Union of Pure and Applied Chemistry and the International Union of Biochemistry*, Elsevier, Amsterdam, 1973.
9. *Biochim. Biophys. Acta* 429, 1–45, 1976.
10. *Enzyme Nomenclature—Recommendations (1978) of the Nomenclature Committee of the International Union of Biochemistry*, Academic Press, New York, 1979.
11. Karlson, P.: 1974. *Naturwissenschaften* 61, 355–360.

Conservation of Primary Structure at the Proteinase-Sensitive Site of Vertebrate Fructose 1,6-Bisphosphatases

B. L. HORECKER, JOHN S. MACGREGOR, and S. PONTREMOLI

B. L. HORECKER was born on October 31, 1914 in Chicago, Illinois, and received the Ph.D. degree in chemistry from the University of Chicago in 1939. During 1941–1959, he worked at the National Institutes of Health, Bethesda, Maryland, where he isolated TPN-cytochrome *c* reductase and identified the intermediates and enzymes of the pentose phosphate pathway. In 1959, he became Chairman of the Department of Microbiology at New York University and in 1963, Chairman of the New Department of Molecular Biology at the Albert Einstein School of Medicine, later becoming associate Dean for Scientific Affairs at that institution. Since 1972, he has been Head of the Laboratory of Molecular Enzymology at the Roche Institute of Molecular Biology. His research has covered the following fields: biological oxidations; hemoglobin derivatives, cytochrome *c* reductase, xanthine oxidase. Carbohydrate metabolism; the pentose phosphate pathway, fermentation pathways of microorganisms, structure and biosynthesis of gram-negative lipopolysaccharide. Enzyme structure, mechanism of action, and regulation; aldolase, transaldolase, and fructose 1,6-bisphosphatase.

Early studies on the properties of mammalian fructose 1,6-bisphosphatase (Fru-P$_2$ase, EC 3.1.3.11), beginning with the first report by Gomori in 1943 (*1*) and continuing until the early 1970's, were carried out with preparations that showed maximum activity in the alkaline pH range. The activity measured at neutral pH was observed to be highly variable, ranging from near zero at pH 7.5 with the partially purified enzyme preparations employed by Gomori, to 30–50% of the maximal activity for the crystalline enzyme isolated from rabbit liver by

Pontremoli and his coworkers (7); similar pH activity curves were reported by other laboratories for purified preparations from rabbit liver (3, 4) and pig kidney (5).

The observed alkaline pH optimum appeared to be incompatible with the proposed role for this enzyme in catalyzing an essential step in gluconeogenesis (6). However, early works by Hers and Kusaka (7), later confirmed by Byrne (8), had demonstrated that in crude extracts of rabbit liver the rate of hydrolysis of $Fru-P_2$ to Fru-6-P was maximal at neutral pH, and Pogell and McGilvery (9) showed that incubation of these extracts with a particulate fraction from rabbit liver, or with papain, resulted in an increase in the activity measured at alkaline pH. In addition, the conditions employed by Gomori for extraction of the enzyme from rabbit liver were shown to cause a selective loss of the neutral activity (8). These observations were confirmed by Nakashima and Horecker (10) who found that incubation of rabbit liver extracts with the heavy particle fraction from the same tissue resulted in a shift in the pH optimum from the neutral to the alkaline range. It was evident that the enzyme with alkaline pH optimum (referred to here as the "alkaline" $Fru-P_2ase$) was generated during its extraction or isolation by proteolytic modification of the native "neutral" $Fru-P_2ase$.

In 1971, several laboratories succeeded in isolating the neutral $Fru-P_2ase$ from a variety of sources, including beef liver (11), rabbit liver (12), and rabbit kidney (13). Neutral $Fru-P_2ases$ have since been isolated from livers of chicken (14, 15), sheep (16, 17), rat (18), mouse (19), and turkey (20), from muscles of rabbit (21), chicken (14, 22), and snake (23), and also from mouse (24) and rabbit (25) intestine. Rabbit liver and kidney $Fru-P_2ases$ have been shown to be immunologically identical (26) and to yield identical tryptic peptide maps (27). They are probably identical proteins coded by the same gene. In contrast, the rabbit muscle enzyme failed to crossreact with antibody against liver $Fru-P_2ase$ (26) and yielded distinctly different peptide maps (27). Thus the muscle $Fru-P_2ase$ must be coded by a different gene. No information is available regarding the structure or genetic origin of the intestinal enzyme.

The availability of pure preparations of the native enzymes from liver and muscle of a variety of species made possible studies on their structure as well as of the molecular basis for the changes in catalytic properties caused by both endogenous and exogenous proteinases. It had become apparent that modification by limited proteolysis was a common property of $Fru-P_2ases$ of vertebrate origin, whether of the liver or muscle type. In order to provide information as to the possible physiological significance of this limited proteolysis, we have sought to determine the extent of conservation of primary structure at the subtilisin-sensitive site of a number of $Fru-P_2ases$. We have found that the primary sequence in this region of the molecule is highly conserved. Since conservation of structure is generally regarded as an indication of essential function, we conclude that proteolytic modification at this site may play a role in either regulation of the activity of the enzyme, or as a first step in its metabolically controlled turnover. This view is reinforced by evidence for the existence of a lysosomal proteinase that catalyzes a modification similar to that catalyzed by subtilisin.

Modification of Fru-P₂ase by Subtilisin

1. Changes in catalytic and allosteric properties

A number of proteinases including papain (28), nagarse, pronase or chymotrypsin (29), and subtilisin (30) have been shown to catalyze limited proteolysis of Fru-P₂ase, resulting in enhanced activity at about pH 9, decreased activity at neutral pH, or both. We selected subtilisin for the studies described here because it appeared to be most specific and to mimic most closely the changes in catalytic properties and structure effected by endogenous proteinases.

Subtilisin has been found to catalyze a similar and specific limited modification

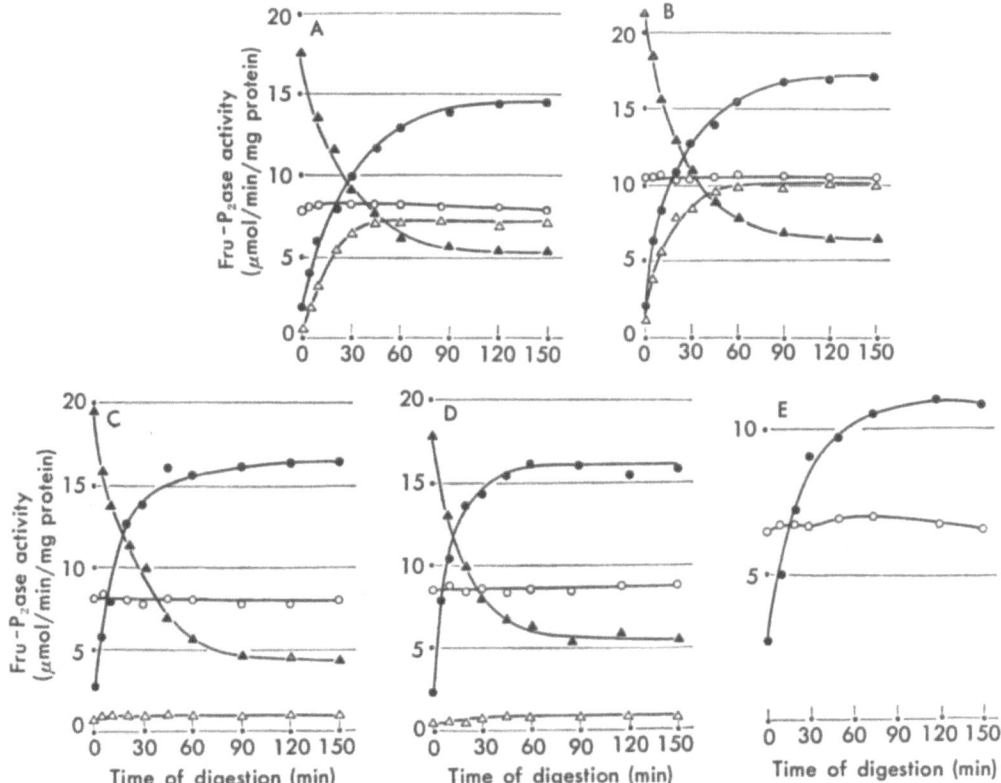

FIG. 1.   Effect of digestion with subtilisin on the catalytic properties of liver and muscle Fru-P₂ases. Digestion was carried out at 22°C with a ratio of subtilisin:Fru-P₂ase of 1:500 (w/w). Assays were carried out as described with additions as indicated. A: rabbit liver. B: rat liver. C: chicken muscle. D: rabbit muscle. E: snake muscle. A–D: pH ▲ 7.5, 40 mM (NH₄)₂ SO₄; ● pH 9.2; ○ pH 7.5; △ pH 7.5, 100 μM AMP. E: ● pH 9.2; ○ 7.5. (From Ph. D. dissertation by John S. MacGregor, in preparation).

of every Fru-P$_2$ase examined, including the enzymes isolated from rabbit liver (*30*), sheep liver (*17*), rat liver (Tejwani and MacGregor, unpublished), pig kidney (*31*), rabbit muscle (MacGregor, unpublished), chicken muscle (*32* and Mac-Gregor, unpublished), and snake muscle (Shi, Xu, and MacGregor, unpublished). In each case digestion with subtilisin resulted in a large (3–7-fold) increase in the catalytic activity measured at pH 9.2. This is illustrated in Fig. 1 for the enzymes from rabbit liver, rat liver, rabbit muscle, chicken muscle, and snake muscle. The catalytic activity at pH 7.5 measured in the absence of NH$_4^+$ (or K$^+$) was not significantly altered, but the activation of NH$_4^+$ was abolished. Digestion with subtilisin also decreased the sensitivity of the liver enzymes to inhibition by AMP, but did not affect inhibition by AMP of the muscle enzymes (Figs. 1 and 2).

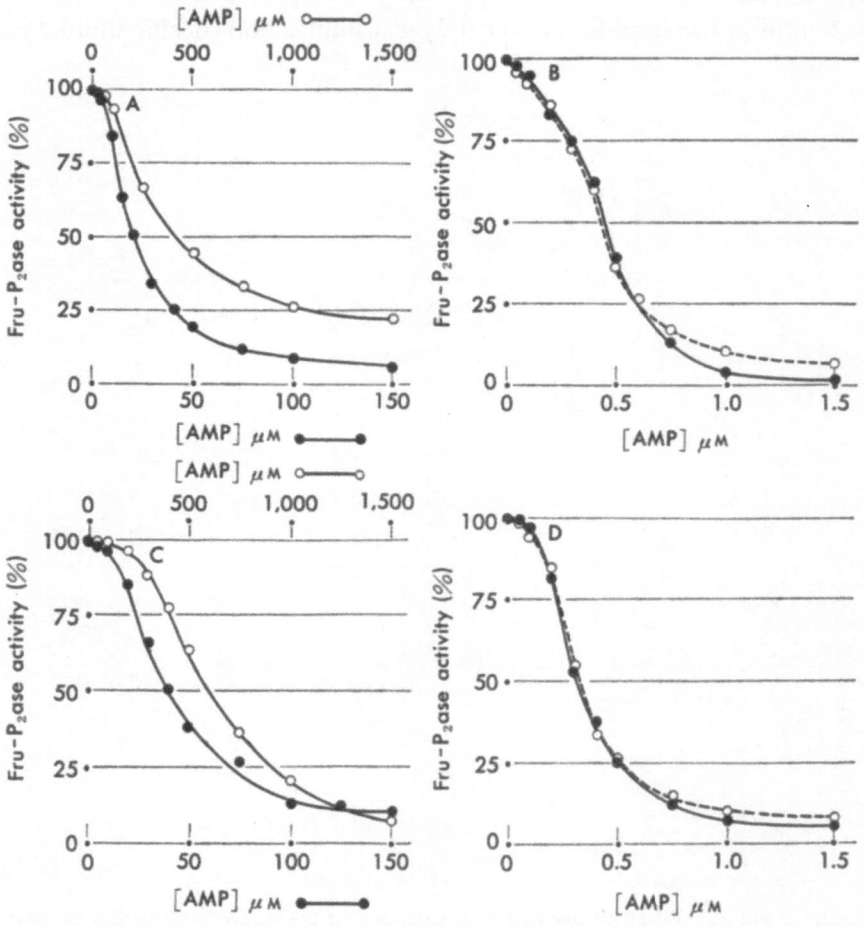

FIG. 2. Effect of digestion with subtilisin on the inhibition by AMP of liver and muscle Fru-P$_2$ases. The conditions of digestion were as in Fig. 1. Samples were analyzed after incubation with subtilisin for 2.5 hr. A: rabbit liver, nK_1=21 μM; sK_1=425 μM. B: rabbit muscle. K_1=0.45 μM. C: rat liver. nK_1=41 μM; sK_1=600 μM. D: chicken muscle. K_1=0.32 μM. (From Ph. D. dissertation by John S. MacGregor, in preparation).

2. Effect of subtilisin on the subunit structure of Fru-P₂ases

The changes in catalytic and allosteric properties described in the preceding section were associated with changes in subunit structure, as revealed by SDS-polyacrylamide gel electrophoresis (Fig. 3). In the case of the enzyme from rabbit liver (*30, 33*), digestion with subtilisin resulted in the conversion of the native subunits ($M_r = 36,000$) to a large fragment with a molecular weight of approximately 29,000, called the S-subunit, and a smaller fragment, molecular weight = 6,500, called the S-peptide. The S-peptide was shown to be derived from the NH_2-terminal portion (*34*). Very similar results to those published earlier were obtained with Fru-P₂ases from rabbit, chicken, and snake muscle (Fig. 3). In the case of Fru-P₂ase from rat liver the pattern was somewhat different; in the first place the native subunit was somewhat larger ($M_r \cong 38,000$) and this was also true for the S-subunit ($M_r \cong 31,000$) generated when the conversion was complete. Also, in the case of rat liver Fru-P₂ase, we detected two species of intermediate molecular weight (Fig. 3A), formed during the early stages of digestion; these were eventually replaced by the 31,000-dalton S-subunit, which was the final product of the limited proteolysis of this enzyme.

Dissociation of the products of digestion by subtilisin did not occur under non-denaturing conditions and the molecular weight and tetrameric structure remained unchanged (*33* and Fig. 4). Thus limited proteolysis by subtilisin may be attributed to the presence of an exposed peptide segment containing the proteinase susceptible bonds (see below), leaving a residual structure resistant to further unfolding and digestion. Although nicking of the susceptible peptide bonds

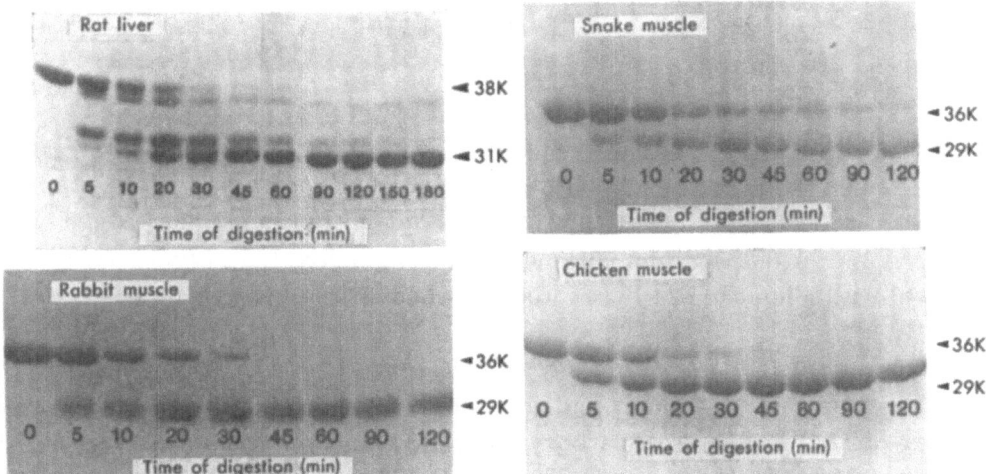

Fig. 3. Changes in subunit structure during digestion with subtilisin. Aliquots from the reaction mixtures in Fig. 1 were removed as indicated, acidified with HCOOH to a final concentration of 20% and lyophilized. SDS-polyacrylamide slab gel electrophoresis was carried out as described (*33*). (From Ph.D. dissertation by John S. MacGregor, in preparation).

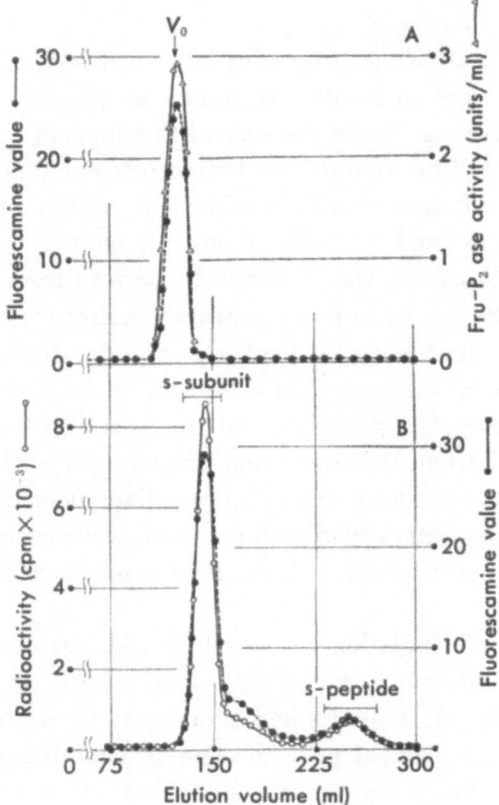

Fig. 4.   Gel filtration of subtilisin-digested rat liver Fru-P_2ase under non-dissociating (A) and dissociating (B) conditions. The conditions for chromatography on Sephadex G-75 were as described (*33*). A: in 0.1 M ammonium acetate, pH 6.5. B: in 9% HCOOH, after reduction and carboxymethylation (*33*). (From Ph.D. dissertation by John S. MacGregor, in preparation).

must result in a change in conformation, as witnessed by the change in catalytic and in some cases also the allosteric properties, this change in conformation does not result in dissociation of the nicked subunits and the molecular weight, which is 144,000 for the native rabbit enzyme, remains unaltered (*33*). Similar results were obtained with all of the Fru-P_2ases studied, whether from liver or muscle.

3.   Sequence analysis of S-peptides and of the NH_2-terminal region of the S-subunits

To separate the S-subunit and the S-peptide, the subtilisin digested proteins were dissociated in 8 M urea, carboxymethylated, and the fragments recovered by chromotography on Sephadex G-75 (Fig. 4). The amino acid sequences of the S-peptides from rabbit liver (*34*), and pig kidney (*31*) have been reported. (Fig. 5), and for the rabbit liver enzyme the sequence has also been determined for a cyanogen bromide peptide containing all of the subtilisin-sensitive bonds (*35*).

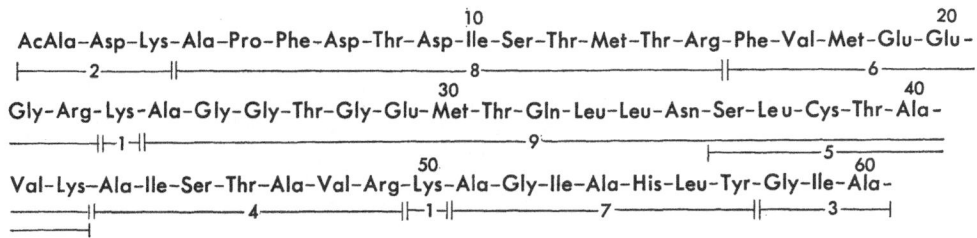

Rabbit liver   AcAla- Asp- Lys -Ala -Pro- Phe-Asp-Thr-Asp- Ile - Ser - Thr - Met - Thr - Arg - Phe - Val - Met - Glu - Glu

Pig kidney   AcThr   Glx   Ala   Asn   Val   Leu

10   20

Gly - Arg - Lys - Ala - Gly - Gly - Thr - Gly - Glu - Met - Thr - Glu - Leu - Leu - Asn - Ser - Leu - Cys - Thr - Ala

Arg

30   40

Val - Lys - Ala - Ile - Ser - Thr - Ala - Val - Arg - Lys - Ala - Gly - Ile - Ala - His - Leu - Tyr - Gly - Ile - Ala

50   60

FIG. 5.   Primary structures of the S-peptides from rabbit liver and pig kidney Fru-P₂ases. The sequences are from El-Dorry *et al.* (*34*) and Marcus *et al.* (*31*).

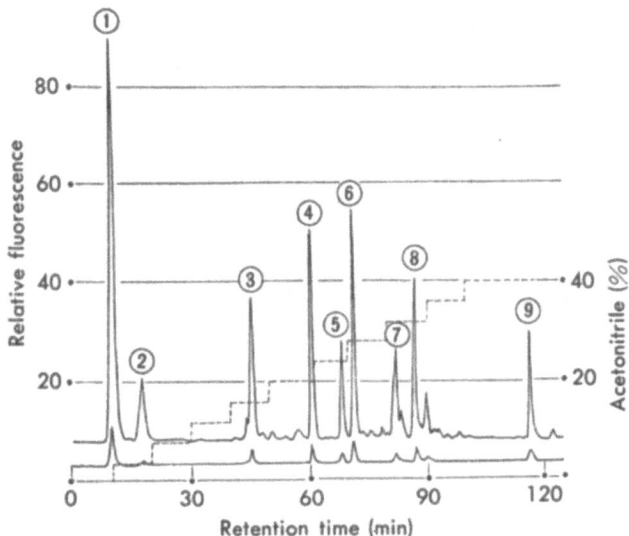

FIG. 6.   Separation of the tryptic peptides derived from the S-peptide of rabbit liver Fru-P₂ase. An aliquot (0.5 mg) of the S-peptide digested for 24 hr at 37°C with a ratio of trypsin:Fru-P₂ase of 1:100 (w/w) was chromatographed using a Roche HPLC Peptide Analyzer (*36*) with a 4.6 × 300 mm, 10 μ Chromegabond C-18 column (EM Industries, Martin, N.J.). Elution was at a flow rate of 22 ml/hr with a gradient of 0–40% acetonitrile in 0.2 M pyridine—1.0 M formic acid buffer, pH 2.8. Fractions corresponding to the peaks shown in the tracing were collected in a fraction collector. (From Ph.D. dissertation by John S. MacGregor, in preparation).

The major sites of cleavage are between Ala[60] and Gly[61] and between Thr[63] and Asn[64], but significant hydrolysis also occurs between Thr[66] and Gly[67], and to a slight extent between Tyr[57] and Gly[58]. The sequence reported for the S-peptide

```
              45        50        55        60        65        70        75
Rabbit liver   A-I-S┌T┬T-A-V-R┬K┐A-G┬I┐A-H-L┬Y┐G-I-A-G-S-T-N-V-T-G-D-Z-V-K-K-L-D-V
Pig kidney     A-I-S┤T│A-V-R┤K│A-G┤I│A-H-L┤Y│G-I-A-?
Rat liver      A-I-S┤S│A-V-R┤E┴A-G┤I│A-H-L┤Y│G-I-A-G-S-T-B-V-T-G-B-Z-V-K-K-L-B
Rabbit muscle  A-I-S┤S┤A-V-R(K)A-G┤L┤A-H-L┤Y│G-I-A-G-S-T-B-V-T-G-B-Z-V-K-K-L-B-V
Chicken muscle A-I-S┤S│A-V-R(K)A-G┤L│A-H-L┤F│G-I-A-?
Snake muscle   A-I-S (S, A, V, R)(K)A-G┤I│A-H-L┤Y│G-I-A-?
```

FIG. 7. Amino acid sequences of liver-and-muscle-type Fru-P$_2$ases in the region adjacent to the site of cleavage by subtilisin. Sequences were based on the result of manual Edman degradation, except for those shown in parentheses, which were deduced from the amino acid composition and expected homologies with the known sequences of the rabbit liver enzyme. The broken lines enclose portions of the sequences that are not fully conserved in all of the enzymes studied. (From Ph.D. dissertation by John S. McGregor, in preparation).

```
               1          5          10         15         20
Rabbit liver    AcA-D-K-A-P┬F┬D┬T-D┬I-S┬T┬M┬T-R┬F-V-M┬E┬E┬G-R-K
Pig kidney      AcT-D-E-A-A┤F│D┤T-N┤I-V┤T┤L┤T-R┤F-V-M┤E│E┤G-R-K
Rat liver       ?  T-D-D-A-P┤F┤E┤T-B┤I-S┤T┤L┤T-R┤F-V-L┤Z│Z┤G-R (K)
Rabbit muscle   (? A, E, K)S-P┤F┤Z┤T-B┤M-L┤T┤L┤T-R┤?
Chicken muscle  (? S, E, K)T-P┤F┤Z┤T (B, M, L┤T┤L┤T-R) F-V-M┤E┤K┤G-R (K)
Snake muscle    ? G-V-K                             Y-L-E┤E┤K┤G-R (K)
```

FIG. 8. Amino acid sequences of the NH$_2$-terminal regions of liver-and-muscle-type Fru-P$_2$ases. See legend at Fig. 7 for explanation. The broken lines enclose portions of the sequences that are not fully conserved in all of the enzymes studied. (From Ph.D. dissertation by John S. MacGregor, in preparation).

from rabbit liver has now been confirmed by isolation of nine tryptic peptides by a single high performance liquid chromatography procedure (Fig. 6). The S-peptides from rabbit liver and pig kidney Fru-P$_2$ases were found to be identical in the region adjacent to the major site of cleavage by subtilisin, but near the blocked NH$_2$-terminus there were substantial differences. The procedure illustrated in Fig. 4 was also employed to determine the sequences near the site of cleavage (Fig. 7) and at the NH$_2$-terminus (Fig. 8) for a number of other Fru-P$_2$ases. Near the protease sensitive site (Fig. 7) residues 43–60 were found to be identical for all Fru-P$_2$ases studied, whether derived from liver or muscle or from mammal, bird or reptile, except for homologous substitutions at positions 46, 53, and 57 and the substitution in the rat liver enzyme of glutamic acid (or glutamine) for lysine at position 50. On the other side of the major cleavage site data are available only for the rabbit liver, rat liver, and rabbit muscle enzymes; the sequences for these enzymes from residues 61 to 74 were found to be identical. Thus, there is almost complete conservation of primary structure in a region including 31 amino acid residues over a wide range of vertebrate classes.

In contrast to the conservation of primary structure near the proteinase-sensitive site, many amino acid substitutions were observed near the NH$_2$-terminus (Fig. 8). Only 9 of the first 23 residues in this region were identical in all of the Fru-P$_2$ases thus far examined.

Bonds susceptible to hydrolysis by subtilisin are located between residues 57 and 67, and this segment must be located on the surface of the protein. Conservation of primary structure in this and adjacent regions of the molecule suggests that it plays an essential role in some function, either in the reaction catalyzed by the enzyme or in some aspect of its regulation. A direct role in catalysis is unlikely, since hydrolysis of peptide bonds in this region does not result in diminished catalytic activity. The essential function may be related to turnover of the protein, or to some unrecognized mechanism of its regulation. This region on the surface of the protein may also be involved in its interaction *in vivo* with other gluconeogenic enzymes; we have recently reported evidence (*37*) for the formation of a complex between rabbit liver Fru-P$_2$ase and rabbit liver aldolase and we have found that complex formation does not occur with Fru-P$_2$ase modified by subtilisin (unpublished observation). However, complex formation involving muscle Fru-P$_2$ase, which shares the subtilisin-sensitive site, has not yet been demonstrated.

4. *Limited proteolysis of Fru-P$_2$ases by lysosomal proteinases*

Rat liver lysosomes contain a proteinase, called Fru-P$_2$ase converting enzyme, that catalyzes the hydrolysis of peptide bond Asn^{64}-Val^{65} in rabbit liver Fru-P$_2$ase (*38*). The effect on catalytic activity is similar to that described for the modification catalyzed by subtilisin. Rat liver cathepsins A, B, C or D did not show this activity. A similar Fru-P$_2$ase converting enzyme is present in rabbit liver lysosomes and cathepsin B from this source also catalyzes the limited proteolysis of rabbit Fru-P$_2$ase (*39*).

Apart from the differences in activities of cathepsin B from the two species, there are also differences in the properties of the converting enzymes. Although both converting enzymes are detected in association with lysosomal membranes, the rat liver lysosomes are fully cryptic (*38*), whereas the modification of rabbit liver Fru-P$_2$ase is observed with intact rabbit liver lysosomes (*39*). Indeed the activity of intact rabbit liver lysosomes is equal to that of the membrane-bound fraction isolated from the disrupted lysosomes. It would appear that modification of Fru-P$_2$ase in rabbit liver can occur without disruption of, or entry of Fru-P$_2$ase into, the lysosomes. This observation, and the finding that the level of the membrane-bound Fru-P$_2$ase in rabbit liver is increased 10-fold during fasting (*40*), suggests that the lysosomal proteinase plays some role in the regulation of this step in gluconeogenesis.

Concluding Remarks

Professor Egami has been a leader in research on the origin of life and the molecular aspects of evolution. The work described in this chapter represents an extension of his pioneering concepts to the study of a particular enzyme; analysis of the structure of this enzyme from a variety of sources has provided insights into its regulation *in vivo* and clues to phylogenetic relationships.

REFERENCES

1. Gomori, G.: 1943. *J. Biol. Chem.* 148, 139–149.
2. Pontremoli, S., Luppis, B., Wood, W.A., Traniello, S., and Horecker, B.L.: 1965. *J. Biol. Chem.* 240, 3464–3472.
3. Pogell, B.M. and McGilvery, R.W.: 1954. *J. Biol. Chem.* 208, 149–157.
4. Mokrasch, L.C. and McGilvery, R.W.: 1956. *J. Biol. Chem.* 221, 909–917.
5. Mendicino, J. and Vasarhely, F.: 1963. *J. Biol. Chem.* 238, 3528–3534.
6. McGilvery, R.W.: 1961. *Fructose-1,6-Diphosphatase and its Role in Gluconeogenesis* (McGilvery, R.W. and Pogell, B.M., eds.) pp. 1–2, American Institute of Biological Sciences, Washington, D.C.
7. Hers, H.G. and Kusaka, T.: 1953. *Biochim. Biophys. Acta* 11, 427–437.
8. Byrne, W.L.: 1961. *Fructose-1,6-Diphosphatase and its Role in Gluconeogenesis* (McGilvery, R.W. and Pogell, B.M., eds.) pp. 89–100, American Institute of Biological Sciences, Washington, D.C.
9. Pogell, B.M. and McGilvery, R.W.: 1952. *J. Biol. Chem.* 197, 293–302.
10. Nakashima, K. and Horecker, B.L.: 1971. *Arch. Biochem. Biophys.* 146, 153–160.
11. Byrne, W.L., Rajagopalan, G.T., Griffin, L.D., Ellis, E.H., Harris, T.M., Hochachka, P., Reid, L., and Geller, A.M.: 1971. *Arch. Biochem. Biophys.* 146, 118–133.
12. Traniello, S., Pontremoli, S., Tashima, Y., and Horecker, B.L.: 1971. *Arch Biochem. Biophys.* 146, 161–166.
13. Tashima, Y., Tholey, G., Drummond, G., Bertrand, H., Rosenberg, J.S., and Horecker, B.L.: 1972. *Arch. Biochem. Biophys.* 149, 118–126.
14. Olson, J.P. and Marquardt, R.R.: 1972. *Biochim. Biophys. Acta* 268, 453–467.
15. Han, P.F., Murthy, V.V., and Johnson, J., Jr.: 1976. *Arch. Biochem. Biophys.* 173, 293–300.
16. Chang, Y.K.Y., Thompson, E.O.P., and Zalitis, J.: 1974. *Proc. Aust. Biochem. Soc. 7, 5.*
17. Zalitis, J.: 1976. *Biochem. Biophys. Res. Commun.* 70, 323–329.
18. Tejwani, G.A., Pedrosa, F.O., Pontremoli, S., and Horecker, B.L.: 1976. *Arch. Biochem. Biophys.* 177, 253–264.
19. Tashima, Y., Mizunuma, H., and Hasegawa, M.: 1979. *J. Biochem.* 86, 1089–1099.
20. Han, P.F., Owen, G.S., and Johnson, J., Jr.: 1975. *Arch. Biochem. Biophys.* 168, 171–179.
21. Black, W.J., Van Tol, A., Fernando, J., and Horecker, B.L.: 1972. *Arch. Biochem. Biophys.* 151, 576–590.
22. Annamalai, A.E., Tsolas, O., and Horecker, B.L.: 1977. *Arch. Biochem. Biophys.* 183, 48–56.
23. Shi Jianping and Xu Genjun: 1981. *Acta Biochim. Biophys. Sin.* 13, 82–90.
24. Mizunuma, H. and Tashima, Y.: 1978. *J. Biochem.* 84, 327–336.
25. Mizunuma, H., Hasegawa, M., and Tashima, Y.: 1980. *Arch. Biochem. Biophys.* 201, 296–303.
26. Enser, M., Shapiro, S., and Horecker, B.L.: 1969. *Arch. Biochem. Biophys.* 129, 377–383.

27. Abrams, B., Sasaki, T., Datta, A., Melloni, E., Pontremoli, S., and Horecker, B.L.: 1975. *Arch. Biochem. Biophys.* 169, 116–125.
28. Pontremoli, S., Melloni, E., and Traniello, S.: 1971. *Arch. Biochem. Biophys.* 147, 762–766.
29. Geller, A.M., Rajagopolan, G.T., Ellis, E.H., and Byrne, W.L.: 1971. *Arch. Biochem. Biophys.* 146, 134–143.
30. Traniello, S., Melloni, E., Pontremoli, S., Sia, C.L., and Horecker, B.L.: 1972. *Arch. Biochem. Biophys.* 149, 222–231.
31. Marcus, F., Edelstein, I., Keim, P.S., and Henrikson, R.L.: 1979. *Fed. Proc.* 38, 507 (Abstr.).
32. Annamalai, A.E.: 1976. Ph.D. Thesis, Rutgers University.
33. Dzugaj, A., Chu, D.K., El-Dorry, H.A., and Horecker, B.L.: 1976. *Biochem. Biophys. Res. Commun.* 70, 638–646.
34. El-Dorry, H.A., Chu, D.K., Dzugaj, A., Botelho, L.H., Pontremoli, S., and Horecker, B.L.: 1977. *Arch. Biochem. Biophys.* 182, 763–773.
35. Botelho, L.H., El-Dorry, H.A., Crivellaro, O., Chu, D.K., Pontremoli, S., and Horecker, B.L.: 1977. *Arch. Biochem. Biophys.* 184, 535–545.
36. Rubinstein, M.: 1979. *Anal. Biochem.* 98, 1–7.
37. MacGregor, J.S., Singh, V.N., Davoust, S., Melloni, E., Pontremoli, S., and Horecker, B.L.: 1980. *Proc. Natl. Acad. Sci. USA* 77, 3889–3892.
38. Lazo, P.S., Tsolas, O., Sun, S.C., Pontremoli, S., and Horecker, B.L.: 1978. *Arch. Biochem. Biophys.* 188, 308–314.
39. Melloni, E., Pontremoli, S., Salamino, F., Sparatore, B., Michetti, M., and Horecker, B.L.: 1981. *Arch. Biochem. Biophys.* 208, 175–183.
40. Melloni, E., Pontremoli, S., Salamino, F., Sparatore, B., Michetti, M., and Horecker, B.L.: 1981. *Proc. Natl. Acad. Sci. USA* 78, 1499–1502.

Some Aspects of the Enzymology of Pancreas and Intestine

P. DESNUELLE

P.A.E. DESNUELLE was born in August 1911 in Lons le Saunier (France) and educated at the University of Lyon (1929–1932) and at the School of Industrial Chemistry in the same city. His Doctor of Sciences degree was prepared with Claude Fromageot and awarded in 1935. A Rockefeller Fellowship in 1935 enabled him to work for one year with Richard Kuhn in the Max-Planck-Institute of Heidelberg. He has been assistant professor in Lyon from 1932–1943, Professor of Biochemistry in Marseille since 1943, Director of the Institute of Biological Chemistry (University of Marseille) during 1953–1968, Director of the Center of Biochemistry and Molecular Biology (Centre National de la Recherche Scientifique) since 1968, and Director of the National Laboratory for Fat Research from 1943–1964. His research covers the following fields: enzymology of the exocrine pancreas (activation of zymogens, heterogeneous biocatalysis induced by lipase), enzymology of the small intestine (mode of insertion and biosynthesis of brush border enzymes), adaptation of pancreatic enzymes to dietary and hormonal stimuli, and technological and biochemical problems of the industry of edible fats.

Under the influence of Claude Fromageot in Paris and Richard Kuhn in Heidelberg, I became an enzymologist more interested in structure-function relationships than in kinetics and theory. I still remember that my first steps in the field were accomplished in 1936 when Richard Kuhn, just after his discovery of lactoflavin, asked me to analyze the non dialyzable component of the "old" yellow enzyme from yeast. The analytical techniques were very crude and hardly specific at that time. The protein nature of enzymes was not yet generally accepted and I felt obliged to crystallize derivatives of some amino acids in the hydrolysates and check their elementary composition and melting point.

A little later, I was attracted by the enzymes secreted by the exocrine pancreas, just because they appeared to be very abundant, reasonably stable, and simple molecules playing an important role in gastroenterology. The book by Northrop and Kunitz (1), read almost by chance, drew my attention to the chemical processes determining the activation of pancreatic zymogens such as chymotrypsinogen and trypsinogen. This work would not have been possible without the help of Moses Kunitz himself who, after a short discussion on the scientific motivations of the project, opened the door of a vast cold room in the Rockefeller University and kindly gave me large amounts of crystalline material prepared several years ago but apparently still very good. Another favorable circumstance was a short stay in Fred Sanger's Laboratory where I got the idea that the FDNB (fluorodinitrobenzene) technique worked out by him for the determination of N-terminal residues in proteins and peptides could also be used for the study of limited proteolysis. In this way, we were able with Dr. Mireille Rovery and others to show that the proteolytic cleavage of a single peptide bound in chymotrypsinogen and trypsinogen was sufficient for converting inactive (or poorly functional) zymogens into fully active enzymes (2–4). The general significance of limited proteolysis in a number of biological processes is now well recognized.

The limited proteolysis inducing trypsinogen activation is of special importance due to the role played by trypsin in the activation of other pancreatic zymogens. Dr. Suzanne Maroux (5) in this Laboratory showed that this activation is not initiated by trypsin, but by an intestinal enzyme called enteropeptidase. This enzyme, which is loosely bound to the duodenal brush border membrane, is built in such a way as to recognize in a strictly specific manner the structure (Asp)4-Lys present in the N-terminal sequence of all trypsinogens just before the Lys-Ile bond cleaved during activation. Thus, enteropeptidase is able to activate trypsinogen very fast and, unlike trypsin, without undesirable side effects leading to "inert" proteins.

Another interesting observation was made by Dr. A. Puigserver on bovine procarboxypeptidase A. This zymogen which is a trimer in the bovine pancreas was first dissociated into constitutive subunits by a nondenaturing technique using dimethylsuccinic acid as an acrylating agent. The advantage of this reagent is that acylation and subsequent deacylation may be carried out at pH near neutrality. Thus, native and fully functional subunits could be separated. The first was found to be the real zymogen of carboxypeptidase A and the second to be a chymotrypsinogen of type C. The third subunit was characterized as a chymotrypsinogen or a proelastase in which the first 17 residues in the N-terminal part of the chain were lacking (6). This component is a real zymogen equipped with a latent active site. But it cannot be converted into a fully functional enzyme. We were very lucky on this occasion because inactive or inactivated enzymes are usually very difficult to characterize. The favorable circumstance here was that the inactive protein formed with active enzymes a complex which could be purified and later dissociated.

Another pancreatic enzyme extensively studied in the laboratory by quite a few people (Louis Sarda, Marie-France Maylié, Catherine Chapus, Robert Verger, Michel Sémériva, and Maurice Charles) is lipase and its cofactor, colipase (7, 8). The most interesting property of lipase is to act very fast on water-insoluble lipid substrates. These substrates do not form aqueous molecular dispersions but emulsified particles containing a considerable number of molecules separated from water by an interface. Lipase recognizes these aggregates much better than isolated monomers. The activation factor at the transition between solution and emulsion is at least 10^3-fold.

The corresponding original process called "heterogeneous biocatalysis" requires as a first step that the enzyme adsorbs to the substrate interface. Lipase adsorbs readily to hydrophobic interfaces. But this adsorption is hindered by bile salts and other amphiphiles present in large amounts in the duodenum during fat digestion. Therefore, the pancreas must synthesize a small protein cofactor called colipase to insure normal digestion. Colipase is able to adsorb to bile salt-coated triglyceride interfaces and then to serve as an anchor for lipase. Colipase is an excellent model for studies of protein-lipid interactions and protein-protein interactions mediated by an organized lipid phase.

The already mentioned activation of trypsinogen by enteropeptidase led Drs. Suzanne Maroux and D. Louvard to investigate intestinal hydrolases (9). These enzymes are entirely different from pancreatic enzymes. Their size is considerably larger. Instead of being dissolved in an extracellular aqueous phase (the pancreatic juice) after exocytosis, they are bound to the enterocyte brush border membrane, thus raising a number of problems concerning their biosynthesis, mode of integration, and function. Investigations were mostly carried out with one of the major brush border enzymes, an aminopeptidase called aminopeptidase N which splits neutral and basic N-terminal residues in peptides originating from the intestinal lumen. Two enzyme forms were solubilized, respectively, by proteolytic digestion of the membrane (the protease form) or incubation with a detergent (the detergent form). The second was found to differ from the first by the presence of a short, mostly hydrophobic sequence (about 30–40 residues) whose role is to penetrate into the lipid matrix and hold the enzyme at the membrane surface.

Moreover, nearly the same number of antigenic sites could be titrated by an ^{125}I-labeled monospecific antibody in the solubilized and membrane-bound enzyme, thus proving that the surface of the molecule is almost completely outside the plane of the membrane. In other words, the brush border hydrolases appear to be amphipathic and highly polarized structures with a short hydrophobic anchor plunging into the membrane and a bulky enzymatically active "head" which protrudes from the external side of the membrane facing the lumen. For thermodynamic reasons, this structure allows free rotation around the anchor and lateral diffusion along the membrane plane. But it excludes the possibility for the strongly hydrophilic "head" to cross the lipid matrix and appear, as suggested by some authors, alternatively on the external and the cytoplasmic side of the membrane. This may have important consequences for our ideas concerning the transport of amino acids and other nutriments across the brush border during intestinal absorption.

Furthermore, the hydrophobic anchor of aminopeptidase was found to be localized at the N-terminal of the enzyme chain. This finding raises the question of how the molecule is integrated into the membrane after its biosynthesis. It has been postulated that membrane proteins penetrate into the cavities of the endoplasmic reticulum with their N-terminal signal sequence forwards and that, in contrast with secretory proteins, they are stopped before discharge by another special sequence which may be assumed to be C-terminal in surface cell constituents. In fact, a number of membrane proteins have been found to be bound by the C-terminal. However, the mode of integration of most brush border hydrolases should not be expected to follow this model. An alternative hypothesis is that the α-NH_3^+ group of the chain does not really enter the lipid matrix and remains on the cytoplasmic side. Then, the growing chain would cross the membrane and form a hairpin loop outside the cell. It will remain hung by the N-terminal after discharge from the ribosome.

In conclusion, it is gratifying to see that enzymology is now at a crossroads at which physicists, physicochemists, chemists, and cell biologists meet for the profit of biochemistry.

REFERENCES

1. Northrop, J.H., Kunitz, M., and Herriott, R.: 1948. *Crystalline Enzymes*, 2nd ed., Columbia Univ. Press, New York.
2. Desnuelle, P.: 1953. *Adv. Enzymol.* 24, 261–317.
3. Desnuelle, P.: 1954. *Annu. Rev. Biochem.* 23, 55–68.
4. Desnuelle, P.: 1960. *The Enzymes*, 2nd ed., Vol. 4 (Boyer, P.D., Lardy, H., and Myrbäck, K., eds.) pp. 93–113 and 119–132, Academic Press, New York.
5. Maroux, S., Baratti, J., and Desnuelle, P.: 1971. *J. Biol. Chem.* 246, 5031–5040.
6. Puigserver, A. and Desnuelle, P.: 1975. *Proc. Natl. Acad. Sci. USA* 72, 2442–2445.
7. Desnuelle, P.: 1972. *The Enzymes*, 3rd ed., Vol. 7 (Boyer, P.D., ed.) pp. 575–616, Academic Press, New York.
8. Sémériva, M. and Desnuelle, P.: 1979. *Adv. Enzymol.* 48, 319–370.
9. Desnuelle, P.: 1979. *Eur. J. Biochem.* 101, 1–11.

The Mechanism of Pepsin Action

JOSEPH S. FRUTON

J.S. FRUTON received his B.A. degree in chemistry (1931) and Ph.D. degree in biochemistry (1934) at Columbia University, and was on the staff of the Rockefeller Institute for Medical Research from 1934 to 1945. He then moved to Yale University and has been Eugene Higgins Professor of Biochemistry there since 1957. From 1951 to 1967 he was Chairman of the Department of Biochemistry and served as Director of Yale's Division of Science from 1959 to 1962. He is a member of the National Academy of Sciences, American Philosophical Society, and the American Academy of Arts and Sciences. In 1944 he received the Eli Lilly Award in biological chemistry, and in 1976 a honorary D.Ss. degree from Rockefeller University. He is the author or co-author of approximately 300 articles, and, with Sofia Simmonds, wrote "General Biochemistry" (1st Ed., 1953; 2nd Ed., 1958), as well as "Molecules and Life: Historical Essays on the Interplay of Chemistry and Biology" (1972). His research interests have been in bioorganic chemistry, with primary emphasis on the specificity and mechanism of enzyme action, peptide chemistry, transpeptidation reaction catalyzed by proteinases, and fluorescence spectroscopy.

Pepsin occupies a significant place in the historical development of biochemistry because on several occasions during the past 150 years the study of this enzyme helped to clarify important problems. It is customary to begin the story of pepsin with the work of Schwann, who gave it its name in 1836, but the credit for its discovery belongs more properly to a Würzburg physician, Johann Nepomuk Eberle (1798–1834). In the last year of his short life, he reported that he had prepared an "artificial gastric juice" by extraction of dried stomach mucosa, and that the physiological digestion of what we now call proteins is a chemical process, because he had demonstrated the dissolution of fibrin by this extract (1). In 1834 also,

Mitscherlich proposed his idea of "contact action" to explain fermentations, and the physiologist Müller recognized that Eberle's finding provided another example of such action. Müller's student Schwann greatly extended Eberle's work, and a few years later another of Müller's students, Adolph Wasmann (1807–1853), described the concentration of the active principle. By 1840, pepsin (along with amylase and emulsin) was well established as the prototype of the so-called "unorganized ferments", to which Kühne gave the name "enzymes" in 1876.

Throughout the latter half of the nineteenth century, there was much debate about the relation of agents like pepsin to the so-called "organized ferments," especially those concerned with the breakdown of sugar to alcohol and CO_2 (2). For Pasteur, such fermentations are correlative with the life of the microorganisms that effect them; along with most leading physiologists (e.g., Nägeli, Pflüger), he considered to be both unnecessary and implausible the proposal of the chemist Moritz Traube that intracellular fermentations and oxidations involve the action of substances analogous to pepsin. Thus, although the discovery of pepsin showed that living cells can secrete into their surroundings catalytic agents that effect chemical breakdown, the concept that such agents are also important in intracellular metabolism was not accepted. Because of the prominence of this debate, Eduard Buchner's report in 1897 that he had prepared a yeast extract that could ferment glucose was widely hailed. Buchner believed that the extract contained an enzyme, which he named zymase, and considered it to be a protein.

The question of the protein nature of enzymes was still open, however, and the study of pepsin figures largely in its resolution. During the 80 years after Wasmann's work, there were numerous attempts to purify pepsin, notably by Brücke and Wittich during the 1870's and by Pekelharing around 1900. Although some of the preparations contained protein, others were claimed to be protein-free. Before 1930, many enzyme chemists, the most famous of whom was Willstätter, believed that enzymes are small active molecules, and that the presence of proteins in enzyme preparations arises from the adsorption of such molecules by colloids. As a consequence, Sumner's claim in 1926 to have isolated urease in the form of a crystalline protein was not generally accepted. A decisive change came after 1930, however, when Northrop (3) described the crystallization of swine pepsin and presented massive evidence to show that its catalytic activity is an intrinsic property of the crystalline protein. The widespread recognition of the protein nature of enzymes that followed Northrop's work had a profound influence on the subsequent development of biochemistry.

In addition to the historical importance of pepsin in relation to protein metabolism, the role of intracellular enzymes, and the protein nature of enzymes, the study of its action played a significant role in the development of ideas about the structure of proteins. For nearly a 100 years after Eberle's discovery of pepsin, it was considered to effect only the dissolution of proteins in an acidic medium without extensive structural change in these substrates. Indeed, during the 1920's, it was believed to act by the deaggregation of non-covalent associations of small peptides or diketopiperazines (*4*). At that time, several noted chemists (Bergmann, Karrer) offered theories of protein structure that invoked such physical aggregation, and X-ray data were cited in support of these views (*5*). Importance was also attached to the fact that no known synthetic peptide had been found to be a substrate for pepsin. This argument against the polypeptide theory of protein structure was removed in 1938 when I showed that Z-Glu-Tyr and its amide are hydrolyzed at the Glu-Tyr bond, albeit slowly (*6*). Subsequently, Baker found that Ac-Phe-Tyr is a better substrate, thus providing the first indication that the action of pepsin on small synthetic substrates is favored by the presence of aromatic amino acid residues on both sides of the sensitive peptide bond.

During the 1960's, I initiated a systematic investigation of the specificity of pepsin by the synthesis of an extensive series of new peptide substrates of varying length and amino acid composition. This work was begun by Inouye, who made peptides of the type Z-His-X-Y-OEt, where X and Y were Phe, Tyr, or Trp; the histidyl residue was inserted to confer solubility on the substrates in aqueous buffers in the pH range 2–5 (*8*). In connection with this work, Inouye showed not only that pepsin exhibits an absolute requirement for the L enantiomer in both the X and Y positions of substrates such as A-Phe-Phe-B, but also that replacement of the L-phenylalanyl residue in the Y position by a β-phenyl-L-lactyl residue gives depsipeptides that are cleaved rapidly by pepsin, thus establishing unequivocally the action of pepsin as an esterase. Also, Inouye introduced the use of the *p*-nitro-L-phenylalanyl residue in the X position of such substrates, and developed a valuable spectrophotometric method for the study of pepsin kinetics. Further work in my laboratory on the specificity of pepsin by Hollands, Trout, and Medzihradszky strengthened the conclusion that, with substrates of the type A-His-X-Y-B, the X-Y bond is cleaved most rapidly when it links two aromatic L-amino acids, and also demonstrated that elongation of the peptide chain of the substrate (as with B=Ala-Ala-OMe) can lead to large increases (up to 700-fold) in the susceptibility of the X-Y bond (*9–11*). Perhaps the most striking result of these kinetic studies was that the differences in the rates of pepsin action were reflected in large changes in k_{cat}, and that the K_m values were relatively invariant. On the assumption that K_m represented a binding constant (later work in our laboratory showed this to be valid), I concluded that the active site of pepsin is a flexible structure and that secondary enzyme-substrate interactions at a distance from catalytic groups may enhance catalysis by the utilization of binding energy to lower the free energy of activation (*12*).

In further work, my associate Sachdev prepared new synthetic substrates in which a pyridyl group, attached to the carboxyl-terminus, served as the site of protonation (*13*). The extensive series of peptide substrates he tested included compounds that were cleaved at a Phe-Phe bond with k_{cat} values differing by 10^3, but with essentially the same K_m values. From the effect of the elongation of substrates of the type A-Phe-Phe-B at the amino and carboxyl ends of the sensitive Phe-Phe unit, we inferred that the active site of pepsin can accommodate at least a heptapeptide; in a fully-extended conformation, this would correspond to a distance of about 25 Å. The results also confirmed our conclusions about the decisive importance of secondary enzyme-substrate interactions in pepsin action, and explained some of the earlier confusion about the specificity of the enzyme drawn from the study of the peptic cleavage of proteins.

The availability of a large series of synthetic peptide substrates for swine pepsin allowed us to examine the specificity of other proteinases which also act optimally in the pH range 2–5 and are therefore named "acid proteinases." These enzymes included rennin (chymosin), several mold proteinases, and the lysosomal cathepsin D; the last had been purified from beef spleen in my laboratory. The results indicated that although their primary specificity with respect to the amino acid residues flanking the sensitive peptide bond is similar to that of pepsin, and their extended active sites have similar dimensions, there are significant differences in secondary specificity (*14*).

These conclusions were drawn before X-ray crystallographic studies had revealed the three-dimensional structure of any of the acid proteinases, or indeed the amino acid sequence of swine pepsin. The X-ray studies (*e.g.*, *15*) have shown the acid proteinases to possess a bilobal structure, with the two catalytic residues (Asp-32 and Asp-215 in the case of pepsin) identified by means of selective chemical modification (*16*) on the two lobes. These are separated by a 30 Å long groove that represents the active site binding region. The agreement between the conclusions we had drawn from kinetic data and the later structural studies was gratifying, and invited the hypothesis that the flexibility we had postulated for the extended active site arises largely from the movement of the lobes with respect to each other. It was also a source of satisfaction that the emphasis we had placed on the flexibility of the active site of pepsin was reflected during the 1970's in the discussion of the dynamics of enzyme action, and in the more widespread recognition that the rigid structures deduced from X-ray crystallographic measurements represent only one of the many conformations an enzyme may assume during the catalytic process in solution.

These newer findings were important in relation to the problem of the intimate chemical mechanism of pepsin action. Like other proteolytic enzymes, pepsin catalyzes transpeptidation reactions, as well as the hydrolysis of amide and ester bonds. For a time it was assumed that the two enzymic carboxyl groups of Asp-32 and Asp-215 effect catalysis by the action of one as a proton donor and the other (in its carboxylate form) as a nucleophile to form a tetrahedral intermediate which can then be converted to either an imino-enzyme or an acyl-enzyme. However, work by Silver (*17*) cast serious doubt on imino-transfer in transpeptidation reactions catalyzed by pepsin, and the question of the catalytic intermediates in the action of pepsin is unclear at present. I have offered the hypothesis that detectable covalently bound acyl-enzyme or imino-enzyme compounds may not be intermediates in pepsin catalysis, but that the sequence of the departure of the two products of hydrolysis may be a reflection of the interaction of each of them with complementary groups in a flexible extended active site (18). According to this hypothesis, such interactions may be coupled in a manner as to allow the nature of one product to influence the rate of departure of the other through the effect it has on the conformational state of the active site. If this hypothesis is correct, the mechanism of pepsin catalysis should be analogous to the acid-base catalysis of non-enzymic reactions, and not involve nucleophilic attack of the carbonyl group of the sensitive bond.

Several methods are now available for the study of the dynamics of conformational changes in proteins, and the one I have used recently for the further study of the mechanism of pepsin action is fluorescence spectroscopy. The initial efforts involved the use of pepsin substrates of the type A-Phe-Phe-B, which carried a dansyl group at the amino terminus, as in Dns-(Gly)$_n$-Phe-Phe-OP4P [$n=0, 1, 2$; OP4P $=3$-(4-pyridyl)propyl-1-oxy)], but we found this probe group to be insufficiently sensitive for our purpose (19). Examination of a variety of other naphthalene sulfonates showed the mansyl group [Mns, 6-(N-methylanilino)-2-naphthalenesulfonyl] to be most suitable, not only because it is essentially non-fluorescent in aqueous solution and fluoresces strongly in non-polar solvents, but also because it is uneffected by reactions in which mansyl chloride reacts with peptides. We found that the active site of pepsin has relatively little intrinsic affinity for the fluorescent probe group when it is present in compounds such as mansylamide or Mns-Gly-Gly-OP4P, but when the mansyl group is attached to a peptide having a Phe-Phe unit, its fluorescence is greatly enhanced (20). It was evident that the mansyl group had been drawn into a region of lower polarity by virtue of the interaction of the Phe-Phe unit of the substrate with the active site of pepsin. Stopped-flow kinetic studies, under conditions of enzyme excess, unequivocally showed that the rate-limiting step in the over-all catalytic process is associated with the transformation of the first detectable enzyme-substrate complex (21). At ordinary temperatures, however, the initial increase in fluorescence upon binding of a mansyl peptide substrate to pepsin is too rapid for measurement by stopped-flow techniques of the rate of conformational change at the active site. For this reason, we are now attempting to study this process at lower temperatures, in the manner that has proved to be valuable in the case of other enzymes (22). We hope that our current studies will make a further contribution to the understanding of the mechanism of pepsin action, and thus to throw more light on the problem of the dynamics of enzyme catalysis in general. In my opinion, research on pepsin will continue to illuminate fundamental problems of biochemistry, as it has done repeatedly since its discovery nearly 150 yaers ago.

REFERENCES

1. Hickel, E.: 1975. *Naturwiss. Rundsch.* 28, 14–18.
2. Fruton, J.S.: 1972. *Molecules and Life*, pp. 22–86, Wiley, New York.
3. Northrop, J.H.: 1930. *J. Gen. Physiol.* 13, 739–766.
4. Oppenheimer, C.: 1926. *Die Fermente und ihre Wirkungen*, 5th ed., Vol. 2, p. 812, Thieme, Leipzig.
5. Fruton, J.S.: 1979. *Ann. N.Y. Acad. Sci.* 325, 1–15.
6. Fruton, J.S. and Bergmann, M.: 1938. *Science* 87, 557.
7. Baker, L.E.: 1951. *J. Biol. Chem.* 193, 809–819.
8. Inouye, K. and Fruton, J.S.: 1967. *Biochemistry* 6, 1765–1777.
9. Hollands, T.R., Voynick, I.M., and Fruton, J.S.: 1969. *Biochemistry* 8, 575–585.
10. Trout, G.E. and Fruton, J.S.: 1969. *Biochemistry* 8, 4183–4190.
11. Medzhiradszky, K., Voynick, I.M., Medzihradszky-Schweiger, H., and Fruton, J.S.: 1970. *Biochemistry* 9, 1154–1162.
12. Fruton, J.S.: 1970. *Adv. Enzymol.* 33, 401–443.
13. Sachdev, G.P. and Fruton, J.S.: 1969. *Biochemistry* 8, 4231–4238.
14. Sampath-Kumar, P.S. and Fruton, J.S.: 1974. *Proc. Natl. Acad. Sci. USA* 71, 1070–1072.
15. Hsu, I., Delbaere, L.T.J., James, M.N.G., and Hofmann, T.: 1977. *Nature* 266, 140–145.
16. Fruton, J.S.: 1974. *Acc. Chem. Res.* 7, 241–246.
17. Silver, M. and Stoddard, M.S.: 1975. *Biochemistry* 14, 614–621.
18. Fruton, J.S.: 1976. *Adv. Enzymol.* 44, 1–36.
19. Sachdev, G.P., Johnston, M.A., and Fruton, J.S.: 1972. *Biochemistry* 11, 1080–1086.
20. Sachdev, G.P., Brownstein, A.D., and Fruton, J.S.: 1973. *J. Biol. Chem.* 248, 6292–6299.
21. Sachdev, G.P. and Fruton, J.S.: 1975. *Proc. Natl. Acad. Sci. USA* 72, 3424–3427.
22. Douzou, P.: 1979. *Q. Rev. Biophys.* 12, 521–569.

Specificity and Cytochemical Localization of Ribonucleases

DAVID SHUGAR

D. SHUGAR received the B.Sc. and Ph.D. degrees in physics, both from McGill University in Montreal. From 1948 to 1950 he was research associate at the Pasteur Institute in Paris, and also the Institut de Biologie Physico-Chimique in Paris. Subsequently he was also research associate from 1950 to 1952 at the University Libre de Bruxelles. Then he worked for some years at the Institute of Hygiene in Warsaw. Presently he is Professor of Biophysics both in this Institute of the Academy of Sciences, and in the Department of Biophysics, University of Warsaw. His general fields of research interest for some years have been the chemistry and physical chemistry of nucleic acids and their constituents, the properties of nucleolytic enzymes, photochemistry of nucleic acids, mechanisms of radiation and chemical mutagenesis, and bacterial transformation.

It is both a pleasure and a privilege to recall, on this occasion, some of our earlier research intimately related to, and in part stimulated by, the well-known investigations of Professor Fujio Egami and his coworkers on ribonucleases T_1 and T_2, as well as a number of other associated fungal and bacterial ribonucleases. RNase T_1, with its specificity for internucleotide linkages adjacent to guanosine residues, is of particular interest not only from the point of view of fundamental research on specificity of RNases, but also because of its widespread use as a tool in many diversified fields. It proved to be decisive in the determination of the first reported sequence of a nucleic acid, that of tRNA[Ala], by Holley and his coworkers, and it is still used in RNA sequence analysis. It is useful for the preparation of oligonucleotides of guanosine. And it is almost a routine tool for digestion of residual RNA as one of the steps in purification of DNA from many sources. Our own laboratory makes use of it almost daily, in particular for purification of transforming DNA.

Some years ago, Dr. Halina Sierakowska and I initiated some work on the cytochemical localization of ribonucleases and, subsequently, phosphodiesterases. Some of this research, principally on nucleases of mammalian origin, has been set forth in detail elsewhere (1, 2). Localization of an enzyme is not only of intrinsic interest; it is frequently helpful in deducing some information about its function in the cell. Here we shall summarize only the principles of the techniques employed, including the types of specific substrates synthesized for this purpose.

One of the useful substrates developed for localization of pancreatic-type ribonucleases was α-naphthyl uridine-3'-phosphate (3). This is completely hydrolyzed by RNase to uridine-2',3'-cyclic phosphate (and eventually to uridine-3'-monophosphate (3'-UMP)) and free naphthol. The relative specificity of this substrate for RNase is enhanced by the resistance of α-naphthyl nucleoside-3'-phosphates to spleen phosphodiesterase (PDase II), which hydrolyzes other esters of nucleoside-3'-phosphates. The liberation of free naphthol accompanying hydrolysis by RNase logically suggested a one-step procedure for RNase localization, based on coupling of the enzymatically liberated naphthol with a suitable diazotate, according to standard azo-dye coupling techniques, to give a precipitate at the site of localization of the enzyme in tissue sections, which are subsequently examined by optical techniques.

The utility of the foregoing substrate was further indicated by the fact that it is hydrolyzed by RNase at a rate about 100-fold greater than uridine-2',3'-cyclic phosphate. Furthermore the V_{max} for this substrate was 80-fold higher than for the corresponding benzyl ester, due to the high leaving tendency of the naphthyl group.

The foregoing substrate, although satisfactory for most purposes, still exhibited slight, but detectable susceptibility to PDase II *in vitro*. Further investigation led to synthesis and selection of 5'-O-benzyluridine-3'-(α-naphthylphosphate), which was fully resistant to PDase II, but still a good substrate for RNase (4).

It was of obvious interest to extend the foregoing to substrates specific for RNase T_2 (5), which shows no base specificity with low-molecular weight substrates and only partial preference towards adenine residues in polyribonucleotides. Because of this lack of base specificity, T_2-like RNases are active towards the α-naphthyl esters of uridine and inosine. These latter are, however, susceptible to other RNases, hence unsuitable as specific substrates for RNase T_2. Subsequent synthesis of adenosine-3'-(α-naphthylphosphate) showed it to be resistant to both pancreatic-type RNase and RNase T_1, hence suitable for cytochemical localization of T_2-type ribonucleases (6). This substrate was subsequently employed to demonstrate the absence of T_2-like activity in a variety of mammalian tissues.

The α-naphthyl esters of both uridine-3'-phosphate and 5'-O-benzyluridine-3'-phosphate have been successfully employed to demonstrate the localization of pancreatic-type RNase activities in mammalian tissues. In the rat, the enzyme was found to occur in the apical portion of acinar cells of the pancreas and salivary glands and, to a small extent, in the lysosomes of kidney proximal tubule epithelial cells (7). No activity was detectable in other rat tissues known to contain heat- and acid-stable alkaline RNases, and which resemble the pancreatic enzyme in specificity towards internucleotide linkages in RNA. This, in turn, pointed to the possibility of the use of these naphthyl esters for delineating subtle differences amongst mammalian alkaline RNases.

A quantitative assay was therefore developed for pancreatic RNase activity in solution and in homogenates, based on extraction of the azo-dye into an organic solvent, followed by its colorimetric estimation. A subsequent study of heat- and acid-stable RNases found in various tissues of the rat and humans (8) revealed the existence of two distinct types of activities: a) RNases of the pancreas, salivary glands, duodenal contents, and a fraction of the serum and urine activities, optimally active at pH 8.5, inhibited by Zn^{2+} and Cu^{2+}, which relatively rapidly hydrolyze uridine-3'-(α-naphthylphosphate); and b) RNases of the liver and spleen, and a fraction of the serum and urine activities, optimally active at pH 7, less sensitive to Zn^{2+} and Cu^{2+}, with negligible activity vs. uridine-3'-(α-naphthylphosphate).

It is also of interest that this ability to distinguish the secretory tissue ribonucleases from the remaining heat- and acid-stable mammalian RNases, together with a detailed study of the properties of these RNases, led to the discovery of an unusual property of human pancreatic RNase, viz. its ability to hydrolyzed double-stranded RNA relatively rapidly. Human pancreatic RNase cleaves double-stranded helical RNA at 2% the rate of single-stranded RNA, as contrasted with relative rates for the enzymes from the whale, rat, and cow of 0.4, 0.03, and 0.003%, respectively (9).

In view of the significant role of double-stranded RNA as an inhibitor of viral replication, via interferon induction and/or inhibition of protein synthesis, the usually high relative activity of human pancreatic RNase towards double-stranded RNA may be of considerable physiological significance.

Last, but not least important, may be the use of synthetic RNase substrates for clinical assays of RNase levels in physiological fluids. Apart from our own studies (8) pointing to the potential utility of serum pancreatic RNase levels in children for detection of cystic fibrosis, there have been a number of reports, which still require confirmation, of changes in serum RNase levels resulting from various malignant disorders (10, 11). The use of the synthetic substrates described above should facilitate such assays and increase their specificity.

REFERENCES

1. Shugar, D. and Sierakowska, H.: 1967. *Prog. Nucleic Acid Res. Mol. Biol.* 7, 369–429.
2. Sierakowska, H. and Shugar, D.: 1977. *Prog. Nucleic Acid Res. Mol. Biol.* 20, 59–130.
3. Kole, R., Sierakowska, H., and Shugar, D.: 1971. *Acta Biochim. Pol.* 18, 187–197.
4. Kole, R., Sierakowska, H., and Shugar, D.: 1971. *Biochem. Biophys. Res. Commun.* 44, 1482–1487.

 5. Sato, S., Uchida, T., and Egami, F.: 1966. *Arch. Biochem. Biophys.* 115, 48–52.
 6. Kole, R., Sierakowska, H., and Shugar, D.: 1972. *Biochim. Biophys. Acta* 289, 323–330.
 7. Zan-Kowalczewska, M., Sierakowska, H., Bardoń, A., and Shugar, D.: 1974. *Biochim. Biophys. Acta* 341, 138–156.
 8. Bardoń, A., Sierakowska, H., and Shugar, D.: 1976. *Clin. Chim. Acta* 67, 231–243.
 9. Bardoń, A., Sierakowska, H., and Shugar, D.: 1976. *Biochim. Biophys. Acta* 438, 461–473.
10. Drake, W.P., Schmukler, M., Pendergrast, W.J., Davis, A.S., Lichtenfeld, J.L., and Mardiney, M.R.: 1975. *J. Natl. Cancer Inst.* 55, 1055–1059.
11. Kottel, R.H., Hoch, S.O., Parsons, R.G., and Hoch, J.A.: 1978. *Br. J. Cancer* 38, 280–286.

Regulation of Ribosome Biosynthesis: Ten Years of Study

MASAYASU NOMURA

M. NOMURA was born in 1927. He graduated from the University of Tokyo in 1951 and received his Ph.D. degree in 1957. After studying in S. Spiegelman's laboratory, University of Illinois, in J.D. Watson's laboratory, Harvard University, and in S. Benzer's laboratory, Purdue University, he became assistant professor, Institute for Protein Research ,Osaka University in 1960. Since 1963, he has held a professorship at the Department of Genetics, University of Wisconsin. He received Japan Academy Award for 1972, and was elected as a member of National Academy of Science of the USA. His research field covers protein and mRNA synthesis, especially biosynthesis, assembly and genetics of ribosomes, and mechanisms of colicin action.

In examining the development of science in certain fields, one finds many factors which influence the speed of development and the probability of success. Sometimes it takes many years to solve a problem, because of failure to ask questions in a proper way, because of lack of suitable technologies, because necessary information is lacking, because generally accepted dogma inhibits understanding, or because the problem simply requires a lot of ground work before crucial experiments can be performed. In contributing an article to this volume, I have decided to present, as a case history, a short historical account of the research on problems related to the regulation of ribosomal gene expression, viewed through my own personal experience.

Bacterial ribosomes consist of two subunits, called 30S and 50S subunits, and the two subunits form 70S ribosomes. We now know that the ribosomes are by and large homogeneous and each 70S ribosome contains 3 rRNA molecules (16S, 23S, and 5S rRNA) and 52 ribosomal proteins (r-proteins). Most r-proteins exist in a single copy per ribosome. We also know that there are no significant amounts of free r-proteins in a cellular pool, and that there is no significant degradation of r-proteins. Thus, the synthesis rates of r-proteins reflect their accumulation rates, which in turn reflect the stoichiometric relationship of all these r-proteins in the ribosome. In addition, the synthesis rates of ribosomes are growth rate dependent; that is, when cells grow faster in nutritionally rich media, cells synthesize more ribosomes to meet the requirement for increased protein synthesis rates. In this respect, the synthesis rates of all r-proteins are coordinately regulated. Therefore, we can ask the following question: what mechanisms ensure the balanced and co-ordinately regulated synthesis rates of as many as 52 r-proteins? As I shall describe below, although such coordinate regulation of r-protein synthesis was suspected for some time, actual experimental evidence to support the formulation of the question became available only rather recently.

In addition to coordination of the synthesis rates of many r-proteins, there is a problem of coordination of rRNA synthesis and r-protein synthesis. Now, we can state that, except under some conditions where cells grow very slowly, exponen-tially growing cells synthesize rRNA in balance with the synthesis rates of r-proteins. Thus one can ask the question how the synthesis rates of rRNA and r-proteins are coordinated? However, the validity of the question itself might be somewhat prob-lematic because, as mentioned above, there are several conditions where rRNA appears to be overproduced and then excess rRNA degraded. Even in "normal" growing cells, published data indicate that rRNA synthesis is somewhat in excess over r-protein synthesis (1). Such data, because of our technical limitations, involve certain errors and could be taken either to support the concept of coordination of rRNA and r-protein synthesis or to indicate that the regulation of r-protein and rRNA is independent and the apparent coordination can be achieved by degrada-tion of components (e.g. rRNA) that are synthesized in excess. As I describe below, we now believe that there actually is coordination of the synthesis rates in ex-ponentially growing cells under "usual" culture conditions.

In 1971, I spent about 6 months in the Department of Molecular Biology at the University of Aarhus, Denmark. I had just recently demonstrated (2) that the 50S ribosomal subunits from *Bacillus stearothermophilus* can be reconstituted from their molecular components, as had been done with *Escherichia coli* 30S ribosomal subunits several years earlier (3). My laboratory was very busy studying the struc-ture and function of ribosomes as well as the mechanism of the ribosome assembly reaction using the newly developed *E. coli* 30S and *B. stearothermophilus* 50S ribo-some reconstitution systems. The availability of the new reconstitution systems inspired many new ideas for experiments, and our work was going well. Therefore, when I received an invitation from Niels Ole Kjeldgaard and Kjeld Marcker to spend a year in Aarhus, I was initially hesitant to take a leave of absence for such a long time. However, I eventually decided that this would be a good time for me to pause and think about my future research plans. I decided to accept the invita-tion—or, rather, to accept half of it—and to spend 6 months in Denmark, far from the daily pressures of my own laboratory. Kjeld Marcker and his coworkers had just moved from Cambridge, England, to a newly created Department in Aarhus University, and were starting several projects related to the gene expression in eukaryotic systems. Switching to eukaryotic systems, as many other molecular biologists did around that time, was a tempting possibility for me, and I was interested in learning techniques used in handling animals and cultured mammalian cells.

However, my short sabbatical stay in Aarhus affected my subsequent research course in a different way than I had originally anticipated. Perhaps I was not so energetic as I had been in my younger days in assimilating new background information and new techniques related to various eukaryotic systems. Perhaps I was not courageous enough to make a big change in my research subjects. I believe that the effect on my career was more because of the influence of daily contacts with Niels Ole Kjeldgaard and his colleagues. Niels Ole had also just moved from Copenhagen, where he, together with Ole Maaløe, spent many years doing careful analysis of biosynthesis of macromolecules in relation to bacterial growth and its control. For example, by analyzing the amounts of rRNA in bacterial cells growing in various different media, they had discovered the growth rate dependent control of ribosome synthesis which I mentioned earlier in this article. The results of their efforts had been compiled and published in 1966 in a classical monograph entitled *Control of Macromolecular Synthesis* (4) which many people working in related fields, including myself, had kept on their bookshelves as an essential reference.

Ole Maaløe's laboratory in Copenhagen had attracted many excellent people who were interested in bacterial physiology, and had been a center of this kind of research for many years. Of course, having worked on ribosomes, I had known the research activity of the Danish school and respected it. However, I was doing different kinds of research which dealt with mainly isolated ribosomes and ribosme assembly *in vitro*. I even had at that time some reluctance to use intact cells, partly because of my experience as a graduate student studying biochemical pathways in bacterial cells and, later, studying protein synthesis. I had seen many controversies on mechanisms deduced from experiments using intact cells and had seen these conflicts resolved by subsequent definitive experiments using cell-free systems. Consequently, I was trying to select, for my work, problems which could be analyzed using biochemical techniques in test tube systems. Another reason for avoiding "physiological" studies to analyze ribosome biosynthesis was my experience with "chloramphenicol particles." In the summer of 1958, soon after my first visit to the United States as a postdoctoral fellow, I worked in Jim Watson's laboratory, then at Harvard University. On Jim's suggestion, I examined the effects of chloramphenicol on ribosomes in growing *E. coli*, and discovered the accumulation of new ribonucleoprotein particles which contained rRNA but smaller amounts of proteins than normal ribosomes (5). This work was done before the development of the methods to separate and analyze the rather insoluble basic r-proteins. Thus without establishing the exact nature of the protein components, we suggested that these incomplete particles might represent precursors for normal ribosomes. In addition, other investigators later interpreted these observations to mean that there are large amounts of free r-proteins in the cellular pool. As mentioned above, this interpretation turned out to be incorrect. The *in vivo* system was messy. Chlorampenicol particles, being heterogeneous, could not be properly purified and characterized. Although there was nothing wrong in our original paper, the observations turned out to be more confusing than enlightening, and we later realized that the particles have nothing to do with normal intermediates. So, I became reluctant to do physiological types of experiments to analyze ribosomal biosynthesis.

Talking with people in Kjeldgaard's laboratory, I realized that the problems of the regulation of ribosome biosynthesis were still there, by then more clearly defined yet unsolved. I found the article just published by Ole Maaløe, "An Analysis of Bacterial Growth" (6) especially interesting and stimulating. For example, to explain apparent growth rate dependency of r-protein synthesis rates, Maaløe proposed the "passive control" model in this article. In addition, to explain coordination of rRNA and r-protein synthesis, he suggested the possibility that one of the r-proteins is an inducer of rRNA synthesis and that this inducer is constantly synthesized and, after functioning as the inducer, removed by incorporation into new ribosomes. Quite often I discussed with Niels Ole these models and suggestions. Thus, by the end of my short stay in Aarhus, I had more or less decided not to switch to eukaryotic systems, but to stay with *E. coli* ribosome research. I was convinced that the regulation problems should be studied using cell-free system, and for that purpose one needed ribosomal genes in the form of isolated DNA. In addition, such DNA would make several new experiments possible, *e.g.*, direct measurements of specific r-protein mRNA and physical mapping of r-protein genes and their promoters. I remember telling Kjeldgaard that I would struggle to isolate r-protein genes from the *E. coli* chromosome region where several r-protein genes such as *str* and *spc* were known to map.

After I returned to Madison from Denmark, we gradually started to do genetics and regulation work. However, most of the people in my laboratory continued to study the structure and function of ribosomes, the *in vitro* ribosome assembly reactions, and the mechanism of action of colicins. The shift in our research emphasis took place only very gradually. Our first effort was to set up a DNA-dependent protein synthesizing system in the expectation that sooner or later we (or other people) would be able to isolate transducing phages carrying r-protein genes and that these would provide DNA templates for the *in vitro* synthesis of r-proteins. Since there was no such DNA available yet, we had to use total *E. coli* DNA as our template. The system was adapted from the two previous systems, one used by Zubay and the other used by Gold and Schweiger. By the fall of 1973, Eberhart Kaltschmidt and Larry Kahan succeeded in this effort (7). It was fortunate that in the course of our research on the reconstitution of 30S ribosomal subunits, we had purified all the 30S r-proteins and prepared antisera against almost every one of these r-proteins. The antisera proved to be very useful in the identification of radioactive r-proteins synthesized in the cell-free system. I should also note that at that time 50S reconstitution studies were being done using *B. stearothermophilus* ribosomes, and hence we were purifying 50S r-proteins from *B. stearothermophilus*, not from *E. coli*. Therefore, we also initiated purification of individual *E. coli* 50S r-proteins and preparation of specific antisera against these proteins, believing that we would need such antisera extensively for the identification of r-protein products in the cell-free *E. coli* system. Bill Strycharz was mainly responsible for this project, and it later turned out to be a worthwhile effort.

Despite my (biased) opinion on experiments using intact cells, we started experiments which were, one might say, similar to those used by the Danish school. These experiments were initiated when Pat Dennis joined our group. Pat Dennis and a very dedicated technical assistant, Laura Sadowski, performed many measurements to answer several questions related to basic phenomenology. For example, growth rate dependent control of the bulk r-proteins was already known, but it was not known whether the synthesis rates of all the individual r-proteins are coordinately regulated. Pat Dennis' experiments done in 1974 (*8*) demonstrated that they are indeed coordinately regulated. From these experimental results, it became possible to formulate the question I mentioned at the beginning of this article: what mechanisms ensure the coordinate and stoichiometric synthesis of most if not all of the 52 r-proteins? Definitive answers have started to emerge only recently, almost 5 years after we first posed the question.

Regarding the apparent coordination of rRNA and r-protein synthesis, people, including us, considered three possibilities. The first possibility was that rRNA synthesis was the primary target of regulatory mechanisms and the regulation of r-protein synthesis was a secondary consequence of the regulation of rRNA synthesis. The second possibility was opposite to the first, namely, that r-protein synthesis was regulated (either directly or "passively" as proposed by Maaløe), and that the regulation of rRNA synthesis was a secondary consequence of the regulation of the r-protein synthesis. The r-protein inducer hypothesis suggested by Maaløe, as I mentioned above, was one such specific example. The third possibility was that both rRNA and r-protein synthesis were regulated, and exact coordination was achieved either by a balance of transcriptional and translational efficiencies inherent in the DNA and mRNA structures themselves, or by degradation of products synthesized in excess, or by both.

Historically, perhaps the first specific model proposed was related to the first possibility. It was proposed that nascent rRNA was the mRNA for r-proteins, and the messenger function stopped immediately after a single translational event because of inhibition by the r-proteins produced or by methylation of the nascent rRNA (*e.g.*, refs. *9, 10*). The model was attractive, and, perhaps for this reason, several papers appeared claiming experimental evidence for this model. Before the discovery of mRNA, rRNA was thought to be the information carrier between genes and proteins. Therefore, the model that nascent rRNA might be the message for r-proteins was a natural extension of the ideas prevalent at that time. Of course, even then the "coding capacity" of rRNA appeared to be insufficient to produce all the r-proteins. However, the possibility existed that there were several different rRNA species. In addition, although the heterogeneity of r-proteins was demonstrated in as early as 1961 by Waller and Harris (*11*), we did not know how many different r-proteins exist in the ribosome. It was only after many years of painstaking effort by several research groups that the size and homogeneity of rRNA, as well as the number of r-proteins, became clearer. Even so, it was not easy to exclude completely a modified form of the model, namely, that rRNA codes for some, not all, r-proteins. Of course, we now know that this model is incorrect and rRNA does not code for r-proteins. However, I have mentioned this old incorrect model, because, as I shall describe below, r-protein gene expression is feedback regulated by r-proteins and is coupled with the ribosome assembly, and there is in a way a similarity between the current model and this old one. The old model could also explain the problem of coordination by coupling the synthesis of r-proteins with ribosome assembly. I should emphasize that at the time when I decided to work on the genetics and regulation of ribosome synthesis, there were various ideas to explain the coordination problems as described above, but definitive tests of various models had to await isolation of rRNA and r-protein genes.

Isolation of r-protein genes was not easy and took a long time. Recombinant DNA techniques were not yet available. Thus we tried to use classical approaches to isolate genes by incorporating them into λ transducing phages. Although the phenomenon of specialized transduction was discovered in 1956 (*12*), isolation of bacterial genes as transducing phages was limited for a long time only to genes located close to the attachment sites of certain temperate phages such as λ and φ80. However, two major developments had taken place by the early 1970's allowing isolation of other genes as λ or φ80 transducing phages. The first was the development of a technique to transpose bacterial genes to other chromosomal locations using temperature sensitive F' factors. With this technique, Beckwith and Signer transposed the *lac* genes to a location close to the φ80 attachment site and then succeeded in the isolation of φ80 transducing phages carrying the *lac* genes (*13*). The second was the discovery of "secondary attachment sites" for λ reported in 1972, which allowed isolation of λ transducing phages carrying genes located close to various secondary attachment sites (*14*).

By the time we decided to isolate r-protein genes, evidence had been accumulated to suggest that there was a r-protein gene cluster in the "*str-spc* region" of *E. coli* chromosome (72 min on the *E. coli* genetic map). Therefore, it was this region which we attempted to isolate as a transducing phage or as part of an F'-plasmid. After we tried several strategies without success, one day (in late 1973, I believe) I was reading an article by Beckwith, Signer, and Epstein published in the 1966 Cold Spring Harbor Symposium volume (*15*). Suddenly a very simple genetic map described in Fig. 2 struck me. They simply stated that F'*lac* factor can be transposed to various chromosomal locations, and without giving any detail, they showed the locations of several transposed F'*lac* factors in the figure. I noticed that one of the locations was near the *str* gene. I immediately thought about inserting *λplac* 5 transducing phage into this position using the *lac* gene homology and hoped that the distance between the transposed *lac* and the *str* genes would be short. Because there were no data shown, but only a rough drawing of the map, I could not judge whether the distance was really sufficiently short to make the isolation of the r-protein genes as transducing phages feasible. We were very fortunate that Jon Beckwith had kept the strain (EC-2), which he kindly provided upon my request, and that the transposed *lac* gene was in fact located close to the target r-protein gene cluster. The actual experiments were carried out mostly by Dick Jaskunas, who was then working as a postdoctoral fellow, and Mary Stroud, an able technical assistant. By the end of 1974, we had several transducing phages, some of which were proven to carry more than a dozen r-protein genes (*16*), and eventually *λfus*2 which carries as many as 27 r-protein genes (*16–18*). As I mentioned above, we had already practiced synthesizing r-proteins *in vitro* using *E. coli* DNA as template and characterizing the protein products. Thus it was not so difficult to demonstrate the synthesis of r-proteins *in vitro* using DNA isolated from those transducing phages. This project was initially carried out by Lasse Lindahl, who had just arrived from Ole Maaløe's laboratory in Copenhagen, and others later joined the project to identify many of the r-protein genes carried by the transducing phages. They were Leonard Post, Janice Zengel, Scott Gilbert, Ann Fallon, and Masayuki Yamamoto. I should add here that after the initial success in isolating r-protein genes from the *str-spc* region, we (*19*), as well as Friesen, Fiil, and their coworkers (*20*), found that *λrif*ᵈ18 isolated originally by Kirshbaum also carried four r-protein genes. Thus, we also started to study genes carried by this transducing phage.

For several years after our success in isolating the transducing phages, we were busy identifying genes, mapping them physically, and identifying their promoters. It was exciting to discover (or isolate for the first time) not only so many r-protein genes, but also other related essential genes, such as genes for elongation factor EF-Tu, EF-G, and EF-Ts, and the gene for RNA polymerase subunit α. The discovery that the elongation factor genes (*21*) and that RNA polymerase subunit genes were cotranscribed with r-protein genes (*22, 23*) was also exciting. It gave an explanation for the known coregulation of EF-G synthesis with r-protein synthesis, but it also posed a new problem: how can it be that r-protein genes and RNA polymerase subunit genes sometimes behave differently even though they are in the same transcription unit? This problem is still, at the time of this writing, under intensive study by several research groups. I should note that the availability of newly discovered restrictiod enzymes and the development of recombinant

DNA techniques certainly accelerated the speed of our work during this time. For example, various DNA fragments could be easily made and used as templates for *in vitro* protein synthesis experiments to physically map various genes; the recombinant DNA techniques were used to subclone smaller segments from the original transducing phage genomes for detailed gene mapping and promoter identification. I also note that during that time, we had a good collaboration with Fred Blattner, who had expertise in recombinant DNA technologies, and Dick Burgess, who had been working on various aspects of *E. coli* RNA polymerase.

From the beginning we considered a model of feedback regulation to explain the coordinate control of many r-proteins and actually suggested such a model to explain certain data (*24*). However, we never tried in our early experiments to test the model seriously. In retrospect, I feel a slight sense of regret. On the other hand, identification of many genes and their promoters, even though it took a great deal of effort, was a necessary step towards our goal, and a detailed map of the organization of these genetic elements was prerequisite to a comprehensive understanding of the regulatory mechanisms. Furthermore, our attention was focused on promoters because the known regulatory mechanisms for gene expression at that time were almost exclusively based on transcriptional control, basically similar to the original operon theory proposed by Jacob and Monod. Therefore, we were certainly eager to identify the promoters, to examine the structure of the promotor regions and to discover any elements involved in the interaction with these promoter regions. Specifically, after it became clear that there were several r-protein operons (even in the major r-protein gene cluster, the *str-spc* region on the chromosome, four operons were identified), we seriously thought about a regulation model based on the assumption that all the promoters controlling r-protein genes had similar structural features, therefore had equal transcriptional activities and equal responses to regulatory signals. It was also thought that the promoters for rRNA operons could share structural features with r-protein promoters and respond to the same signals (such as the signals for stringent control), but be of different strength from r-protein promoters, since the rRNA gene products are not amplified by translation.

Around that time, the rapid DNA sequencing method was developed by Maxam and Gilbert (*25*) and it became possible to test the above possibilities. Thus, one of my graduate students, Leonard Post, started sequencing the r-protein promoters. He was extremely efficient and sequenced five r-protein promoters out of the six that had been identified in the *rif* and *str-spc* regions of *E. coli* chromosome. Although Post's sequencing work made significant contributions in several respects (*e.g.* resolving ambiguous gene orders and revealing an interesting bias in codon usage frequencies in r-protein genes; ref. *26*), comparison of the five r-protein promoter sequences (and rRNA promoter sequences) failed to reveal any common features which distinguish these r-protein promoters from other un-

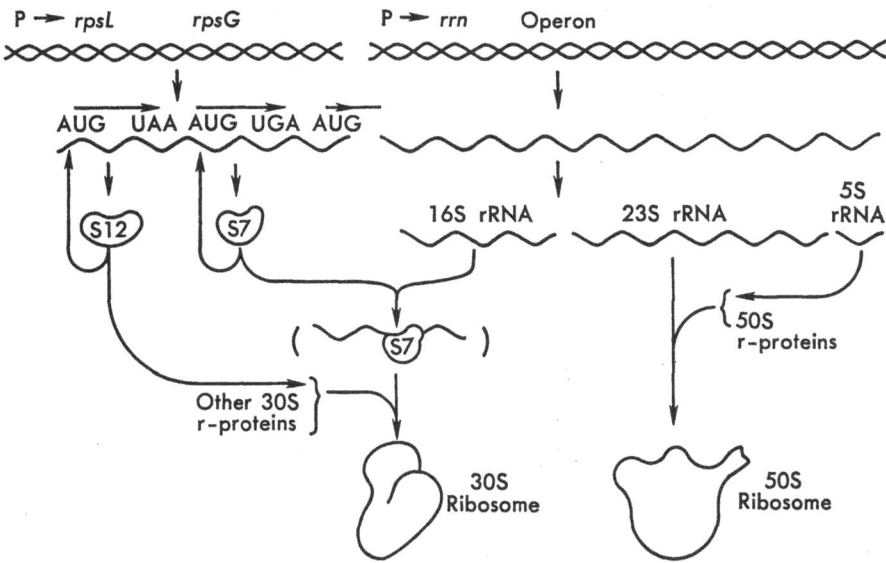

FIG. 1.   Feedback regulation model of ribosomal protein synthesis in *E. coli*. One of many r-protein operons, the *str* operon, is shown as an example. This operon contains the promoter (P), the genes for r-protein S12 (*rpsL*), S7 (*rpsG*), elongation factors EF-G (*fusA*), and EF-Tu (*tufA*). Only the first two genes are shown in the figure. The operon is transcribed to produce a ploycistronic mRNA. The mRNA is then translated to produce r-proteins S12 and S7. As long as rRNA is produced from rRNA operons (*rrn* operon) and ribosome assembly continues, free S7 and S12 are removed and the translation continues. However, when r-proteins are overproduced relative to rRNA, the excess r-proteins interact with their own mRNA and prevent further translation. S7 is known to inhibit the synthesis of S7 (and EF-G, the product of the next cistron) without inhibiting the synthesis of S12 (*32*). Similarly, S12 presumably feedback regulates its own synthesis at the level of translation. S7 is known to make a specific complex with 16S rRNA *in vitro*. It was found that there is a structural homology between the S7 binding site in 16S rRNA and the mRNA region between S12 and S7 cistrons where S7 presumably acts as a "translational repressor." The 16S-S7 complex is not necessarily an intermediate in the ribosome assembly, and hence, is shown in parenthesis in the figure. In some other instances, one r-protein feedback regulates the synthesis of more than one r-protein. For example, L1 (but not L11) feedback regulates the synthesis of L11 and L1, both of which are encoded in a single operon, the L11 operon (*30, 36*).

related conventional promoters (*cf.*, ref. *27*). Thus the answer to the question of the mechanism of coordinate regulation had to be sought in other ways.

I continued to consider the possibility that ribosome synthesis might be regulated by some kind of feedback mechanisms, and around 1976, when Ann Fallon joined our group, we started systematic studies on gene dosage effects in r-protein synthesis to test this idea. Sue Jinks and Geneva Strycharz also participated in this project. Our approach was very indirect; we did not yet try to test the feedback model directly using *in vitro* systems. We examined the effects of increased gene dosages on r-protein synthesis rates and on mRNA synthesis rates in strains merodiploid for a group of r-protein genes (in the *str-spc* region) or in strains carrying these extra r-protein genes on a multi-copy plasmid. We found that the rates of transcription of the r-protein genes increased in proportion to the increase in gene dosage, but that the rates of r-protein synthesis did not increase relative to the synthesis rates of other r-proteins whose genes existed in a single copy. (Around that time, Geyl and Böck in Regensburg, Germany, also analyzed gene dosage effects on r-protein synthesis, and obtained similar results (28), but they did not analyze the effects on mRNA synthesis nor did they emphasize any specific model.) After the initial results were obtained, I immediately thought about a translational feedback regulation model; I thought that the experimental results were best explained by a model which assumes that some r-proteins are feedback inhibitors and inactivate their own mRNA, and that as long as the assembly of ribosomes removes r-proteins, the corresponding mRNA escapes from the inhibition and continues to direct synthesis of r-proteins (Fig. 1). I thought that the model was very attractive in its simplicity. However, the results and the model were different from the general belief at that time that r-protein genes are transcriptionally controlled (e.g., ref. 1). Therefore, we repeated somewhat tedious experiments many times. I suppose that the classical transcriptional control model so well studied in the lac operon and other similar systems had exerted so much influence on people's thoughts and created an (unproven) dogma of the transcriptional control for r-protein gene expression. I remember that I was vacillating for a long time between the temptation of publishing the new elegant (to me) model and the fear of proposing a heresy which could be completely wrong. However, I gradually became confident of our data and the model, and in early 1979, finally published a paper (29) proposing the translational feedback regulation model explained above. In addition, I decided, for the first time, to test the model directly and seriously using the DNA-dependent in vitro protein synthesizing system which we had developed over many years, in hopes of using it exactly for this kind of experiment.

John Yates, a graduate student who was then working on a different project, agreed to test the model. We had almost all the r-proteins purified and stored in a freezer, and various DNA molecules carrying r-protein genes were also there. It was exciting to discover, soon after John started the experiments, that some r-proteins had in fact specific inhibitory effects on the synthesis of some r-proteins. The inhibited proteins were always in the same operon as the added r-protein; the specificity was clear and dramatic. The effects of these inhibitory r-proteins were demonstrated to be at the level of translation, rather than transcription, as predicted from the gene dosage experiments. Ribosomal proteins S4, S8, and L1 were immediately identified as translational repressors in this way (30), followed by identification of S7 and L4 (31, 32). Independently, Fukuda in Japan (33), and Weissbach and his coworkers in the Roche Institute (34), showed that L10 inhibits its own synthesis from λrif^{d}18 transducing phage DNA.

The feedback regulation model proved by our in vitro experiments, was also confirmed by other kinds of experiments. One powerful approach was used by

Lasse Lindahl and Janice Zengel, both of whom had left my laboratory by then and settled at the University of Rochester. They fused several r-protein genes to the *lac* operon promoter so that they could flood a cell at will with an excess of a particular protein. They found that overproduction of certain r-proteins inhibits the synthesis of a group of r-proteins (*35*). Dennis Dean, a postdoctoral fellow in my laboratory, used a similar approach and demonstrated that the repressor r-proteins identified *in vitro* (S4, S8, and L1) in fact function as (translational) repressors *in vivo* (*36, 37*). Lindahl and Zengel's team also identified L4 as a (translational) repressor *in vivo* (*38*).

Subsequent experiments in this story are not to be told in this article; they are still under intensive study and rapid progress is being made. I should just note that the feedback regulation model proposed by us has been proven to be basically correct, and that major principles and general schemes of the regulation have largely emerged on a solid experimental basis. Some new concepts have also been proposed or established. For example, r-protein operons are sometimes subdivided into units of regulation and each unit has a gene coding for a repressor r-protein which regulates the whole unit by acting at or near the beginning of the first cistron of that regulatory unit at the translational step. The target site on mRNA for a repressor r-protein presumably has a structural homology with the binding site on rRNA for that r-protein and the feedback regulation involves a competition between rRNA and r-protein mRNA for repressor r-proteins (*e.g.*, ref. *32*).

It was exciting and very satisfying to see definite proof of the translational feedback regulation model and all the progress achieved within a short time after the formulation of the model, giving answers to the questions which we once thought would be very difficult to solve. The work took almost nine years after I first decided to study the regulation problem and six years since we succeeded in isolating transducing phages carrying r-protein genes. The rapid progress we are making now owes thanks to the efforts made by many coworkers during these many years. Yet, I often wonder why we did not perform some of the simple *in vitro* experiments earlier, when we had all the necessary tools at our disposal—the transducing phages that made r-proteins *in vitro* and a freezer containing individual purified r-proteins. We did not lack hypotheses: the feedback regulation model, though it was a transcriptional feedback regulation model, I conceived and discussed often in published articles. But we never seriously attempted testing the model using the *in vitro* protein synthesizing system. Only after I was convinced by the gene dosage experiments and had published the specific model, did I become serious about proving the model by *in vitro* experiments. Perhaps we were too busy identifying and mapping the genes (which were certainly important and formed the basis for the rapid progress we are making now on the regulation problems). Perhaps we were too prejudiced in favor of transcriptional regulation, spending our time in identifying promoters, sequencing them, and performing *in vitro* transcription experiments (which were again important and nothing to be regretted). Perhaps I was not sufficiently optimistic or confident to be able to hope that the problem had such a basically simple solution. Having found the basic mechanisms responsible for the coordination of so many genes in making such a complex structure as the ribosome, I am impressed by the clever and pretty (to me) mechanism Nature has evolved for this purpose. Of course, details and actual mechanics are always complex and I expect more struggles and efforts before we completely understand this system. Nevertheless, I feel it is worthwhile now to tell the story of this research to young students (and to myself) as a case history to indicate that beneath apparently formidable biological complexities, there are often rather simple principles hidden, waiting to be discovered.

Acknowledgment

As I indicated in this article, our ribosome research was a collaborative effort and I thank my present, as well as my previous, coworkers, some of whose names are mentioned in the article. I also acknowledge the National Science Foundation and the National Institutes of Health, United States Public Health Service, and the College of Agriculture and Life Sciences, University of Wisconsin, which supported research described in this article.

REFERENCES

1. Gausing, K.: 1977. *J. Mol. Biol.* 115, 335–354.
2. Nomura, M. and Erdmann, V.A.: 1970. *Nature* 228, 744–748.
3. Traub, P. and Nomura, M.: 1968. *Proc. Natl. Acad. Sci. USA* 59, 777–784.
4. Maaløe, O. and Kjeldgaard, N.O.: 1966. *Control of Macromolecular Synthesis*, W.A. Benjamin, Inc., New York.
5. Nomura, M. and Watson, J.D.: 1959. *J. Mol. Biol.* 1, 204–217.
6. Maaløe, O.: 1969. *Dev. Biol.* (Suppl.) 3, 33–58.
7. Kaltschmidt, E., Kahan, L., and Nomura, M.: 1974. *Proc. Natl. Acad. Sci. USA* 71, 446–450.
8. Dennis, P.P.: 1974. *J. Mol. Biol.* 89, 223–232.
9. Roberts, R.B.: 1965. *J. Theor. Biol.* 8, 49–53.
10. Nakada, D.: 1965. *J. Mol. Biol.* 12, 695–725.
11. Waller, J.-P. and Harris, J.I.: 1961. *Proc. Natl. Acad. Sci. USA* 47, 18–23.
12. Morse, M.L., Lederberg, E.M., and Lederberg, J.: 1956. *Genetics* 41, 142–156.
13. Beckwith, J.R. and Signer, E.R.: 1966. *J. Mol. Biol.* 19, 254–265.
14. Shimada, K., Weisberg, R.A., and Gottesman, M.E.: 1972. *J. Mol. Biol.* 63, 483–503.
15. Beckwith, J.R., Signer, E.R., and Epstein, W.: 1966. *Cold Spring Harbor Symp. Quant. Biol.* 31, 393–401.
16. Jaskunas, S.R., Lindahl, L., and Nomura, M.: 1975. *Proc. Natl. Acad. Sci. USA* 72, 6–10.
17. Jaskunas, S.R., Fallon, A.M., and Nomura, M.: 1977. *J. Biol. Chem.* 252, 7323–7336.

18. Lindahl, L., Post, L., Zengel, J., Gilbert, S.F., Strycharz, W.A., and Nomura, M.: 1977. *J. Biol. Chem.* 252, 7365–7383.
19. Lindahl, L., Jaskunas, S.R., Dennis, P.P., and Nomura, M.: 1975. *Proc. Natl. Acad. Sci. USA* 72, 2743–2747.
20. Watson, R.J., Parker, J., Fiil, N.P., Flaks, J.G., and Friesen, J.D.: 1975. *Proc. Natl. Acad. Sci. USA* 72, 2765–2769.
21. Jaskunas, S.R., Lindahl, L., Nomura, M., and Burgess, R.R.: 1975. *Nature* 257, 458–462.
22. Jaskunas, S.R., Burgess, R.R., and Nomura, M.: 1975. *Proc. Natl. Acad. Sci. USA* 72, 5036–5040.
23. Yamamoto, M. and Nomura, M.: 1978. *Proc. Natl. Acad. Sci. USA* 75, 3891–3895.
24. Dennis, P.P. and Nomura, M.: 1975. *Nature* 255, 460–465.
25. Maxam, A.M. and Gilbert, W.: 1977. *Proc. Natl. Acad. Sci. USA* 74, 560–564.
26. Post, L.E., Strycharz, G.D., Nomura, M., Lewis, H., and Dennis, P.P.: 1979. *Proc. Natl. Acad. Sci. USA* 76, 1697–1701.
27. Post, L.E., Arfsten, A.E., Davis, G.R., and Nomura, M.: 1980. *J. Biol. Chem.* 255, 4653–4659.
28. Geyl, D. and Böck, A.: 1977. *Mol. Gen. Genet.* 154, 327–334.
29. Fallon, A.M., Jinks, C.S., Strycharz, G.D., and Nomura, M.: 1979. *Proc. Natl. Acad. Sci. USA* 76, 3411–3415.
30. Yates, J.L., Arfsten, A., and Nomura, M.: 1980. *Proc. Natl. Acad. Sci. USA* 77, 1837–1841.
31. Yates, J.L. and Nomura, M.: 1980. *Cell* 21, 517–522.
32. Nomura, M., Yates, J.L., Dean, D., and Post, L.E.: 1980. *Proc. Natl. Acad. Sci. USA* 77, 7084–7088.
33. Fukuda, R.: 1980. *Mol. Gen. Genet.* 178, 483–486.
34. Brot, N., Caldwell, P., and Weissbach, H.: 1980. *Proc. Natl. Acad. Sci. USA* 77, 2592–2595.
35. Lindahl, L. and Zengel, J.: 1979. *Proc. Natl. Acad. Sci. USA* 76, 6542–6546.
36. Dean, D. and Nomura, M.: 1980. *Proc. Natl. Acad. Sci. USA* 77, 3590–3594.
37. Dean, D., Yates, J.L., and Nomura, M.: 1981. *Nature*, 289, 89–91.
38. Zengel, J., Mueckel, D., and Lindahl, L.: 1980. *Cell* 21, 523–535.

Is Ageing a Biochemical Problem?—On the Accuracy of Protein Biosynthesis

FRIEDRICH CRAMER

F. CRAMER was born in Breslau on September 20, 1923. He studied chemistry at the University of Breslau and the University of Heidelberg and received Dr.rer.nat. in 1949. He became associated with Chemisches Institut der Universität Heidelberg as a teaching assistant (=assistant professor) in 1949. In 1953–1954, he studied at the University of Cambridge in the institute directed by A.R. Todd and started to work on nucleic acids. In 1959, he promoted to a professorship at the Technische Hochschule Darmstadt. He moved to Max-Planck-Institut für Experimentelle Medizin, Göttingen, in 1963, where he is the Department Head of the Abteilung Chemie. He is elected as a member of EMBO (1967), Chairman of the Biological-Medical Section of the Max-Planck-Gesellschaft (1976), member of the Polish Academy of Science, member of the Göttingen Akademie der Wissenschaften, and Honorary Fellow of the American Society of Biological Chemists.

What Is Ageing?

Our life is embedded between birth and death. Physically it is initiated with the fertilization of the female egg and ceases with the breakdown of the metabolic functions in the central nervous system. These physical borderlines can be defined. It is the scientific aspect which can be treated only in such an article. We all are aware that the existential, metaphysical part of the problem is at least equally important for our lives. Science, however, can solve its problems only by abstraction and narrowing down the problems until they become solvable and treatable by science. Thus I shall discuss ageing as a biochemical problem.

Ageing in terms of biochemistry is the gradual decrease and the final cessation

FIG. 1.   Albrecht Dürer (1514 A.D.): portrait of the painter's mother at the age of 63 years.

of certain cellular and organic functions. One can study this in the portrait of the mother of the famous German painter Albrecht Dürer (1471–1528) (Fig. 1). In this portrait one can see at least three typical phenomena of ageing. The old woman has lost her teeth, which can be seen from the form of the mouth and lips. Bacteria normally living in the oral cavity have finally destroyed the worn down dental enamel. Time and the disorganization of a symbiosis with bacteria have led to this phenomenon of ageing. Secondly we can conclude from the protruding right eyeball that the old woman suffered from a glaucoma, a pathological increase in the internal eye pressure, which finally leads to the destruction of the retina and to blindness. This phenomenon of ageing is a loss of regulating capacity of the organ eye. Thirdly we see the typical alterations of the skin which come with age. The skin becomes wrinkled and brittle. This change in the senile skin is called senile elastosis and it is a change of the keratins in the connective tissue. The composition of the proteins in the skin has been altered. Therefore this must have something to do with protein biosynthesis.

Is Life Expectancy Programmed?

The average life expectancy seems to be programmed for each biological species. It can, however, be altered by environmental influence. Thus medical progress in our time has elongated the life span of man. The life expectancy for various species in years is as follows: fly (0.077), mouse (3–3.5), rat (3–3.5), rabbit (5–7), guinea-pig (8), cat (9–10), fox (10), squirrel (10–12), dog (10–12), ant (10–

15), frog (10–15), sheep (10–15), goat (12–15), wolf (12–15), herring (16), chicken (20), tiger (20), lion (20–25), cow (20–25), primate monkey (20–30), pig (20–30), camel (40–50), horse (40–50), crocodile (50), carp (50–60), falcon (60–70), crow (60–70), man (70–74), Galapagos turtle (100–150), elephant (150–200). One can easily recognize that longevity is by no means connected with size or weight of the individuals, although in general large animals live longer.

What is the maximal life span of Man? One can calculate a theoretical maximal life expectancy under the assumption that all known diseases do not abbreviate the life of a particular individual such as carcinoma, arteriosclerosis, failures of kidney, heart or particular organs *etc.* With such a theoretical calculation one arrives at a maximal age of 120 years for man. This seems to be in fact the maximal age which is reported, if such reports are correct. There are certain forms of ageing which apparently are precisely programmed. The capacity of the ear to hear very high frequencies already begins to decrease after the 15th year. At the age of almost precisely 46 years the elasticity of the eye lens begins to decrease leading to far-sightedness. There exists the so-called Werner-syndrome (Progeria), an autosomal recessively inherited disease. Persons with this disease are very old people at the age of 15 to 20 and die without an observable organic disease in their early twenties. This disease is considered to be the result of a disturbed and accelerated ageing programme of the particular individual.

Is There a Molecular Basis of Ageing?

Life functions as a network system of interacting macromolecules. In this scientific concept there must necessarily be a molecular basis for ageing. Cells produce macromolecules, especially the nucleic acids of the cellular nucleus, and the proteins which are synthesized according to the programme laid down in the DNA. These proteins in turn are not only structural elements of the cell, but also carriers of catalytic functions as enzymes. Some of them are important in the reproduction of the cell, *e.g.* in protein biosynthesis, nucleic acid replication or messenger synthesis. If these enzymes do not fulfill their synthetic functions with high fidelity they will produce proteins which function even less well. By feedback of errors one can finally arrive at an error catastrophy. In 1963, Orgel proposed an error catastrophy as a possible explanation for ageing (*1*). The correctness of Orgel's hypothesis has so far not been demonstrated. It is, however, clear in principle that macromolecular biosynthesis cannot be infinitely precise. It is therefore very important to know the true error rate of protein biosynthesis.

R. Loftfield was the first to consider the problem of fidelity of protein biosynthesis in mechanistic terms (*2*). It is clear that synthesis of macromolecules requires a much higher fidelity than the precision by which low molecular weight compounds are metabolized. Assuming an error rate of 1 : 1,000 for a normal enzymatic reaction, the same error rate for the synthesis of a protein with a chain length of a thousand amino acids would mean that on the average every protein is wrong. The problem is even more severe with DNA synthesis, but in DNA synthesis nature has developed a repair and control system which in a proofreading manner eliminates errors (3). Such proofreading enzymes are as yet unknown for proteins. The errors in proteins can certainly not be measured by analytical procedures such as sequencing, because sequencing is much less precise than 1%. Loftfield postulated an error of less than 1 in 3,000. In later studies this figure was verified by different methods, e.g. in the synthesis of bacterial flagellin (4). Flagellin does not contain cysteine. When flagellin biosynthesis was carried out in the presence of high a concentration of ^{35}S-cystein, a small amount of radioactivity could be observed in the flagellin produced. From this it was calculated that the error is less than 1 in 10,000, taking in account the limits of the analytical procedure.

How Are Amino Acids Recognized?

A mino acids are the building blocks for proteins. They must be incorporated into the proteins in a specific way. The sequence is guided by the arrangement of the triplets in the messenger, to which the tRNAs fit with their anticodons. The crucial step in the incorporation of amino acids is the correct correlation of the amino acid to its particular tRNA. Thus the aminoacylation of tRNA is the key reaction in the fidelity of protein biosynthesis. We have studied the mechanism of this reaction in great detail (5). We have asked the question of how enzymes can distinguish between two very similar amino acids like valine and isoleucine with the required precision. Linus Pauling had predicted that no enzyme in nature could distinguish between valine and isoleucine better than 1% if the distinction is regulated by a simple adsorption—desorption process in the enzyme cavity (6). One

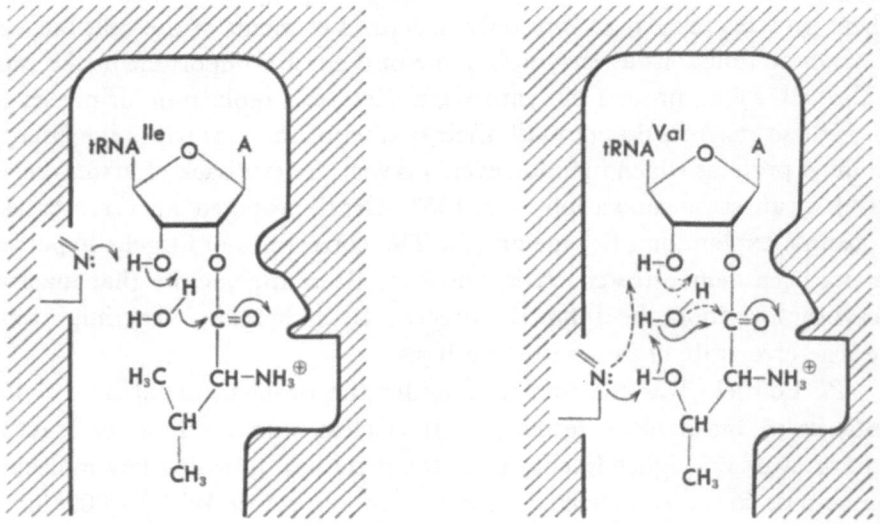

FIG. 2.   Mechanism of proofreading by isoleucyl- and valyl-tRNA synthetases.

can theoretically calculate the difference in binding energy between isoleucine and valine in the isoleucine cavity of the enzyme. From this one can derive a difference in ΔG, which amounts to an error rate of 1% (7). In fact we have found that amino-acyl-tRNA synthetases do have a proofreading mechanism (5). The proofreading requires the nonaccepting hydroxyl group at the tRNA terminus which is involved in a hydrolytic process directed primarily against the wrong amino acid (8). Its mechanism is depicted in Fig. 2.

How large is the contribution of the proofreading to the overall fidelity of the aminoacylation reaction? When using a modified tRNA, tRNA^{Phe}–C–C–NA, carrying a 3' amino group at the accepting end, one can prevent proofreading since the amino group adjacent to the accepting hydroxyl serves as a trap for the amino-acids. Thus we could establish that the distinction between isoleucine and valine in the first enzymatic selection process is one error in 225. This is in close agreement with the predicted value of Pauling. It is clear, that such a fidelity would never be good enough for macromolecular synthesis. The fidelity of the second part of the proofreading process was determined to be one error in 800. The overall fidelity is the product of the two, $225 \times 800 = 180,000$. Thus the total error which occurs in the distinction between isoleucine and valine in yeast aminoacyl-tRNA synthetase is one in 180,000. This is to our knowledge the first exact figure on the fidelity of protein biosynthesis.

Is the Error Rate in Protein Biosynthesis a Natural Constant?

The observed error rate of one error in 180,000 incorporations of isoleucine might be a special case for the distinction between valine and isoleucine in yeast. On the other hand, the evolution of a proofreading system which minimizes the errors is not an easy task for nature and has certainly been induced by evolutionary pressure. Also the proofreading costs much energy and therefore will only be carried out in the highly economical systems of nature to an extent which is absolutely necessary. Thus one can assume that the above figure is not accidental in evolution and not an isolated case for the distinction between isoleucine and valine. One can use the following argument: it would be useless for nature to develop a highly sophisticated system in the distinction between a particular pair of amino acids when in the incorporation of other amino acids the error rate is much higher. Evolutionary pressure must have the tendency to equalize the error rates for the incorporation of all amino acids. There are, of course, certain mistakes which may have graver consequences, *e.g.* when a neutral amino acids is replaced by a nega-tively or positively charged one. It might, therefore, be necessary to use certain weighting factors for certain errors.

In general, however, from the preceding arguments I would like to argue that the error rate of one in 200,000 for protein biosynthesis is a constant which has been elaborated during evolution as a compromise between the necessity to maintain life on the one hand and to have to save energy on the other. I would like to propose that this fidelity is a biological constant. This should also apply to the fidelity of the ribosomal system about which much less is as yet known. It has been proposed that also at the ribosomal level a proofreading step is involved, since the simple physical interaction between codon and anticodon could never explain the high fidelity of the alignment of aminoacyl tRNAs at the ribosome (9).

This Ageing in Error Catastrophy?

We certainly do not yet have enough data to decide whether the cause of ageing is the decline of the biopolymers. First one should know more exact figures on other proofreading aminoacyl-tRNA synthetases. We are presently collecting such data. Since ageing is a decline of phenotype (old-aged fathers generate babies as young as young fathers do) and since by the definitions of molecular biology the phenotype is the sum of the structure and assembly of proteins, ageing must be defined as decline of proteins. There might certainly be some key proteins which are essential or perhaps even serve as clocks for ageing. From purely logical arguments, however, one cannot avoid the conclusion that ageing and errors in protein biosynthesis are connected. They might not be directly connected since even the most primitive organism is not a pool or test tube of substances and reactions but rather a highly complex network system with hierarchies and interdependencies. For such network systems entirely different rules are valid. The network at one location or time might be insensitive to some influence and at another location or time break down completely under the same influence (10). Although we have collected the first data on the error rate in protein biosynthesis, it might still take some effort until we understand the phenomenon of ageing.

REFERENCES

1. Orgel, L.: 1963. *Proc. Natl. Acad. Sci. USA* 49, 517–521.
2. Loftfield, R.B.: 1972. *Prog. Nucleic Acid. Res. Mol. Biol.* 12, 87–128.
3. Kornberg, A.: 1980. *DNA Replication*, W.H. Freeman & Co., San Francisco.
4. Edelmann, P. and Gallant, J.: 1977. *Proc. Natl. Acad. Sci. USA* 74, 3396–3398.
5. Cramer, F., v.d. Haar, F., and Igloi, G.: 1979. *Transfer RNA: Structure, Properties and Recognition* (Schimmel, P., Söll, D., and Abelson, J., eds.) pp. 267–279, Cold Spring Harbor Laboratory, New York.
6. Pauling, L.: 1958. *Festschrift Arthur Stoll*, pp. 597–602, Birkhäuser Verlag, Basel.
7. DeMaeyer, L.C.M.: 1976. *Ber. Bunsen Ges.* 80, 1189–1196.
8. v.d. Haar, F. and Cramer, F.: 1975. *FEBS Lett.* 56, 215–217.
9. Thompson, R.C. and Stone, P.J.: 1977. *Proc. Natl. Acad. Sci. USA* 74, 198.
10. Prigogne, I.: 1979. *Vom Sein zum Werden—Zeit und Komplexität in den Naturwissenschaften*, p. 186, Piper Verlag, München.

tRNA-Like Structures: Possible Role of Ancestral tRNAs in Viral RNA Genome Formation

FRANÇOIS CHAPEVILLE

F. CHAPEVILLE is Professor of Biochemistry, Paris University (VII) and Director of the "Institut de Recherche en Biologie Moléculaire"—CNRS, Université Paris VII. From 1953 to 1960 he worked at the French Atomic Commision, and studied sulphur metabolism; he discovered several enzymes and the reduction of inorganic sulphate by bird yolk sac endodermal cells. From 1960 to 1962 he worked with F. Lipmann at the Rockefeller University and published the experimental proof of Crick's adaptor hypothesis. Using chemically modified aminoacyl-tRNAs with his collaborators he contributed to the study of polypeptide chain initiation and elongation, discovered the peptidyl-tRNA hydrolase, and the tRNA-like structures in viral RNA genomes. Since 1970, his laboratory has been largely involved in the study of animal DNA polymerases, DNA ligases, DNA terminal transferase, and terminator codon suppression in animal cells. These last ten years he was invited as Visiting Professor by several foreign Universities including the University of Tokyo in 1980.

Nos connaissances sur l'origine des virus et leur évolution comme celles sur les organites normaux de la cellule sont très limitées. Il semble de plus en plus certain que l'apparition de la cellule eucaryotique résulte de la combinaison et de la fusion de plusieurs systèmes autoréplicatifs simples devenus vite indépendants mais incapables ni de s'autorépliquer en dehors de la cellule, ni de sortir d'une cellule et passer dans une autre. Selon l'hypothèse d'Atenburg, au cours de l'évolution, une réacquisition partielle d'indépendance primitive par un organite cellulaire et sa capacité de passer d'une cellule à une autre pourraient être à l'origine des virus. S'il en est ainsi, il serait logique qu'au moins une partie des signaux utilisés dans la régulation de la réplication et la traduction du matériel génétique de ces systèmes capables d'infecter les cellules soit similaire à ceux utilisés par la cellule elle-même dans ces processus.

Au cours de l'évolution, des modifications structurales d'un signal de nature polynuclèotidique ont pu conduire au changement ou élargissement de la significa- tion originale de ce signal et l'acquisition d'un spécificitévirale. Si ces modifications structurales n'ont pas été très importantes, on devrait retrouver dans certains signaux utilisés actuellement par la cellule et par le virus les traces d'ancêtres communs. Des données récentes sur les structures de différents enzymes protéoly- tiques et d'autres protéines, sur la structure de leurs gènes et leur diversification précédée par des duplications d'un gène ancestral, suggèrent que des structures polynucléotidiques autres que celles qui sont traduites en protéines ont subi le même type d'évolution.

En 1970, notre laboratoire en collaboration avec celui du Dr H. Duranton (1, 2) a mis en évidence la présence dans le génome d'un virus végétal à RNA d'une structure reconnue par plusieurs enzymes spécifiques des tRNA, structure que nous avons appelée tRNA-like. Ce même phénomène a été ensuite trouvé dans d'autres virus. Plus récemment, dans les virus oncogènes à RNA où un tRNA spécifique cellulaire sert d'initiateur de la synthèse du DNA-RNA dépendante, on a trouvé qu'environ un tiers de la structure complémentaire du tRNA initiateur existe dans le génome viral (3). Si le rôle de cette structure est bien connu, celui des structures tRNA-like n'a pas encore été élucidé. Celles-ci interviennent sans doute, soit directe- ment comme signaux spécifiques des virus dans un des processus qui permettent leur reproduction, soit indirectement en interférant avec le système cellulaire pour diminuer ou empêcher son fonctionnement au profit du développement viral.

RNA du Virus de la Mosaique Jaune du Navet (VMJN)

Le RNA génomique de ce virus de masse moléculaire de 2×10^6 daltons, incubé en présence d'un mélange d'acides aminés-^{14}C, d'ATP, et d'un extrait d'*E.coli* débarassé d'acides nucléiques fixe un seul acide aminé, la valine (1, 2). Une molé- cule de valine est fixée par molécule de RNA génomique. Les conditions dans lesquelles s'opère cette fixation et la cinétique de la réaction sont similaires à celles de l'estérification du tRNA^{Val}. L'examen de la liaison qui unit la valine au RNA viral a montré qu'à tout point de vue elle est analogue à la liaison ester du valyl- RNA. Après la digestion du Val-RNA viral par la RNAse pancréatique, la valyl- 3'-adénosine a été isolée. Le traitement des extraits bactériens a permis cependant d'isoler des fractions qui estérifiaient bien le tRNA^{Val} mais pas le RNA viral. La valyl-tRNA synthétase purifiée seule se révéla être aussi inactive vis-à-vis du RNA viral mais en présence d'un autre enzyme spécifique des tRNA, la tRNA nucléo- tidyl-transférase, elle catalysait son estérification (2, 4). Ce résultat confirmé par d'autres expériences a démontré que le RNA génomique du VMJN avant de fixer la valine devait être complété par l'addition terminale d'un résidu d'adénylate.

L'analyse des oligonucléotides estérifiés par la valine et obtenus après digestion par la RNAse T_1 soit à partir des valyl-tRNA^{Val} isolés des feuilles de la plante-hôte (chou chinois), soit à partir du valyl-tRNA de VMJN, a montré qu'ils sont différents et que par conséquent ils ne pouvaient s'agir ni d'un artefact dû à une contamination du RNA viral par un tRNA, ni même d'une ligation enzymatique hypothétique d'un tRNA cellulaire sur le RNA viral.

L'étude récente des paramètres cinétiques en présence de la valyl-tRNA synthétase purifiée (constantes de vitesse et de Michaélis) de la valylation du RNA viral a montré qu'ils sont du même ordre que ceux observés lors de l'estérification du tRNA^{Val} de levure (5). Ces données ont conduit à la conclusion que la valylation du RNA viral diffère de la valylation incorrecte observée dans le cas du tRNA^{Phe} ou du tRNA^{Met} de levure par la valyl-tRNA synthétase homologue, et en conséquence qu'il ne peut s'agir d'une aminoacylation fortuite. Le traitement du RNA de VMJN par la RNAse P d'*E. coli* un enzyme impliqué dans le processus de maturation des précurseurs de tRNA, a conduit à la libération d'un fragment de RNA de 40,000 daltons environ (4,5 S) provenant de l'extrémité 3' du RNA viral (6). Ce fragment possède les mêmes capacités d'aminoacylation que le RNA viral de haut poids moléculaire. Ce résultat a montré que les structures reconnues dans le RNA viral par la valyl-tRNA synthétase ne sont pas dispersées le long de la molécule du RNA viral mais sont localisées dans la région 3' représentant moins de 2% de la molécule.

On dispose de peu d'informations sur l'aminoacylation du RNA de VMJN *in vivo*. Si celle-ci se produit, elle ne peut être que transitoire ou concerner seulement une faible partie des molécules puisque le RNA du virion non seulement n'est pas estérifié mais il lui manque l'adénosine 3' terminal. Récemment, des indications ont été obtenues sur la réalité de cette deuxième possibilité. L'injection du RNA viral dans les oocytes de Xenope conduit à une fixation de la valine (7).

Aminoacylation D'autres RNA Viraux de Plantes

Depuis la découverte de l'aminoacylation du RNA de VMJN on a montré qu'au moins 10 autres génomes de virus végétaux appartenant à quatre groupes différents peuvent être estérifiés par divers acides aminés. La liste de ces virus et leur spécificité pour un acide aminé est donnée dans le Tableau I. (voir revue réf. 8). A l'exception des tobamovirus où deux représentants examinés acceptent deux acides aminés différents, les RNA des virus de même groupe acceptent le même acide aminé. Dans le cas des virus à composants multiples (virus de la mosaïque du concombre contenant des molécules différentes de RNA de 1,1 à $0,3 \times 10^6$ daltons) chaque molécule de RNA accepte le même acideaminé. Dans tous les cas où cela a été étudié, les RNA viraux sont estérifiés en présence des aminoacyl-tRNA synthétases isolées de la plante hôte. Dans certains cas, l'enzyme bactérien est actif, dans d'autres il doit être d'origine eucaryotique.

TABLEAU 1.   Esterification des RNA Viraux par les Acides Amines

Groupe	Virus	Acide aminé
Bromovirus	Virus de la mosaïque du brome	Tyrosine
	Virus de la marbrure de la fève	Tyrosine
	Virus de la marbrure chlorotique	Tyrosine
	de *Vigna Sinensis*	Tyrosine
Cucumovirus	Virus de la mosaïque du concombre	Tyrosine
Tobamovirus	Virus de la mosaïque du tabac	Histidine
	Virus de la mosaïque du tabac	
	souche "Cowpea"	Valine
Tymovirus	Virus de la mosaïque jaune du navet	Valine
	Virus de la mosaïque de l'aubergine	Valine
	Virus de la mosaïque du "concombre sauvage"	Valine
	Virus de la mosaïque janue du cacao	Valine
Cardiovirus	Virus de l'encéphalomyocardite	Sérine
	Virus de Mengo	Histidine

Deux cas d'aminoacylation des RNA viraux d'origine animale ont été rapportés, l'un concernant l'estérification par l'histidine du RNA de Mengovirus, l'autre par la sérine du RNA du virus de l'encéphalomyocardite (*8*). Ces observations sont surprenantes puisque les deux RNA viraux possèdent à leur extrémité 3′ une prolongation par le poly A. Une estérification suppose dont une fragmentation préalable du génome et le dégagement d'une extrémité reconnue par l'aminoacyl-tRNA synthétase.

Reconnaissance des RNA Viraux Par D'autres Enzymes Specifiques des tRNA

Le fait que le RNA du virus réagit avec une seule aminoacyl-tRNA synthétase a conduit a rechercher si d'autres enzymes spécifiques des tRNA qui reconnaissent dans l'ensemble de ces molécules un ou plusieurs motifs communs sont aussi capables de réagir avec le RNA viral. Les résultats précédemment mentionnés, obtenus avec le RNA de VMJN et les préparations purifiées de valyl-tRNA synthétase ont montré que la tRNA nucléotidyl-transférase était nécessaire à l'estérification. Les analyses ont montré que cet enzyme catalyse l'addition terminale d'un résidu adénylate accepteur de la valine (*2, 4*). D'autres enzymes se sont révélés être aussi actifs. La peptidyl-tRNA hydrolase, qui hydrolyse la liaison ester lorsque le groupement aminé de l'aminoacide est substitué, hydrolyse le N-acétyl-valyl-RNA de VMJN (*2*). Il en est de même pour deux autres enzymes spécifiques des tRNA, la tRNA cytosine méthylase (*9*) et la RNAse P mentionnée précédemment.

Le fragment de RNA libéré en présence de ce dernier enzyme est long de 112 nulcéotides (*6*). Il a été aussi démontré qu'en présence de GTP le Val-RNA de VMJN forme un complexe avec le facteur d'élongation EF-1 dont les propriétés sont semblabes à celui formé avec les aminoacyl-tRNA (*10, 11*). Dans certaines de ces investigations lorsque les enzymes d'origine autre qu'*E. coli* ont été essayés, ils se sont révélés également actifs. L'ensemble de ces résultats a montré que le RNA viral semble posséder une structure dans laquelle les groupements fonctionnels essentiels pour la reconnaissance par les enzymes spécifiques des tRNA sont répartis dans l'espace de la même façon que dans les tRNA. Le système ribosomal n'est cependant pas capable d'utiliser l'aminoacide estérifié par le RNA viral pour la synthèse des protéines; cela a été observé dans le cas d'au moins trois RNA viraux chargés par les acides aminés différents. De nouvelles études en présence du système ribosomal pourraient révéler l'existence d'une compétition entre un aminoacyl-tRNA et le RNA viral estérifié par le même acide aminé. Possédant l'extrémité-CCA estérifiée, par l'analogie aux fragments 3′ terminaux des tRNA, l'aminoacyl-RNA viral pourrait être utilisé comme accepteur de la chaîne peptidique au niveau du site A. Une telle réaction conduirait probablement à la libération de la chaîne attachée au RNA viral de la même façon que cela se produit avec la puromycine ou des fragments aminoacyl 3′ terminaux des tRNA.

Structure des Regions Acceptrices D'aminoacides des RNA Viraux

L e fait que plusieurs enzymes spécifiques des tRNA reconnaissent la partie 3′ terminale de certains RNA viraux était a priori en faveur de l'existence d'une analogie de structure entre ces molécules. Les données obtenues sur la séquence des nucléotides dans trois RNA viraux au niveau de la région acceptrice (Fig. 1) n'ont cependant pas confirmé cette hypothèse (*12–14*). En effet, si dans le cas du RNA de VMJN, une faible similarité à un tRNA peut être décelée, dans les deux

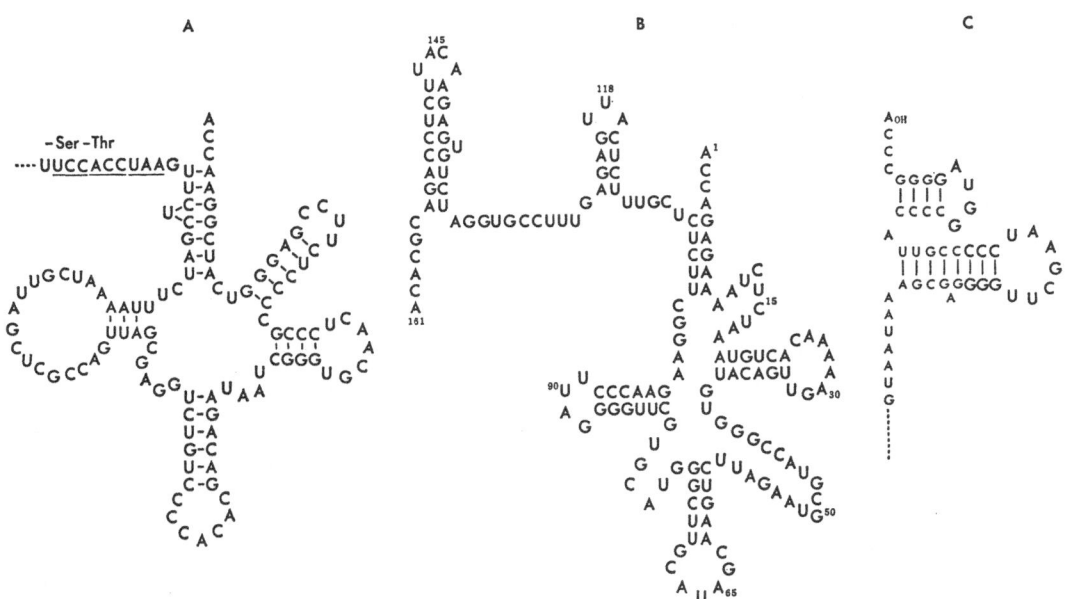

Fig. 1. Structure primaire et secondaire possible des régions acceptrices d'aminoacides de trois RNA viraux. A: RNA du virus de la mosaïque jaune du navet. B: RNA du virus de la mosaïque du brome. C: RNA du virus de la mosaïque du tabac.

autres cas il n'en est rien. Cela est surprenant et pose le problème de la reconnaissance de ces régions par les enzymes spécifiques des tRNA. Il est cependant très probable qu'en dépit des apparences les structures tertiaires de ces régions possèdent des points communs avec celles des tRNA. Les informations sur ce point seraient précieuses dans l'étude des interactions protéines-acides nucléiques. Ces régions des RNA viraux ne possèdent pas de nucléotides modifiés. Le fait qu'elles sont reconnues par les enzymes spécifiques des tRNA montre que les modifications des nucléotides ne doit pas intervenir d'une façon importante dans la réaction de ces enzymes avec les tRNA. Par ailleurs, dans le génome viral ces régions, du fait qu'elles sont reconnues par des enzymes spécifiques des tRNA ne doivent pas former des associations fortes avec d'autres parties de la chaîne polynucléotidique.

tRNA Ancestraux et al Formation des Genomes Viraux

L'élucidation du rôle des structures tRNA-like dans les génomes viraux apportera sans doute des éléments intéressants dans l'étude de l'évolution des virus. A l'heure actuelle, seules des spéculations sont permises. Tôt dans l'évolution les structures qui ressemblent à des tRNA modernes ont joué probablement un rôle important dans la formation des premiers organismes et dans leur diversification. Certaines de ces molécules ont donné naissance aux tRNA qui jouent un rôle fondamental dans la traduction des gènes en protéines ou assument d'autres fonctions importantes. L'utilisation de certains tRNA comme le tRNA de tryptophane ou de proline comme amorces de la synthèse des DNA au cours de la transcription reverse ainsi que celle d'un $tRNA^{Gly}$ spécifique pour la synthèse des peptidoglycanes illustre l'importance de la diversification fonctionnelle de ces molécules. On ignore pourquoi elles ont été choisies pour remplir autant de fonctions disparates.

D'autres molécules ancestrales des tRNA ont probablement joué un rôle dans la formation de divers génomes à RNA dans lesquels on les trouve maintenant et où elles sont utilisées comme signaux de régulation dans la réplication et dans l'expression de ces génomes. L'évolution de ces molécules ancestrales n'a pas suivi nécessairement le même chemin que celle des molécules qui ont donné naissance aux tRNA. Comme seulement une partie des structures présentes dans les tRNA modernes seraient nécessaires pour ces fonctions régulatrices spécifiques du virus, on doit s'attendre à l'existence de grandes diversités dans la ressemblance entre la structure de ces signaux et les tRNA. Cette diversité résulte des modicafitions nécessaires ou permissives des structures initiales pour devenir des régulateurs efficaces, et elle a été influencée par l'ensemble des processus aboutissant à l'apparition du génome viral tel qu' il se présente maintenant. Il est probable que l'évolution de ces structures se soit arrêtée depuis longtemps puisque dans les fragments des génomes à composants multiples qui possèdent une structure tRNA-like, elle est la même dans chaque fragment.

REFERENCES

1. Pinck, M., Yot, P., Chapeville, F., and Duranton, H.: 1970. *Nature* 226, 954–956.
2. Yot, P., Pinck, M., Haenni, A.L., Duranton, H.M., and Chapeville, F.: 1970. *Proc. Natl. Acad. Sci. USA* 67, 1345–1352.
3. Haseltine, W.A., Maxam, A.M., and Gilbert, W.: 1977. *Proc. Natl. Acad. Sci. USA* 74, 989–992.
4. Litvak, S., Tarrago-Litvak, L., and Chapeville, F.: 1973. *J. Virol.* 11, 238–242.
5. Giégé, R., Briand, J.P., Mengual, R., Ebel, J.P., and Hirth, L.: 1978. *Eur. J. Biochem.* 84, 251–256.
6. Prochiantz, A. and Haenni, A.L.: 1973. *Nartue New Biol.* 241, 168–170.
7. Joshi, S., Haenni, A.L., Hubert, E., Huez, G., and Marbaix, G.: 1978. *Nature* 275, 339–341.
8. Haenni, A.L. and Chapeville, F.: 1980. *Transfer RNA: Biological Aspects* (Söll, D., Abelson, J.N., and Schimmel, P.R., eds.) pp. 539–556, Cold Spring Harbor Laboratory, New York.
9. Haenni, A.L., Benicourt, C., Teixeira, S., Prochiantz, A., and Chapeville, F.: 1975. *FEBS Proc. Meet.* 39, 121–131.
10. Litvak, S., Tarrago, A., Tarrago-Litvak, L., and Allende, J.E.: 1973. *Nature New Biol.* 241, 88–90.
11. Haenni, A.L., Prochiantz, A., and Yot, P.: 1974. *Lipmann Symposium: Energy Regulation and Biosynthesis in Molecular Biology* (Richter, D., ed.) pp. 264–276, de Gruyter, Berlin.
12. Siberklang, M., Prochiantz, A., Haenni, A.L., and RajBhandary, U.L.: 1977. *Eur. J. Biochem.* 72, 465–478.
13. Dasgupta, R. and Kaesberg, P.: 1977. *Proc. Natl. Acad. Sci. USA* 74, 4900–4904.
14. Guilley, H., Jonard, G., and Hirth, L.: 1975. *Proc. Natl. Acad. Sci. USA* 72, 864–868.

¹H NMR Study on a Short DNA Helix in Aqueous Solution

LOU-SING KAN, DORIS M. CHENG, PAUL S. MILLER,
ELDON E. LEUTZINGER, KRISHNA JAYARAMAN,
and PAUL O. P. TS'O

P.O.P. Ts'o is Professor and Director of the Division of Biophysics, School of Hygiene and Public Health, the Johns Hopkins University since 1973. He was previously professor (1967–1973) and associate professor (1962–1967) of biophysical chemistry at Johns Hopkins, and a staff member (1955–1962) at the California Institute of Technology, where he received his Ph.D. degree. His research interests include nucleic acid chemistry and biology, the application of nuclear magnetic resonance to biochemical research, chemical and viral carcinogenesis, interferon research, and cellular research on aging and differentiation. He has published about 180 scientific and review papers, and is the editor of the following books: "Basic Principles in Nucleic Acid Chemistry, Vol. I and II," "The Molecular Biology of the Mammalian Genetic Apparatus, Vol. I and II," "Polycyclic hydrocarbons and Cancer, Vol. I and II," and "Carcinogenesis: Fundamental Mechanisms and Environmental Effects." He has served on the editorial boards of "Biochemistry" and "Biophysical Journal," "Molecular Pharmacology," "Biochimica et Biophysica Acta," "Cancer Review," and "Journal of Environmental Health Science," "Biopolymers," and "Cancer Research." Currently, he has served as a member of the NIH Study Section A on Biophysics and Biophysical Chemistry, a member of European Expert Committee on Biophysics (UNESCO), and a member of the Clearinghouse on Environmental Carcinogens, National Cancer Institute.

One direct and effective way to study the conformation of DNA helix, as well as the mechanisms of DNA-protein (or with other agents) in aqueous solution at the atomic level of resolutions, is to apply NMR spectroscopy on short DNA helix with defined sequence. Previously, we have described extensive ¹H NMR studies on a self-complementary ribooligonucleotide, r-(A-A-G-C-U-U), which forms a stable duplex in aqueous solution (1, 2). These studies revealed that this short duplex exists in an A'-type conformation and thus serves as a good model for helical RNA.

In this short communication, we report some preliminary results of ^{1}H NMR studies on a chemically synthesized decadeoxyribonucleotide having a self-complementary sequence: d-(C-C-A-A-G-C-T-T-G-G). This decamer helix is expected to form one full turn of DNA helix in solution. In addition, this decamer contains a sequence which can be recognized by several restriction endonucleases such as *Alu* I (recognizes -A-G-C-T), *Hind* III, and *Hsu* I (both recognize -A-A-G-C-T-T-).

Synthesis of d-(C-C-A-A-G-C-T-T-G-G)

The decanucleotide was prepared by chemical synthesis using phosphotriester approach (*3*). The overall synthetic scheme is shown in Fig. 1. The detailed steps of the procedures was published elsewhere (*4*).

The protected decamer was deblocked by treating with 1 M pyridine-2-aldoximate tetramethylguanidinium salt in dioxane solution. A total of 1,318 O.D. (A_{257}) (representing 21% of yield) of d-(C-C-A-A-G-C-T-T-G-G) were obtained. The purity of this chemical synthesized decamer is higher than 95% by high pressure liquid chromatography (HPLC) results(*4*).

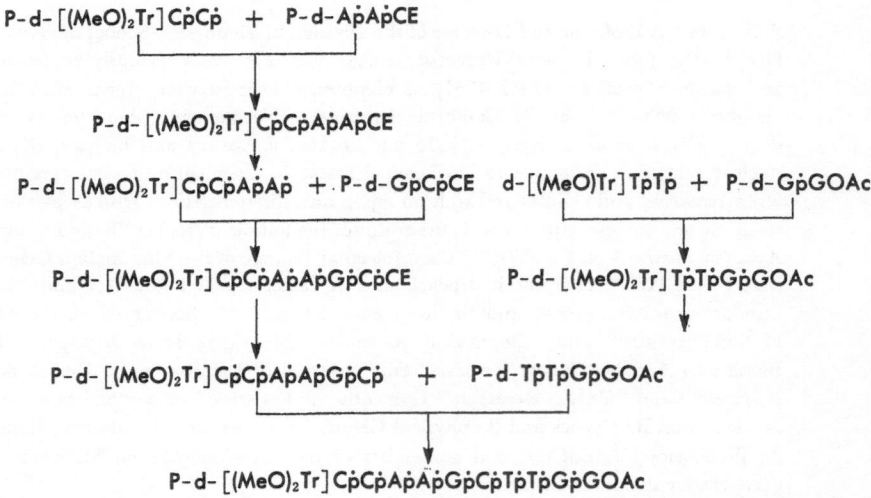

FIG. 1. The synthetic scheme for the preparation of protected oligodeoxyribonucleotides. The letter P before the oligonucleotide indicates the base groups are protected: *i.e.*, C=bzC; A=bzA; G=ibuG; T=T. The symbol p̄ indicates a *p*-chlorophenylphosphotriester internucleotide linkage and (MeO)₂ Tr is dimethoxytrityl group.

Characterization of d-(C-C-A-A-G-C-T-T-G-G)

The decamer was completely digested to mononucleotides (-side) with snake venom phosphodiesterase. The ratio of nucleotides in the hydrolysate is

1.00:1.95:1.98:1.98:3.01 of dC:dpC:dpA:pT:dpG and the calculated ratio should be 1:2:2:2:3.

The decamer was also hydrolized by *Alu* I restriction endonuclease (P.L. Biochemicals) to yield two pentanucleotides, d-(C-C-A-A-G) and d-(pC-T-T-G-G) in the expected 1:1 ratio. Further treatment with bacterial alkaline phosphatase gave two pentanucleotides, d-(C-C-A-A-G) and d-(C-T-T-G-G) whose chromatographic mobilities were identical to those of authentic samples (Leutzinger *et al.*, unpublished data).

However, the decamer is not hydrolyzed by *Hind* III restriction endonuclease at 0°C or as high as 37°C. It will be important to investigate whether this decamer can be bound by the enzyme.

UV and CD Spectra of d-(C-C-A-A-G-C-T-T-G-G)

The melting profile of the decamer measured in 0.13 M NaCl-phosphate buffer (pH 7.0) at strand concentration of 0.013 mM is shown in Fig. 2. The CD spectra of the decamer at 19 and 80°C are shown in Fig. 3. Under the condition of these experiments, the melting temperature of the decamer helix is 47°C. The shape of the CD spectrum of the decamer at low temperature is qualitatively similar to that of DNA, suggesting that the duplex may have a B-type geometry.

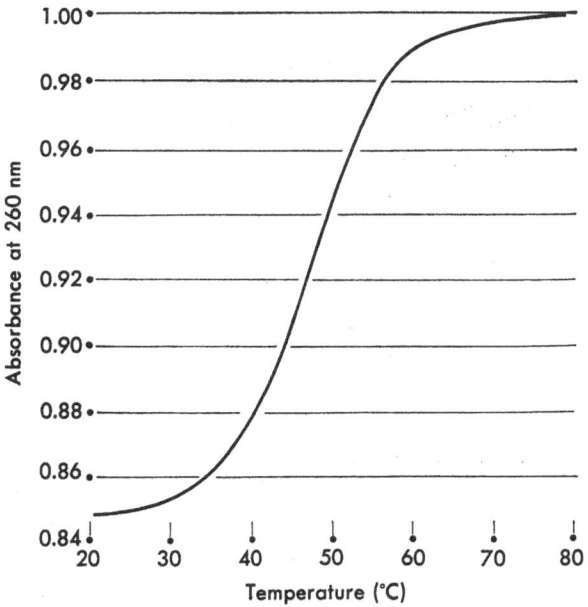

Fig. 2. The UV melting profile of d-(C-C-A-A-G-C-T-T-G-G) at a strand concentration of 13 μM in 0.10 M sodium chloride, 0.01 M sodium phosphate, 0.10 mM ethylenediamine tetraacetate at pH 7.0.

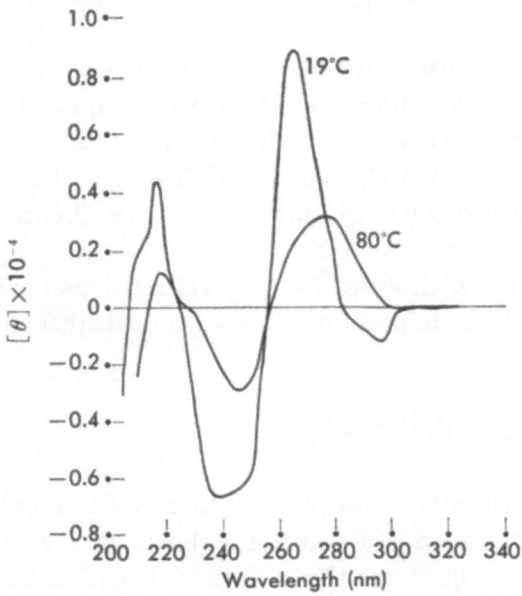

FIG. 3.   The effect of temperature on the CD spectrum of 13 μM d-(C-C-A-A-G-C-T-T-G-G) in 0.10 M sodium chloride, 0.01 M sodium phosphate, 0.10 mM ethylenediamine tetraacetate at pH 7.0.

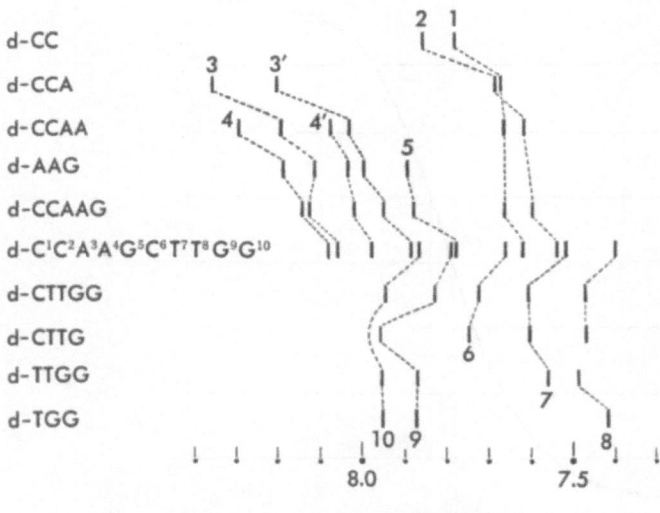

Chemical shift, ppm

FIG. 4.   The scheme of sequential incremental assignment of d-(C-C-A-A-G-C-T-T-G-G) at high temperature where 1, C^1-H_6; 2, C^2-H_6; 3, A^3-H_8; 3', A^3-H_2; 4, A^4-H_8; 4', A^4-H_2; 5, G^5-H_8; 6, C^6-H_6; 7, T^7-H_6; 8, T^8-H_6; 9, G^9-H_8; and 10, G^{10}-H_8.

¹H NMR Study on d-(C-C-A-A-G-C-T-T-G-G)

One of the big advantages of ¹H NMR spectroscopy on nucleic acids is the high information content of the spectrum, *i.e.* signals are obtained from the protons of each nucleotide residue in the decamer. For this reason, the unambiguous assignment of all proton resonances in the NMR spectrum becomes an essential requirement. In the past, we have used the information from spectral patterns, the area under peaks, spin-lattice relaxation time, effect of paramagnetic ions, and homo- and heterodecoupling techniques to assign the ¹H NMR spectra of oligonucleotides up to three nucleotidyl units in length (*5, 6*). For longer oligomers, in addition to the techniques mentioned above, we have adopted the so-called sequential incremental method (*1*) as illustrated in Fig. 4. In this case, the base

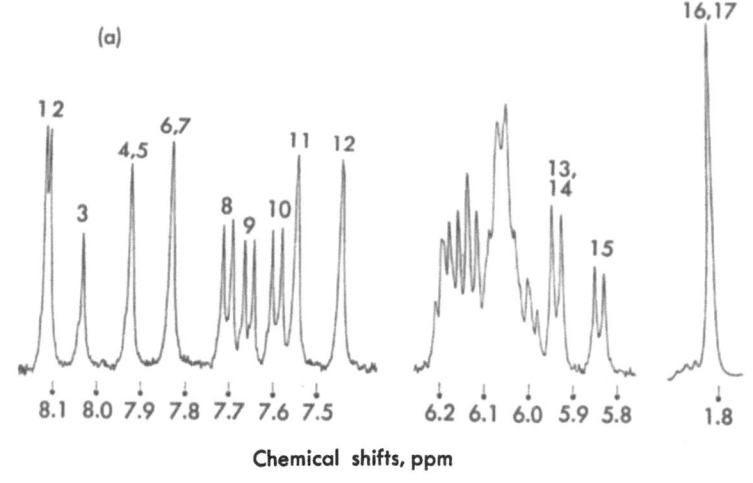

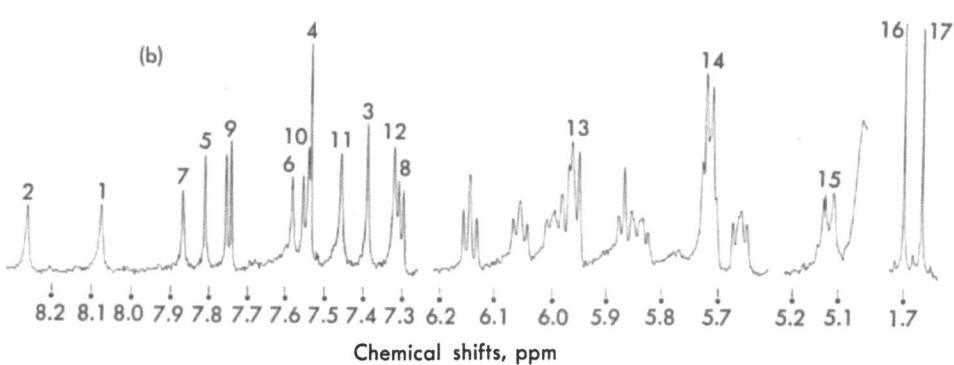

FIG. 5.   ¹H NMR spectra of 6.3 mM d-(C¹-C²-A³-A⁴-G⁵-C⁶-T⁷-T⁸-G⁹-G¹⁰) in 1×10^{-4}M EDTA, 0.01 M phosphate buffer, pH 7.0: (a) 360 MHz spectrum at 90°C; (b) 600 MHz spectrum at 20°C. The assignments of the base proton resonances are: 1, A⁴-H₈; 2, A³-H₈; 3, A⁴-H₂; 4, A³-H₂; 5, G¹⁰-H₈; 6, G⁵-H₈; 7, G⁹-H₈; 8, C⁶-H₆; 9, C¹-H₆; 10, C²-H₆; 11, T⁷-H₆; 12, T⁸-H₆; 13, C¹-H₅; 14, C²-H₅; 15, C⁶-H₅; 16, T⁷-CH₃; and 17, T⁸-CH₃.

proton resonances of the decamer are compared systematically to its nine smaller fragments at high temperature. (And these fragments can be readily obtained during the course of synthesis of the decamer). As shown in Fig. 4, the identities of all base proton resonances of the decamer can be assigned with high confidence (see also Fig. 5a).

The identification of all base proton resonances of the decamer at low temperature (at helical state) (Fig. 5b) can be achieved by carefully tracing the signals from high temperature to low temperature (data not shown). Figure 5b shows a 600 MHz ^{1}H NMR spectrum (obtained from Carnegie-Mellon University, Pittsburgh, Pennsylvania) at 20°C, together with all the assignment of the base proton resonances.

In summary, we have employed a phosphotriester synthetic procedure to prepare a decadeoxyribonucleotide (as well as the important fragments) having a defined sequence (CCAAGCTTGG) in high purity and large quantities. The sequence of this decamer can be recognized by several restriction endonucleases. Optical studies indicate that this decamer can form a double-stranded helix, probably in B-form, in aqueous solution with melting temperature around 50°C depending on ionic strength and strand concentration. In the ^{1}H NMR study, all base proton resonances of this decamer can be identified at a wide temperature range, including both single-stranded and helical states (Fig. 5). These results indicate that we are able to obtain conformational information at atomic level of this short DNA helix in aqueous solution, which may serve as a model system for a much longer DNA.

REFERENCES

1. Borer, P.N., Kan, L.S., and Ts'o, P.O.P.: 1975. *Biochemistry* 14, 4847–4863.
2. Kan, L.S., Borer, P.N., and Ts'o, P.O.P.: 1975. *Biochemistry* 14, 4864–4869.
3. Itakura, K., Katagiri, N., Bahl, C.P., Wightman, R.H., and Narang, S.A.: 1975. *J. Am. Chem. Soc.* 97, 7327–7332.
4. Miller, P.S., Cheng, D.M., Dreon, N., Jayaraman, K., Kan, L.S., Leutzinger, E.E., Pulford, S.M., and Ts'o, P.O.P.: 1980. *Biochemistry* 19, 4688–4698.
5. Ts'o, P.O.P., Kondo, N.S., Schweizer, M.P., and Hollis, D.P.: 1969. *Biochemistry* 8, 997–1029.
6. Kan, L.S., Barrett, J.C., and Ts'o, P.O.P.: 1973. *Biopolymers* 12, 2409–2421.

How Different Is Z from B?

BERNARD PULLMAN

B. PULLMAN was born in 1919. He studied at the Sorbonne, became Licencié-ès-Sciences in 1946 and Docteur-ès-Sciences in 1948. From 1946 to 1954, he belonged to the National Research Council. Since 1954, he has been Professor of Quantum Chemistry at the Sorbonne. Since 1959, he is the Director of the Department of Quantum Biochemistry at the Institut de Biologie Physico-Chimique of Paris and since 1963, he is the Director of this important Institute. He is a member of the French, Hungarian and GDR Academian of Sciences and of the Royal Academy of Pharmacy of Spain; Past President and Honorary President of the International Academy of Quantum Molecular Sciences; Honorary President of the International Society of Quantum Biology; Doctor Honoris Causa of the Universities of Liège (Belgium), Uppsala (Sweden), and Madrid (Spain); Honorary Member of the Italian Society of Cancerology, the Union of Medical Sciences of Roumania, the Chemical Society of Austria and the Biochemical Society of Japan; He is the recipient of the Essec Prize of the French Ligue against cancer, of the Louis Bonneau Prize and of the Charles Louis de Saulses de Freycinet Prize of the French Academy of Sciences, of the Stas Medal of the Chemical Society of Belgium, of the Sir Yagdish Bose Medal of the Indian Academy of Medical Physics, of a Medal of the Israel Academy of Sciences and Humanities, and of the award of the International Society of Quantum Biology (USA).

The readers of this volume will of course have guessed that we have no intention of dealing here with the graphical difference between two letters of the Roman alphabet. These who follow the excitation curve of scientific discussion in modern molecular biology must, moreover, have already realized that what is in stake here are Z- *versus* B-DNAs.

Truly, Z is not the first competitor to B and everybody has heard about A-, C-, D-, V-DNA, *etc...* What makes Z so particular with respect to all the other DNAs are some spectacular differences of structure, which render it, so far, unique.

The discovery in its present form comes from a recent crystallographic study of the self-complementary hexamer d(CpGpCpGpCpG) by Rich and colleagues (1, 2) which opened the road to a family of new DNA double helices, termed Z-DNAs. These Z-DNAs differ markedly from the classical B-DNA structure, most notably, because they form *left-handed* helices. Further, their repeat unit is a dinucleotide, rather than a mononucleotide as in B-DNA; their constituent nucleotides, dG and dC, differ in their sugar conformations, 3' endo and 2' endo respectively, their glycosidic torsion angles, *syn* and *anti* respectively; in addition, the two phosphates, GpC and CpG differ in the phosphodiester torsional angles.

Similar left-handed helices have also been found in crystal studies of related tetranucleotides (3, 4), always with the alternating base sequence $\left(\begin{smallmatrix} G\text{-}C \\ C\text{-}G \end{smallmatrix}\right)$ and Arnott *et al.* (5) have presented evidence for Z-DNA conformations in fibers of poly(dG.dT). poly(dA.dC). Sarma *et al.* (6) brought conclusive evidence for the persistence of the Z-DNA structure of poly(dG.dC). poly(dG.dC) in solution.

An important question concerns the possible biological significance of such left-handed helices. It has been suggested and shown to be geometrically feasible that sections with a Z-DNA conformation may be interspersed within long B-DNA double helices (1, 2). Should such a possibility be realized in nature the next question concerns, naturally, the consequences for the biochemical and biological behaviour of DNA of the local presence of such sections.

The answer to this question will be determined by the differences in gross and fine structure between the Z and B helices. The gross geometrical differences are given by the X-ray crystallographic results. The differences in fine structures, by which we understand in particular the *electronic* and *steric* properties involved in the biochemical reactivity, may be evaluated by theoretical procedures among which the quantum-mechanical ones are the most appropriate.

We have computed two such properties relevant to this problem the *electrostatic molecular potential* at the reactive sites in particular of the purine and pyrimidines basis and the *steric accessibility* to them. The significance and the technique

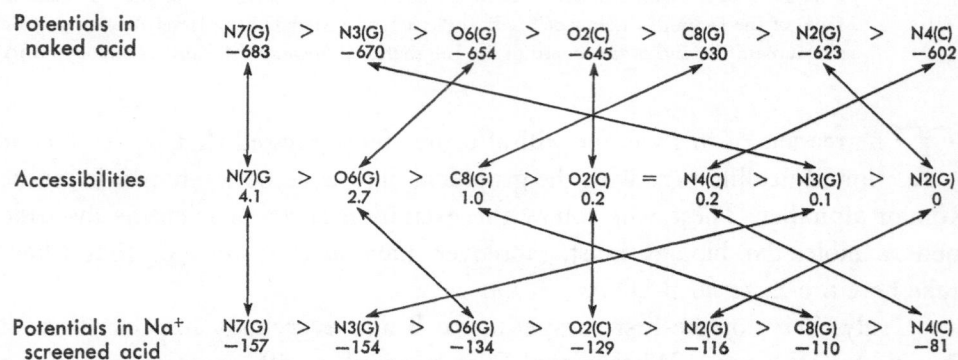

FIG. 1.   Potentials (in kcal/mol) and accessibilities (in Å²) of reactive sites in a poly (dG.dC) model of B-DNA.

of calculating these quantities has been discussed in previous publications. (For electrostatic potential see *e.g.* Refs. *7, 8*; for the steric accessibility, Refs. *9, 10*). The minima of the electrostatic potential are a measure of the intrinsic affinity of the different reactive centers for electrophiles and have been shown (*11, 12*) to correlate satisfactorily with a large number of relevant experimental data. An appropriate steric accessibility is, of course, a prerequisite for a reaction to occur. (The computed accessibilities refer to a sphere of 1.2. Å² radius, shown to conveniently represent a water molecule and indicating an upper limit for accessibility toward more complex attacking species.)

Figure 1 presents results of a computation of potentials (in kcal/mole) and accessibilities (in Å²) at reactive sites of the bases for a complete helical turn of a poly (dG.dC) model of B-DNA (*10, 11*). The turn involves 22 phosphates and as many sugars and bases. The potentials were evaluated both for the free acid and for the acids acreened by a Na⁺ ion at each phosphate (as a model of a complete screening).

Figure 2 presents the same results for Z-DNA (*13*). As a complete helical turn involves in this case 24 phosphates, sugars, bases, we have for the sake of comparison with B-DNA computed also the potentials for a truncated helix in which two nucleotide units have been suppressed. The effect is significant only for the free acid.

Figure 3 carries out a comparison between the accessibilities to the reactive sites of the bases in B- and Z-DNAs and Table 1 presents the same comparison for the potentials.

The figures are self-explanatory. The main conclusions to be drawn from them seem to be following:

1) The accessibilities to the reactive sites run nearly parallel in both acids. The most accessible sites are in both cases located on the guanine moiety and concern in particular its N7, O6, and C8. Altogether, the accessibility to these most reactive sites is, however, greater in Z-DNA than in B-DNA.

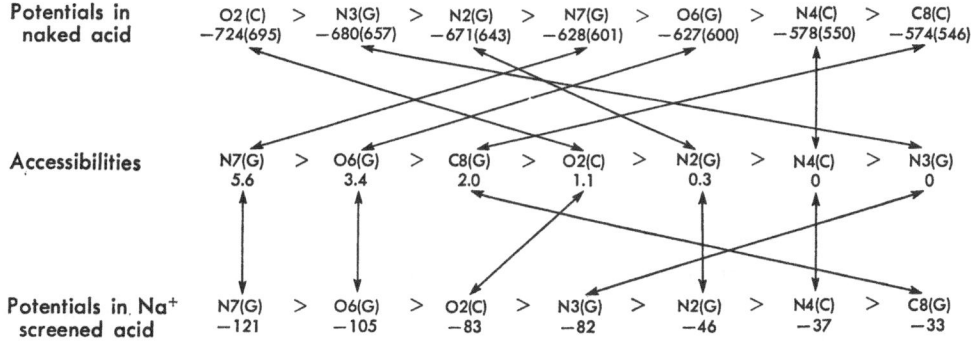

FIG. 2.   Potentials (in kcal/mol) and accessibilities (in Å²) of reactive sites in Z-DNA. Complete helix; the potentials in parenthesis for the naked acid refer to a truncated helix: see text.

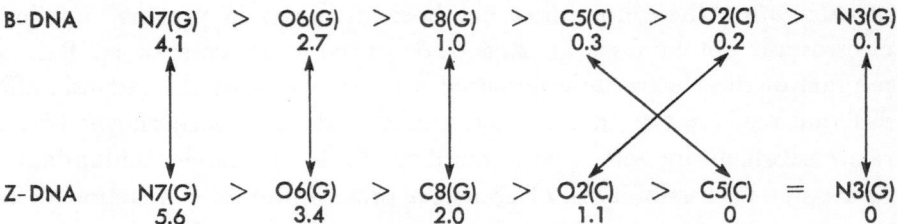

FIG. 3.   Comparison of accessibilities (in Å²) to reactive sites in B- and Z-DNAs.

TABLE 1.   Potentials in B-DNA and Z-DNA (in kcal/mol)

Site	Unscreened			Na⁺ screened		
	B-DNA	Z-DNA		B-DNA	Z-DNA	
		Helical turn	Truncated helix		Helical turn	Truncated helix
N7 (G)	−683 (1)	−628 (4)	−101 (4)	−157 (1)	−121 (1)	−121 (1)
N3 (G)	−670 (2)	−686 (2)	−657 (2)	−154 (2)	−82 (4)	−81 (4)
O6 (G)	−654 (3)	−627 (5)	−600 (5)	−134 (3)	−105 (2)	−105 (2)
O2 (C)	−645 (4)	−724 (1)	−695 (1)	−129 (4)	−83 (3)	−81 (3)
C8 (G)	−630 (5)	−574 (7)	−546 (7)	−110 (6)	−33 (7)	−32 (7)
N2 (G)	−623 (6)	−671 (3)	−643 (3)	−116 (5)	−46 (5)	−45 (5)
N4 (C)	−602 (7)	−578 (6)	−540 (6)	−81 (7)	−37 (6)	−37 (6)

Numbers in parenthesis indicate, in each column, the order of diminishing potentials.

2)   In the naked acids, the deepest electrostatic potentials are located on N7, N3, and O6 of guanine in B-DNA and on O2 of cytosine, N3 and N2 of guanine in Z-DNA. N3 of guanine having a very small accessibility in both compounds, it may be concluded that the deepest potentials at accessible sites are thus located in the major groove in B-DNA and in the minor one in Z-DNA.

3)   The screening by Na⁺ ions leaves unchanged this results for B-DNA. It produces, however, a drastic transformation in Z-DNA, in which the deepest potentials now occur on its convex side (there is no major groove in Z-DNA), at N7 and O6 of guanine, thus at sites similar to those of B-DNA.

4)   The relative values of potentials at similar sites of B- and Z-DNAs depend on whether we consider the screened or unscreened species and whether we use a complete or a truncated helix for Z-DNA. For the unscreened acids, the deepest potential of Z-DNA (at O2(C)) is slightly deeper than the deepest potential of B-DNA (at N7 (G)). In the screened acids the deepest potential situated in all cases at N7(G), is deeper in B-DNA than in Z-DNA.

It has been shown (*11, 12, 14*) that the electrostatic molecular potentials of the nucleic acids are particularly significant for the interpretation of the reactivity of these acids and their constituents with chemical carcinogens. The results on the structure of Z-DNA suggest that the interaction with some of these species like *e.g.* the metabolic ultimate carcinogens of aromatic hydrocarbons, which attack preferentially the exocyclic amino group of guanine, could occur more easily in the (unscreened) Z-DNA than in B-DNA. Z-DNA segments if present in a B-DNA chain, could thus represent potentially vulnerable sites for such attacks.

Acknowledgments

We wish to thank Professor A Rich, R.E. Dickerson and R.H. Sarma for communication of results prior to publication, in particular Prof. Rich for the crystallographic coordinates of Z-DNA. Thanks are also due to the National Foundation for Cancer Research for its support of this work.

REFERENCES

1. Wang, A.H-J., Quigley, G.J., Kolpak, F.J., Crawford, J.L., Van Boom, J.H., Van der Marel, G., and Rich A.: 1979. *Nature* 282, 860–866.
2. Wang, A.H-J., Quigley, G.J., Kolpak, F.J., Van der Marel, G., Van Boom, J.H., and Rich, A.: 1980. *Science* 211, 171–176.
3. Drew, H.R., Takano, T., Tanaka, S., Itakura, K., and Dickerson, R.E.: 1980. *Nature* 286, 567–573.
4. Crawford, J.L., Kolpak, F.J., Wang, A.H-J., Quigley, G.J., Van Boom, J.H., Van der Marel, G., and Rich, A.: 1980. *Proc. Natl. Acad. Sci. USA*, in press.
5. Arnott, S., Chandrasekaran, R., Birdsall, D.C., Leslie, A.G.W., and Ratliff, R.L.: 1980. *Nature* 283, 743–745.
6. Mitra, C.K., Sarma, R.H., and Sarma, R.H.: 1980. *Biochemistry*, in press.
7. Pullman, A.: 1976. Mécanismes d'Altération et de Réparation du DNA, Relations avec la Mutagénèse et la Cancérogénèse Chimique. *Colloq. CNRS Paris* 256, 103–113.
8. Pullman, A. and Pullman, B.: 1980. *Molecular Electrostatic Potentials in Chemistry and Biochemistry* (Politzer, P. and Truhlar, D.G., eds.), Plenum Press, New York, in press.
9. Lavery, R., Pullman, A., and Pullman, B.: 1980. *Theor. Chim. Acta*, in press.
10. Lavery, R., Pullman, A., and Pullman, B.: 1980. *Biochim. Biophys. Acta*, in press.
11. Pullman, B., Perahia, D., and Cauchy, D.: 1979. *Nucleic Acids Res.* 6, 3821–3830.
12. Pullman, A. and Pullman, B.: 1980. *Int. J. Quant. Chem.* (Quant. Biol. Symp. 7) 275–260.
13. Zakrzewska, K., Lavery, R., Pullman, A., and Pullman, B.: 1980. *Nucleic Acids Res.* 8, 3917–3932.
14. Pullman, B. and Pullman, A.: 1980. *Carcinogenesis, Fundamental Aspects and Environmental Effects* (Pullman, B., Ts'O, P.O.P., and Gelboin, H., eds.) pp. 55–66, Proceedings of the 13th Jerusalem Symposium in Quantum Chemistry and Biochemistry, Reidel Publishing Co., Dordrecht, Holland.

Cleavage with Restriction Enzymes of DNA from 39 Species of Plants and Animals

A. LIMA-DE-FARIA, MARGARETH ISAKSSON, EVA OLSSON, and JOHAN ESSEN-MÖLLER

A. LIMA-DE-FARIA was born at Cantanhede in Portugal on July 4, 1921. He graduated from the University of Lisbon (biology) in Portugal in 1945 and obtained the Ph.D. degree in genetics from the University of Lund in Sweden in 1956. In 1957 he became a docent of genetics at the University of Lund, and in 1964 a research docent. In 1969 he was Designated Professor of Molecular Cytogenetics and Director of Institute of Molecular Cytogenetics. He was a Visiting Professor at the Department of Anatomy, Duke University, Durham, N.C. in 1964 and a Visiting Professor of the University of Edinburgh from 1967 to 1968. He was decorated Knight of the Northern Star and Great Official Order Santiago. Medals from the University of Helsinki and Oporto. In 1961 he received the Oscar II prize. He is a member of the Swedish Roral Fysiografic Society and of the Nordic Society of Cell Biology. In 1969 he edited the Handbook of Molecular Cytology.

Restriction endonucleases are enzymes which cleave DNA. A group of these enzymes recognize specific sequences of base pairs and cleave the DNA molecule at specific sites (1–3).

To understand the molecular organization of the eukaryotic chromosome it is necessary to know the frequency of these segments in the chromosome complement and their distribution pattern.

For this purpose high molecular weight nuclear DNA was extracted from: 1 protozoan, 11 species of plants, 1 insect, 1 fish and 2 bird species, 10 rodents, 10 whales and other mammals including humans. Three restriction enzymes were employed: EcoR1, HpaII, and HaeIII. Four molecular weight markers were used:

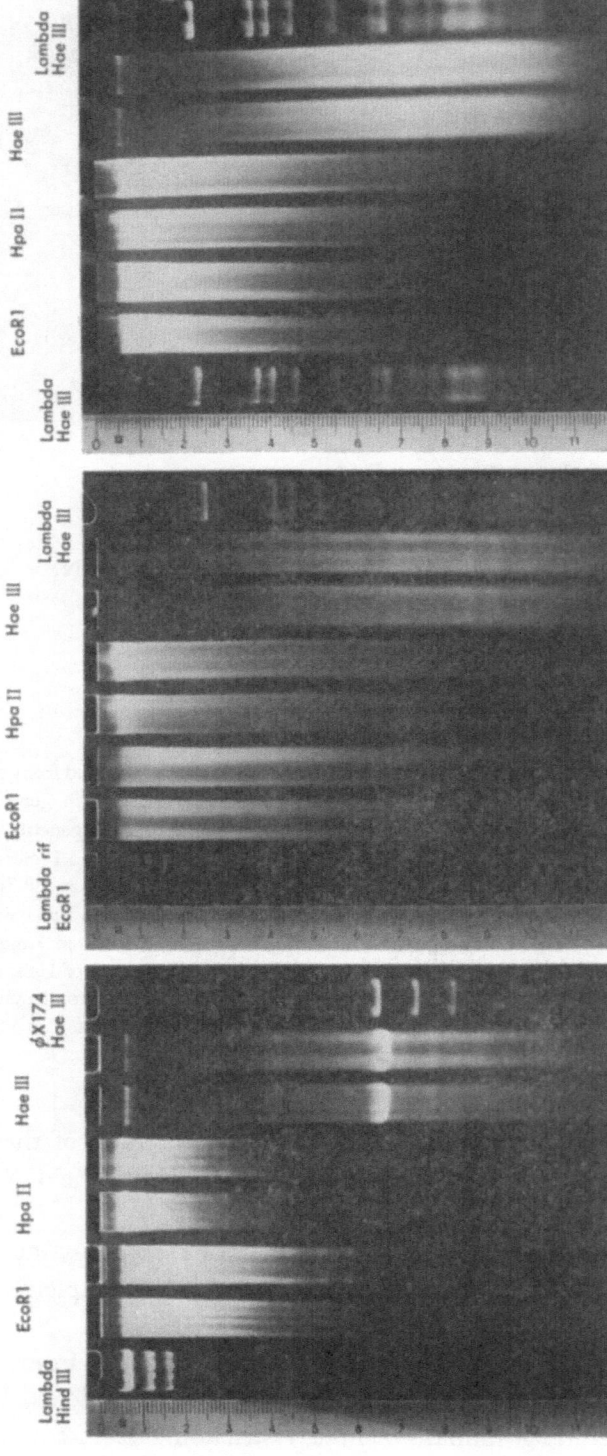

FIG. 1.   (Left) Nuclear DNA of *Balaenoptera acutorostrata* (whale) treated with the restriction enzymes EcoR1, HpaII, and HaeIII. After digestion the DNA was run on agarose gels. Two wells were used for each enzyme to show that the pattern of cleavage is reproducible. Lambda DNA was cleaved with HindIII as a molecular weight marker (left side, only one well). On the right side can be seen φ X174 DNA cleaved with HaeIII (one well). All three enzyme treatments show distinct bands. One of the bands present in HaeIII is very large. HpaII and HaeIII cleave CCGG and GGCC respectively, but they produce quite different general patterns. HpaII forms a gradient of mainly large sequences in the range of 11.0×10^3 to 600 base pairs, whereas HaeIII is characterized by mainly small segments in the range of 3.4×10^3 to 200 base pairs.

FIG. 2.   (Middle) Nuclear DNA of *Hordeum vulgare* (barley) treated with the same enzymes as in Fig. 1. Molecular weight markers are in this case Lambda rif DNA cleaved with EcoR1 and Lambda DNA cleaved with HaeIII. Note the difference between the HpaII and Hae III molecular patterns and the presence of a large number of bands in the HaeIII enzyme treatment.

FIG. 3.   (Right) Nuclear DNA of *Acomys cahirinus* (spiny mouse) treated in the same way as Fig. 1. Lambda DNA cleaved with HaeIII is found on each side. The EcoR1 pattern shows a light band but the HpaII and HaeIII patterns do not display distinct bands. The difference between HpaII and HaeIII in the general distribution of the molecules is the same as in Figs. 1 and 2.

Lambda DNA+EcoR1, Lambda rif DNA+EcoR1, Lambda DNA+HindIII, and ϕX174+HaeIII. After enzyme incubation the DNA segments obtained were separated in agarose gels by means of electrophoresis. The detailed description of the methods is to be found in another report (4).

The following results were obtained: 1) Each enzyme produces a pattern of cleavage characteristic of each DNA. 2) Different enzymes produce different patterns when the same DNA is digested. 3) The cleaved DNA appears as molecules of different size and often forms sharp bands. 4) HpaII and HaeIII cleave DNA segments which are CCGG and GGCC respectively but they produce quite different patterns with the same DNA. HpaII leads to the formation of large DNA segments with molecular weights ranging mainly between 11.0×10^3 and 600 base pairs, whereas HaeIII consistently gives a quite different picture with small DNA segments mainly between 3.4×10^3 and 200 base pairs.

The HaeIII pattern is characteristic of all eukaryotes which we have so far investigated from plants to humans with the exception of the protozoan *Tetrahymena pigmentosa*.

Acknowledgment

This work was supported by grants from the Swedish Natural Science Research Council.

REFERENCES

1. Roberts, R.J.: 1976. *Critical Reviews in Biochemistry* (Fasman, G.D., ed.) pp. 123–164, CRC Press, Boca Raton, Fla.
2. Smith, H.O.: 1979. *Science* 205, 455–462.
3. Arber, W.: 1979. *Science* 205, 361–365.
4. Lima-de-Faria, A., Isaksson, M., and Olsson, E.: 1980. *Hereditas* 92, 267–273.

Chromosome Replication in *Bacillus subtilis* and Differentiation of Rat Nervous System

NOBORU SUEOKA

N. SUEOKA was born on April 12, 1929, in Kyoto, Japan and graduated from Kyoto University in 1953 and received a M.S. degree in 1955. He received a Ph.D. degree in biochemical genetics in 1959 from California Institute of Technology. He discovered the correlation between base composition and density of DNA in 1960 when he was a postdoctoral fellow at Harvard University. He was assistant professor in the Department of Microbiology, University of Illinois at Urbana, Ill. (1961–1962), associate professor and Professor in the Department of Biology and Biochemistry, Princeton University (1962–1972), and has been Professor in the Department of Molecular, Cellular and Developmental Biology, University of Colorado, Boulder (1972–present). He has been a fellow of the American Academy of Arts and Sciences since 1969. His research covers biometrical genetics of wheat morphogenesis, biochemical genetics of *Neurospora crassa*, molecular biology of meiosis and mitosis in *Chlamydomonas reinhardi*, compositional analyses of DNA, DNA replication in *Bacillus subtilis*, tRNA and protein synthesis, and rat neural cell differentiation.

M an learns a lot from books about life, both truths and illusions. None is, however, more acute and instructive than his own experience. In contrast, over the years I have learned that pomposity and trickery are the most despicable acts in science, and dishonesty the ugliest. The creation of genuinely new concepts in your own mind is most beautiful, and the contribution that comes out of it from your own effort or with your colleagues is most satisfying.

To celebrate Dr. Egami's 70th birthday, we are told to narrate our thoughts and aspirations. Since the value of a man is mainly in the minds of others and I have not quite reached the age for reminiscence, I will not attempt to analyze my past, just describe my current thoughts and hopes for the future.

My current research interests are in two specific areas: DNA replication in

Bacillus subtilis and mammalian neural-cell differentiation. I shall describe our current activities and hopes of achievement in these areas.

DNA Replication in Bacillus subtilis

This is a long-lasting project started at the University of Illinois, Urbana, in the Department of Microbiology when I got my first faculty appointment in 1962. Since then, a number of my colleagues have contributed to this work: Hiroshi Yoshikawa, Aideen M. O'Sullivan, Michio and Akiko Oishi, Lincoln Arceneaux, Robert Armstrong, Tatsuo Matsushita, Roger Kennet, Robert Bishop, Chip Quinn, Seigo Ohi, Elliot Kaufman, Sumi Imada, Jim Beeson, Scott Winston, and many others who assisted with the research. Since 1968, we have been studying the role of membrane in the control of replication of the bacterial chromosome. The original concept of Jacob, Brenner, and Cuzin (1) of replicons and membranes now has a good chance to be right. Experimentally, first Ganesan and Leferberg (2) in *B. subtilis* and Smith and Hanawalt (3) in *Escherichia coli* found evidence for the replication fork association with membrane. Sueoka and Quinn (4) found that the origin and the terminus of replication as well as the replication fork are enriched in the membrane. A number of laboratories have confirmed the membrane association of bacterial chromosome at both the origin and the terminus, although none has offered any solid chemical evidence. Recently, H. Yoshikawa's laboratory in Kanazawa (5) found a new DNA protein complex (S-complex) occupying a region close to the origin as a possible membrane attachment site. In turn, Winston and Sueoka (6) found that two initiation genes (7), dnaBI and dnaBII, are responsible for the membrane association of the replication origin area of the *B. subtilis* chromosome. Using temperature sensitive mutants of these two genes, we can now demonstrate dissociation and re-association of the origin carrying DNA and certain plasmids *in vitro*. Now the essentiality of the membrane in initiation seems to be certain, and future research quite exciting in view of the elucidation of the mechanism of initiation of the bacterial chromosome and its regulation. Currently, we are engaging in several research projects on chromosomal replication, both genetic and biochemical. *In vitro* initiation of replication of bacterial origin is one of the coveted goals. Dr. Teruo Tanaka from the Mitsubishi-Kasei Institute of Life Sciences, has just joined our laboratory and is analyzing the structure of the origin of the *B. subtilis* chromosome and the interaction of the origin and proteins, an essential part of our research.

Neural-Cell Differentiation

This project was started in 1970, while I was at Princeton University. The objective here is to contribute to our understanding of the genetic and biochemical basis of neural-cell specificity. The establishment of a number of neural cell lines and analysis of their cell surface proteins was the original plan. Dr. Masaru Imada (currently Assistant Professor at the Denver Medical Center, University of Colorado) and I induced neurotumors in rats and established several clonal cell lines from a peripheral neurotumor, RT4 (8). To analyze the cell surface proteins, Dr. Imada developed a new two-dimensional gel electrophoresis system, which separates cell surface proteins by their affinities to ionic and non-ionic detergents as well as molecular weight (9, 10). Using this technique, cell surface proteins of RT4 cell lines have been analyzed. We have also detected cell surface proteins which may be necessary for cell spreading in various mammalian cell lines (11).

The cell lines we established from RT4 revealed an interesting feature. Among 4 morphologically different cell types, one (D) showed a glial property (cytoplasmic presence of S100 protein) and two (B and E) showed a neuronal property (voltage dependent Na⁺ influx), whereas the fourth type (AC) showed both glial and neuronal properties (8, 12).

Moreover, the AC type converts to other cell types (B, D, and E) at an approximate frequency of 10^{-6} (*cell-type conversion*). Thus, this cell type has the property of a stem cell. One interesting feature of this system is the fact that several independently derived cell lines from AC of one type (D, B or E) always show the same biochemical property, *i.e.*, presence of S100 protein (D) or voltage dependent Na⁺-influx (B and E) (12). This phenomenon is defined as *conversion coupling*. Tumorigenic properties are also coupled. In this case, cell types with S100 protein (AC and D) are tumorigenic and those without S100 (B and E) are non-tumorigenic (13).

Our current objective is to analyze individual cells of the stem cell type (AC) for the expression of glial and neuronal properties. For this, immunocytochemical methods are most promising. Dr. Y. Tomozawa is working out the conditions for immunofluorescent staining of our cell lines, and Mrs. T. Chow and I are isolating hybridomas for monoclonal antibodies against proteins unique to neuronal and glial cells. If we find that both glial and neuronal properties are expressed in individual cells of the stem cell type, we will have every incentive to look for such cells in developing nervous tissues, for example, in sympathetic ganglia. With this in mind, a graduate student (Kurt Droms) is following the development of rat lumber sympathetic ganglions, L4 and L5.

My hope is that we can make basic contributions to our understanding of stem-cell differentiation with its symmetric feature of differentiation and multi-gene regulation correlated with conversion coupling among neuronal proteins and among glial proteins.

Genomic Expression in Nervous Tissue

Dr. Dona Chikaraishi joined our laboratory in 1975 and in 1976 started working on the extent of genomic expression of various rat tissues. There had been several reports that mammalian brains had the highest (~30%) genomic expression of any tissue. First, we confirmed this and further discovered that nuclear RNA complexity of other tissues like liver (~20%) and kidney (~10%) are included in the complexity of brain nuclear RNA (*14*). Moreover, the complexity of kidney nuclear RNA (10%) is included in the liver nuclear RNA complexity. This nesting feature of the genomic expression fired my imagination, and made me formulate the following hypothesis (germ layer hypothesis):

> The hypothesis is that the extent of gene transcription manifested in the complexity of nuclear RNA is highest in oocytes and that irreversible restrictions of transcription occur sequentially and stepwise, leading to primary germ layers during early stages of development; and that the restricted features of the primary germ layers—ectoderm least restricted, endoderm further restricted, and mesoderm most restricted—are retained in developing tissues as well as adult tissues (*15*).

The first prediction of the hypothesis is that different tissues of the same germ layer origin should have similar nuclear RNA complexity. To test this hypothesis, I spent a summer examining the nuclear RNA complexity of another ectodermal tissue, mammary gland. For technical reasons I used mammary tumor. To my excitement, the complexity of this tissue revealed a complexity value (~25%) almost but not quite as high as that of brain. Later, a graduate student, Murray Brilliant collected a sufficient amount of rat normal mammary gland and confirmed that it has the same complexity value as that of mammary tumor. To date, by the efforts of Shelley Beckmann, Jim Beeson, Murray Brilliant, and Dona Chikaraishi, we have accumulated the data which are essentially consistent with the prediction of the hypothesis.

The second prediction is that once irreversible restriction of transcription occurs in clearly developmental stage, the cells should not change the nuclear RNA complexity. This is being tested by Murray Brilliant.

The third prediction is that any messenger RNA sequence in kidney (mesodermal) and liver (endodermal) should be found in brain nuclear RNA, whereas there must be mRNA sequence in brain which are not found in the nucleus of liver or kidney.

To pursue this line of questions, Jim Beeson and Yoichi Gondo are currently cloning DNA fragments which are exclusively expressed in the brain. By using this cloned DNA, we hope to examine the genomic expression more rigorously in the future. Particularly interesting is the high complexity of brain RNA, both nuclear RNA (~30%) (*14*) and cytoplasmic RNA (~20%) (*16*). Our results so far favor the model that in the brain the high complexity of the nuclear RNA is true for all ectodermal cells (including neurons and glial cells) but cytoplasmic RNA complexity are the result of summing up of different cell types, each of which may have lower cytoplasmic RNA complexity than that of the total cytoplasmic RNA of the brain.

It is amazing that nature offers us so much to learn and to explore, even with many new discoveries made every year, and each detail is so logical and simple in the end. I am most of the time encouraged, enthused, and inspired and yet sometimes I get anxious, particularly when I realize how little one can do in his lifetime. It has been fortunate that I had a chance to study various aspects of biology, with a wide variety of organisms ranging from bacteriophage to wheat and to rat. Maybe, or at least I hope, I can last as a scientist to my 60's, 70's, or 80's, as several predecessors have demonstrated possible. Certainly, Dr. F. Egami is one of them, having proven to be original over the years, having achieved so much and still remaining active.

REFERENCES

1. Jacob, F., Brenner, S., and Cuzin, F.: 1963. *Cold Spring Harbor Symp. Quant. Biol.* 28, 329–336.
2. Ganesan, A.T. and Lederberg, J.: 1965. *Biochem. Biophys. Res. Commun.* 18, 824–835.
3. Smith, D.W. and Hanawalt, P.C.: 1967. *Biochim. Biophys. Acta* 149, 519–531.
4. Sueoka, N. and Quinn, W.G.: 1968. *Cold Spring Harbor Symp. Quaut. Biol.* 33, 695–705.
5. Yamaguchi, K. and Yoshikawa, H.: 1977. *J. Mol. Biol.* 110, 219–253.
6. Winston, S. and Sueoka, N.: 1980. *Proc. Natl. Acad. Sci. USA* 77, 2834–2838.
7. Imada, S., Carroll, L.E., and Sueoka, N.: 1980. *Genetics* 94, 809–823.
8. Imada, M. and Sueoka, N.: 1978. *Dev. Biol.* 66, 97–108.
9. Imada, M., Hsieh, P., and Sueoka, N.: 1978. *Biochim. Biophys. Acta* 507, 459–469.
10. Imada, M. and Sueoka, N.: 1980. *Biochim. Biophys. Acta* 625, 179–192.
11. Hsieh, P. and Sueoka, N.: 1980. *J. Cell Biol.* 86, 866–873.
12. Tomozawa, Y. and Sueoka, N.: 1978. *Proc. Natl. Acad. Sci. USA* 75, 6305–6309.
13. Imada, M., Sueoka, N., and Rifkin, D.B.: 1978. *Dev. Biol.* 66, 109–116.
14. Chikaraishi, D.M., Deeb, S.S., and Sueoka, N.: 1978. *Cell* 13, 111–120.
15. Sueoka, N., Chikaraishi, D.M., Deeb, S.S., Hsieh, P., Imada, M., and Tomozawa, Y.: 1978. Proceedings of the XIV International Congress of Genetics, Moscow, in press.
16. Chikaraishi, D.M.: 1979. *Biochemistry* 15, 3249–3256.

Observations on Enzyme Systems Involved in the Biodegradation of Anionic Surfactants

KENNETH S. DODGSON

K. S. DODGSON was born on February 15, 1922 in Blackburn, Lancs, U.K. and educated at the University of Liverpool (1940–1942 and 1945–1949). He was awarded Doctor of Philosophy degree in 1949 and Sc.D. degree in 1961. He was elected to Fellowship of the Institute of Biology in 1971. Since 1963 he has occupied the Chair of Biochemistry in University College, University of Wales, Cardiff, U.K., and since 1979 has served as Visiting Professor to the Department of Microbiology of the University of Georgia. He served as Honorary Meetings Secretary (1964–1968) and Honorary Committee Secretary (1968–1970) of the Biochemical Society and was a member of the British National Committee for Biochemistry during the period 1967–1972. His research has been particularly concerned with the biochemistry and enzymology of sulphur-containing compounds and has involved incursions into the following fields: bio-organic chemistry, analytical biochemistry, metabolism of xenobiotics, biochemistry of connective tissue components, radiation chemistry, lung biochemistry, cellular and tissue effects of toxic industrial dusts, biochemistry of fibrinogen and the fibrinogen-fibrin transition, biodegradation of surfactants, mucopolysaccharase enzymes and sulphatases, phosphosulphatases and sulphotransferases.

My own interest in sulphate esters and sulphatases has certainly been influenced by Professor Egami's earlier researches and I therefore deem it appropriate to outline some of the recent work in the field of sulphated detergents with which I and colleagues have been involved.

The sulphate esters of long-chain primary alcohols, secondary alcohols and alkylethoxylates all feature as viable commercial surfactants with characteristics suited to particular purposes. Detergent formulations based on these sulphate esters must be non-toxic under normal use and biodegradable when discharged into the environment. In the early days of their use such detergents were subject to extensive tests for toxicity but, surprisingly, little attention was paid to the ways in which animals and microorganisms metabolized them or to the enzyme systems involved. Indeed, it was not until 1969 that the first detailed report on the metabolic fate in mammals of an alkyl sulphate ester was made (1). An element of serendipity was associated with this particular investigation, and the rationale that led my colleagues and I to undertake the work is worth recounting.

Briefly, we were sceptical of a hypothesis (2) that related biliary excretion of xenobiotic metabolites to a minimum molecular weight of about 325. It seemed to us that biliary excretion was more likely to depend on the ability of the metabolite to serve as a surfactant rather than on its molecular weight and it might therefore be possible to select compounds of molecular weight significantly less than 325 that nevertheless might undergo biliary excretion. Accordingly, we decided to study the route of excretion in the rat of dodecyl sulphate (molecular weight of the free acid =266). We expected this compound to be excreted unchanged in the bile but, to our considerable discomfit, it was extensively metabolized (1). Moreover, the major product, butyric acid 4-sulphate, and the minor product, SO_4^{2-} ions, were both excreted in urine. The liver was the major site of the metabolism, which involved ω-oxidation of the alkyl chain followed by β-oxidation.

The finding that SO_4^{2-} ions were produced led to some temporary excitement, as we believed that we had stumbled at last on an alkylsulphatase enzyme. It was not to be! We were able to show (3) that these ions appeared as a result of the spontaneous loss of the ester sulphate group of butyric acid 4-sulphate under pH conditions in which the carboxyl group was fully dissociated. The other product of this transformation, γ-butyrolactone, is a potent central nervous system depressant, but did not accumulate in sufficient quantity to be hazardous.

These various findings were so intriguing that the biliary excretion problem was left in abeyance whilst my colleagues went on to show that an ω-/β-oxidation sequence operated in the metabolism of other primary alkyl sulphates (4), alkylethoxy sulphates (5) and primary alkyl sulphonates (6).

Meanwhile, Dr. W.J. Payne and his colleagues at the University of Georgia (7–9), as well as others (10, 11) had begun to study the ways in which microorganisms biodegraded primary and secondary alkyl sulfate surfactants and had concluded that degradation of each type was initiated by enzymic removal of the ester sulphate group. The subsequent metabolism of the liberated alcohols was considered to be initiated by an alcohol dehydrogenase-type enzyme. Enzymic liberation of sulphate from these sulphate esters implied the involvement of alkylsulphatases—a type of enzyme the existence of which had not been established with any degree of certainty.

By a chain of fortuitous circumstances collaboration began between our own group and the scientists in Georgia. Essentially we provided the enzymological skills and Drs. Payne and J.W. Fitzgerald provided the microbiological expertise. The present purpose is to provide a brief summary of some results of this collaboration and to point to some of the novel enzyme systems that have emerged in consequence.

Investigations have concentrated on two microorganisms isolated from soil. One of these, a strain of *Comamonas terrigena*, possesses two constitutive secondary alkylsulphatases that have been designated as CS1 and CS2 on the basis of electrophoretic mobility on gels (*12*). Enzyme activity on gels is detected by incubation with "decan 5-sulphate" (erroneously thought to be pure), and the liberated and insoluble alcohol appears as a white band (*13*). The other organism, designated *Pseudomonas* C12B, has the capacity to produce no less than five alkylsulphatases (*14*). When grown on nutrient broth the bacterium possesses a primary alkylsulphatase (P1, detected on gels using dodecyl sulphate as substrate) and two secondary alkylsulphatases (S1 and S2). Addition of a primary alkyl sulphate (*e.g.* dodecylsulphate) to the broth leads to the appearance of a further primary alkylsulphatase (P2), whilst substitution of Oronite (a mixture of C_{10}-C_{20} secondary alkyl sulphates) for the primary alkyl sulphate results in the production of all four enzymes plus an additional secondary alkylsulphatase (S3).

The Secondary Alkylsulphatases

An intriguing question—why the multiplicity of sulphatases?—was posed as a result of these studies, particularly with respect to *Pseudomonas* C12B. This question has now been answered so far as the secondary enzymes are concerned. The breakthrough actually came from studies (*15*) on the two *Comamonas* enzymes. One of these enzymes (CS2) was purified to homogeneity and the other was partially purified. Space does not permit a full account of the subsequent course of events but it suffices to say that investigations (*16*) on the time-course of the enzymic reaction of these enzyme towards "decan 5-sulphate," together with analytical studies, showed the substrate to be heterogeneous. Indeed, 70% of the material consisted of racemic decan 2-sulphate. Hitherto we had prepared secondary alkyl sulphates by treating alcohols with sulphuric or chlorosulphonic acids at low temperatures, a process that had apparently led to the simultaneous shift of the ester sulphate group along the alkyl chain. This work, together with enzymological studies employing pure preparations of D- and L-alkan 2-sulphates, revealed that one of the two alkylsulphatases (CS1) was specific for L-alkan 2-sulphates of chain length greater than C_5, whilst the other (CS2) was specific for the corresponding D-isomers. These enzymes therefore provided the first examples of sulphatases exhibiting stereospecificity as well as specificity towards positional isomers. Kinetic studies (*17*) on the pure CS2 enzyme have established the importance of the alkyl chain in the binding of substrate to enzyme.

An understanding of the roles of the three secondary alkylsulphatases of

Pseudomonas C12B soon followed. The S1 enzyme was specific for D-alkan 2-sulphates and the S2 enzyme for the corresponding L-isomers (*18*), whilst the S3 enzyme turned out to exhibit remarkable specificity (*19, 20*). Symmetrical alkyl sulphates such as nonan 5-sulphate and undecan 6-sulphate are good substrates and are rapidly hydrolysed to completion. When the sulphate group is more remote from the centre of the alkyl chain, the enzyme is less active and begins to show a preference for L-isomers. For example, L- and D-forms of octan 4-sulphate are hydrolysed at roughly the same rate, whereas D-octan 3-sulphate is hydrolysed at a much slower rate than the L-isomer. D-Octan 2-sulphate is not a substrate but the L-isomer is, albeit a poor one. An explanation of the specificity has been advanced (*20*) and is based on a three-point interaction between enzyme and substrate, with a binding site for the sulphate group, a site accomodating short alkyl chains and one accomodating longer alkyl chains.

Collectively, the S1, S2, and S3 enzymes exhibit a range of stereo- and positional-specificity that should enable the degradation of most of the components of a commercial secondary alkyl sulphate detergent to be initiated, although any D-alkan 3-sulphates present would be relatively resistant to attack.

The Primary Alkylsulphatases

The significance of the ability of *Pseudomonas* C12B to produce two primary alkylsulphatases is obscure. The inducible P2 enzyme has a rather transitory existence and appears in greatest amounts at mid-logarithmic growth phase (*21*). This enzyme has been purified and kinetic, specificity and inhibition studies have been reported (*22*). Primary alkyl sulphates (C_4-C_{14}) are substrates, whilst secondary alkyl sulphates and alkanesulphonates are competitive inhibitors. The alkyl chain plays a key role in the binding of both substrates and inhibitors.

Little is yet known about the P1 enzyme but initial work on a partially purified preparation suggests a substrate specificity very similar to that of the P2 enzyme. Reasons for the co-existence of two enzymes, both of apparently similar specificity, are currently being sought.

Induction of the S3 and P2 Alkylsulphatases of Pseudomonas C12B

Induction of each of these enzymes presents remarkable features, Induction of S3 by Oronite was shown (*14*) to depend on the presence in the detergent of both secondary alkyl sulphates and the parent alcohols. It was further established that any combination of alkyl sulphate and alcohol from tetradecan 2-sulphate, hexadecan 2-sulphate, tetradecan 2-ol and hexadecan 2-ol would serve to induce the enzyme. Subsequent work, not yet published, shows that D- and L-forms of these compounds are equally effective as inducers. Why an enzyme that exhibits little activity towards L-alkan 2-sulphates and none towards D-alkan 2-sulphates should be induced by, for example, tetradecan 2-sulphate and tetradecan 2-ol is still not clear. Meanwhile, this system appears to be the first reported example of enzyme induction requiring a combination of two inducer components. Two other such systems have since been reported (23, 24).

Both primary alkyl sulphates and alkanesulphonates can serve as inducers for the P2 primary alkylsulphatase (25). Kinetic studies with alkylsulphates are complicated because the compounds are also substrates, but the alkanesulphonates are gratuitous inducers and are not metabolized by the bacterium under the experimental conditions employed. All alkanesulphonates tested (C_4–C_{12}) showed some ability to induce the enzyme in resting cell suspensions and experiments were designed to investigate the effects of increasing concentrations of some of these inducers. Increasing amounts of enzyme were produced as inducer concentration was increased but with alkanesulphonates of chain length C_8 or greater a pronounced inhibitory effect, followed by a further increase in enzyme production, was noted as concentrations were increased beyond 0.2 mM. From the initial parts of the various curves it was possible to plot $K_{inducer}$ values (analogous to K_m) for each inducer and these decreased as chain length increased. A plot of log $K_{inducer}$ against the number of carbons in the alkyl chains, gave a linear relationship characteristic of an homologous series.

The multiphasic character of the inducer concentration/specific enzyme activity curves for alkanesulphonates of chain length C_8 or greater is very puzzling. However, we hold the view that the curves reflect the involvement of two different inducer transport mechanisms rather than a phenomenon at gene level. The Michaelis-Menten-type of relationship observed at low inducer concentrations and the subsequent inhibitory phase could reflect an effective transport system that is inhibited as the inducer concentration increases. Restoration of induction when inducer concentration is further increased may then reflect the activity of a second and less efficient transport system that becomes apparent only when the other system is inhibited.

Alkylsulphatases or Alkyltransferases?

A further surprising and novel property of primary and secondary alkylsulphatases emerged from studies designed to establish which bond of the C–O–S ester sulphate linkage was cleaved by the enzymes. Based on studies on arylsulphatases (26) it had been tacitly assumed that all sulphatases operated by fission of the O–S bond. However, we have now shown (20, 27, 28) that the S1, S3, and P2 enzymes of *Pseudomonas* C12B and the CS2 enzyme of *C. terrigena* all cleave the C–O bond. Moreover, in the cases of the secondary alkylsulphatases, the process is accompanied by inversion of configuration so that alcohols are produced that are of opposite sign to those originally used to prepare the substrates.

The Enzyme Nomenclature Section of the IUPAC-IUB Commission on Biochemical Nomenclature recognizes six different classes of hydrolytic enzymes acting on ester bonds of the type alcohol-O-acid. The sulphatases (EC 3.1.6) feature in the classification. A literature search indicates that the only mode of action observed for these various hydrolases involves cleavage of the O-acid rather than the alcohol-O bond. The discovery of alkylsulphatase enzymes that attack the alcohol-O bond therefore adds a new dimension to the field of ester hydrolases. Whether, in the fullness of time these enzymes emerge as alkyltransferases rather than alkylsulphatases, remains a matter of conjecture.

Alcohol Dehydrogenase Activity in C. terrigena

Yet one further novel finding has recently emerged from these various biodegradation studies. It occurred to the writer that the existence in these detergent-degrading bacteria of enzymes exhibiting stereospecificity might be accompanied by the presence of stereospecific alcohol dehydrogenases. This indeed seems to be the case in the one example so far studied. Preliminary studies (29) revealed the presence in C. terrigena of two NAD-dependent enzymes, one active towards primary alcohols and secondary alcohols of the D-configuration and the other active towards secondary alcohols of the L-configuration. The latter enzyme appears to provide the first example of an alcohol dehydrogenase exhibiting a marked preference for the L-forms of secondary alcohols. This particular enzyme has now been purified to homogeneity and it is hoped to report on its properties in due course. Meanwhile, it emerges as a tetrameric enzyme of molecular weight approximately 130,000 that seems to operate by a random order mechanism.

Meanwhile, it will be recalled that the alkylsulphatase complement of C. terrigena is by no means as complex as that of Pseudomonas C12B. Clearly it will now be most interesting to study alcohol dehydrogenase activity in the latter organism.

Role of Bacterial Alkylsulphatases

Unpublished work in progress in our laboratory is revealing that many other soil microorganisms possess alkylsulphatase enzymes. One such organism (a pseudomonad) possesses no less than six distinct and apparently constitutive alkylsulphatases, three of which are active towards primary alkyl sulphates and three towards secondary alkyl sulphates. This organism also has the capacity to liberate sulphate from 2-(2,4-dichlorophenoxy)ethyl sulphate (Crag herbicide). The possibility that alkylsulphatase enzymes have some major role to play in the physiology of some soil bacteria therefore begins to seem likely. Current indications are that alkylsulphatases are located in the periplasmic region of the bacterial cell (see, for example Refs. 21, 30, 31) and that, unlike several other bacterial and fungal sulphatases, they are not solely concerned with the acquisition of sulphur for growth. The question arises as to why these enzymes should exist at all, especially in constitutive form, when detergent formulations containing potential substrates have featured in the environment for such a short time-span. It seems not unlikely that several naturally occurring compounds could serve as substrates (and, in some cases, as inducers) for some or all of these enzymes. Examples of such compounds include the chlorinated and non-chlorinated alkyl diol disulphates of *Ochromonas danica* and certain other unicellular algae (*32*), the alkyl mono- and di-sulphates of the brown seaweeds (*33*), the bile alcohol sulphates of fish and amphibian biles (*34*), the C-20 and C-21 sulphate ester end-products of steroid metabolism (*35*), and the short-chain alkyl sulphates present in developing avian eggs (*36*). Soil microorganisms must have encountered these various esters from time immemorial.

Experiments with such potential substrates/inducers are now planned. Meanwhile the bacterial alkylsulphatases and associated enzymes have emerged as fascinating enzymes with unusual properties that throw new light on the molecular aspects of detergent biodegradation and provide a fascinating area of research for the enzymologist.

Acknowledgment

It is a pleasure to record my thanks for the co-operation and companionship of the colleagues whose names are associated with this work.

REFERENCES

1. Denner, W.H.B., Olavesen, A.H., Powell, G.M., and Dodgson, K.S.: 1969. *Biochem. J.* 111, 43–51.
2. Millburn, P., Smith, R.L., and Williams, R.T.: 1967. *Biochem. J.* 105, 1275–1281.
3. Ottery, J., Olavesen, A.H., and Dodgson, K.S.: 1970. *Life Sci.* 9, 1335–1340.
4. Burke, B., Olavesen, A.H., Curtis, C.G., and Powell, G.M.: 1975. *Xenobiotica* 5, 573–584.
5. Taylor, A.J., Powell, G.M., Howes, D., Black, J.G., and Olavesen, A.H.: 1978. *Biochem. J.* 174, 405–412.
6. Taylor, A.J., Olavesen, A.H., Black, J.G., and Howes, D.: 1978. *Toxicol. Appl. Pharmacol.* 45, 105–117.
7. Payne, W.J. and Feisal, V.E.: 1963. *Appl. Microbiol.* 11, 339–344.
8. Williams, J. and Payne, W.J.: 1964. *Appl. Microbiol.* 12, 360–362.
9. Payne, W.J. and Painter, B.G.: 1971. *Microbios* 3, 199–206.
10. Hsu, Y-C.: 1965. *Nature* 207, 385–388.
11. Lijmbach, G.W.M. and Brinkhuis, E.: 1973. *Antonie van Leeuwenhoek* 39, 415–423.
12. Fitzgerald, J.W., Dodgson, K.S., and Matcham, G.W.J.: 1975. *Biochem. J.* 149, 477–480.
13. Payne, W.J., Fitzgerald, J.W., and Dodgson, K.S.: 1974. *Appl. Microbiol.* 27, 154–158.
14. Dodgson, K.S., Fitzgerald, J.W., and Payne, W.J.: 1974. *Biochem. J.* 138, 53–62.
15. Matcham, G.W.J., Dodgson, K.S., and Fitzgerald, J.W.: 1977. *Biochem. J.* 167, 723–729.

16. Matcham, G.W.J. and Dodgson, K.S.: 1977. *Biochem. J.* 167, 717–722.
17. Barrett, C.H., Dodgson, K.S., and White, G.F.: 1980. *Biochem. J.* 191, 467–473.
18. Bartholomew, B., Dodgson, K.S., and Gorham, S.D.: 1978. *Biochem. J.* 169, 659–667.
19. Matcham, G.W.J., Bartholomew, B., Dodgson, K.S., Fitzgerald, J.W., and Payne, W.J.: 1977. *FEMS Microbiol. Lett.* 1, 197–200.
20. Shaw, D.J., Dodgson, K.S., and White, G.F.: 1980. *Biochem. J.* 187, 181–190.
21. Fitzgerald, J.W. and Laslie, W.W.: 1975. *Can. J. Microbiol.* 21, 59–68.
22. Cloves, J.M., Dodgson, K.S., White, G.F., and Fitzgerald, J.W.: 1980. *Biochem. J.* 185, 23–31.
23. Blackwell, C.M. and Turner, J.M.: 1978. *Biochem. J.* 176, 751–757.
24. Mandrand-Berthelot, M-A., Novel, G., and Novel, M.: 1977. *Biochimie* 59, 163–170.
25. Cloves, J.M., Dodgson, K.S., White, G.F., and Fitzgerald, J.W.: 1980. *Biochem. J.* 185, 13–21.
26. Spencer, B.: 1978. *Biochem. J.* 69, 155–159.
27. Bartholomew, B., Dodgson, K.S., Matcham, G.W.J., Shaw, D.J., and White, G.F.: 1977. *Biochem. J.* 165, 575–580.
28. Cloves, J.M., Dodgson, K.S., Games, D.E., Shaw, D.J., and White, G.F.: 1977. *Biochem. J.* 167, 843–846.
29. Barrett, C.H., Dodgson, K.S., White, G.F., and Payne, W.J.: 1980. *Biochem. J.* 187, 703–709.
30. Fitzgerald, J.W. and Laslie, W.W.: 1974. *Biochem. Soc. Trans.* 2, 1072–1073.
31. Dodgson, K.S., Fitzgerald, J.W., and Payne, W.J.: 1978. *FEMS Microbiol. Lett.* 3, 115–117.
32. Haines, T.M.: 1973. *Annu. Rev. Microbiol.* 27, 403–411.
33. Liem, P.Q. and Laur, M-H.: 1976. *Biochimie* 58, 1381–1396.
34. Haslewood, G.A.D.: 1978. *The Biological Importance of Bile Salts*, North-Holland Amsterdam.
35. Hadd, H.E. and Blickenstaff, R.T.: 1969. *Conjugates of Steroid Hormones*, Academic Press, New York.
36. Yagi, T.: 1966. *J. Biochem.* 59, 495–500.

Biosynthesis of Proteoglycans

ALBERT DORFMAN

A. DORFMAN was born in Chicago on July 6, 1916 and received his early education in the Chicago Public Schools. His professional education was entirely at the University of Chicago from which institution he received an S.B. degree in 1936, a Ph.D. degree in 1939, and an M.D. degree in 1944. His early scientific interests were in bacterial metabolism and from 1939–1943 he was a research associate in Biochemistry at the University of Chicago. From 1946 to 1948 he was a member of the United States Army and Chief of Biochemistry at the Army Medical Department Research and Graduate School in Washington, D.C. He returned to the University of Chicago in 1948 as assistant professor of Pediatrics and since 1957 he has been Professor of Pediatrics and Biochemistry. From 1957 to 1972 he served as Director of the La Rabida University of Chicago Institute and from 1962 to 1972 he was Chairman of the Department of Pediatrics. Since 1967 he has been Director of the Joseph P. Kennedy, Jr. Mental Retardation Research Center and the Richard T. Crane Distinguished Service Professor of Pediatrics and Biochemistry. Since 1948, his research has been primarily concerned with the biochemistry of connective tissues with particular emphasis on the biosynthesis of glycosaminoglycans, the etiology of mucopolysaccharidoses, and the developmental biology of cartilage. He is a member of numerous professional societies including the National Academy of Sciences (USA).

A major accomplishment of modern biochemistry has been the elucidation of the biosynthesis of macromolecules. In the case of nucleic acids and proteins, we know now that the specific structure is determined by template mechanisms which have their ultimate origin in the structure of the genome. However, if the macromolecule additionally contains carbohydrate, the structure is also determined by the specificity of the enzymes involved in the formation of glycoside bonds.

More than thirty years ago my laboratory set out on a program to define the mechanism of biosynthesis of mucopolysaccharides, now called glycosaminoglycans.

What seemed at that time a relatively straightforward problem has led to increasingly complex and interesting questions of cell biology. This problem has led us to utilize the new knowledge and techniques of molecular biology, biochemistry, and cell biology. Although many important questions have been answered, new questions have constantly arisen. It is worthwhile to delineate the problem from the point of view of current thinking, summarize our current knowledge and finally to pose remaining unsolved problems.

Initially, we were concerned only with the formation of glycoside bonds and concentrated on the synthesis of hyaluronic acid and chondroitin sulfate. The biosynthesis of hyaluronic acid by cell-free preparations of Group A streptococci was demonstrated and that of chondroitin sulfate by cell-free preparations of embryonic chick cartilage (1, 2). However, it has become apparent that more complete comprehension of proteoglycan biosynthesis requires consideration of protein synthesis and the integration of this synthesis with the modifications of the protein, including the addition of carbohydrate side chains.

Recent work has concentrated on a consideration of the biosynthesis of cartilage chondroitin sulfate proteoglycan, the proposed structure of which is illustrated in Fig. 1 (3). It is this substance which imparts the unique physical and biological properties to cartilage. This complex substance is composed of a protein core to which are attached chondroitin sulfate and keratan sulfate chains. It has also been shown that mannose-containing oligosaccharides are present (4). In order to manufacture a single subunit, approximately 25,000 covalent bonds must be formed. This estimate does not include the reactions necessary for the synthesis

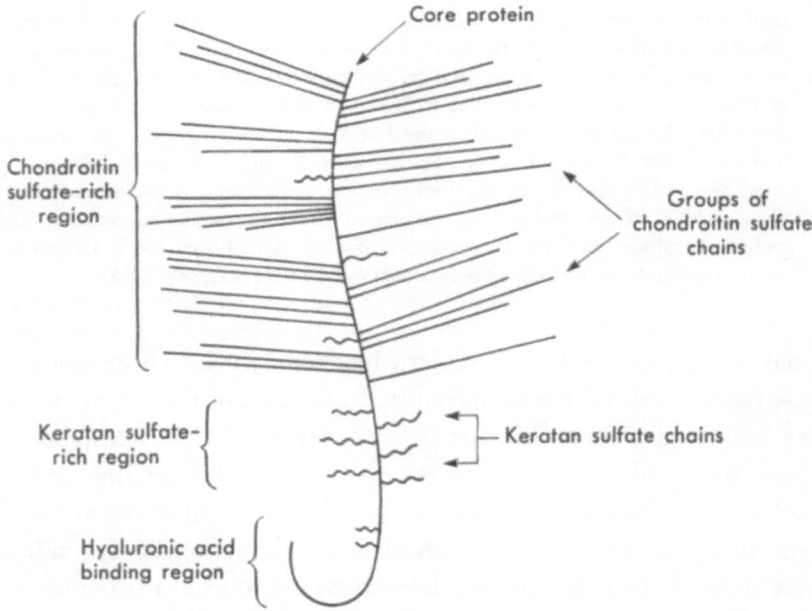

FIG. 1.   Structure of a proteoglycan monomer (reproduced from Ref. 3).

of such precursors as sugar nucleotides, tRNAs and mRNA. Nevertheless, in a matter of minutes, a differentiated cartilage cell carries out and coordinates all of these reactions. Additionally, the proteoglycan subunit then interacts with two other chondrocyte products, link protein and hyaluronic acid, to produce the aggregate which characterizes cartilage.

On the basis of currently accepted biochemical concepts, one may outline the processes that are necessary to achieve this impressive feat.

1) The cell must synthesize and maintain adequate levels of immediate precursors, including specific nucleotides, amino acids, and tRNAs.

2) The cell must synthesize the appropriate specific enzymes such as glycosyltransferases, the many enzymes involved in protein synthesis and the requisite cofactors as well as ribosomes.

3) The gene coding for the core protein must be activated or derepressed and transcribed to form the appropriate precursor mRNA.

4) The mRNA must be processed by a variety of steps involving removal of introns, capping and polyadenylation.

5) The mRNA must be transported to the cytoplasm and combined with ribosomes.

6) The mRNA must be translated to form a preprotein which is probably processed as it proceeds from the ribosome through the membrane of the endoplasmic reticulum to the interior of the endoplasmic reticulum.

7) Polysaccharide chains of a least three types must be added to the appropriate portion of the peptide and polysaccharide formation must be completed in the Golgi apparatus.

8) The completed molecule must be conveyed from Golgi vessicles to the exterior of the cell for excretion of its contents to the matrix.

9) The proteoglycan subunit must interact with hyaluronic acid and link protein to form appropriate aggregates in the matrix.

Given this set of probable steps, we may examine the current state of our specific knowledge of the biosynthesis of proteoglycan.

Recently with William Upholt and Barbara Vertel (5), a cell-free system for the biosynthesis of core protein utilizing mRNA obtained from chick embryonic sternae or differentiated limb bud culture has been developed. In either wheat germ or rabbit reticulocyte systems, a specific protein of approximately 340,000 daltons is formed. This large molecular weight protein is not made by mRNAs obtained from calvaria, liver or undifferentiated limb buds and only to a limited extent by cultures treated with the thymidine analogue, bromodeoxyuridine. The product has been identified by clonal antibodies. These antibodies were obtained from a clone developed as a result of fusion of spleen cells from a rat immunized with hyaluronidase-treated core protein and a strain of mouse myeloma cells (6). In this same cell-free system type II collagen is also formed (5).

In a separate series of studies, we have sought to define the mechanisms of

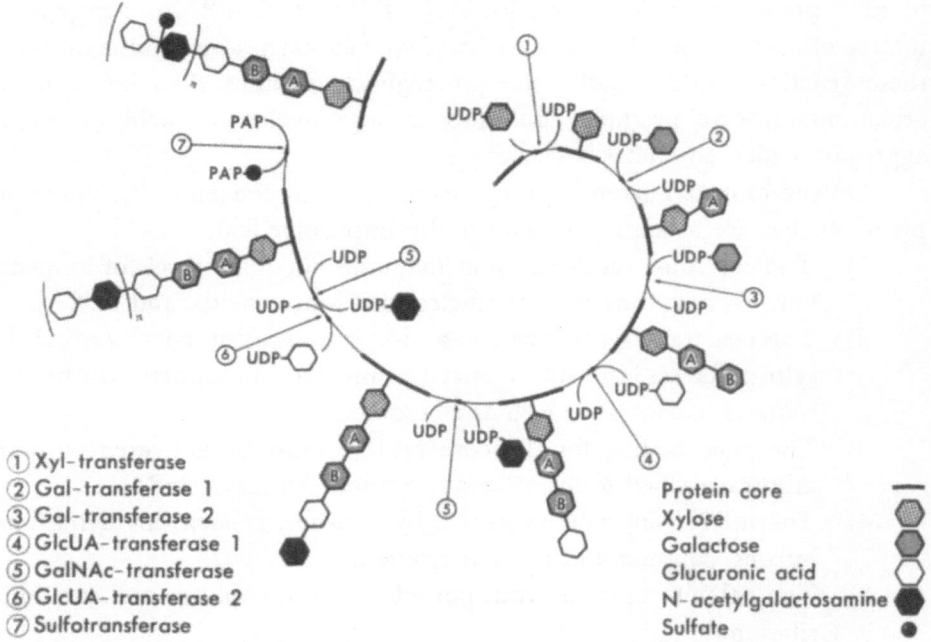

① Xyl-transferase
② Gal-transferase 1
③ Gal-transferase 2
④ GlcUA-transferase 1
⑤ GalNAc-transferase
⑥ GlcUA-transferase 2
⑦ Sulfotransferase

Protein core
Xylose
Galactose
Glucuronic acid
N-acetylgalactosamine
Sulfate

Fig. 2.   Pathway of biosynthesis of cartilage chondroitin sulfate proteoglycan.

initiation of the covalently bound chondroitin sulfate chains which are known to be linked to serine residues *via* a linkage region composed of glucuronosyl-galactosyl-galactosyl-xylose (7). It became of special interest to determine whether the initial linkage sugar, xylose, was added before or after completion of the peptides. Previous studies on synthesis of glycoproteins and gastric mucin had indicated the addition of carbohydrate to nascent chains (8, 9). With Kazuyuki Sugahara, it has been possible to obtain tRNA peptides from polysomes derived from differentiated limb bud cultures. Such peptides were treated with alkaline borotritide to release any carbohydrate chains containing xylitol. Following acid hydrolysis [^{3}H]xylitol was demonstrated by gas liquid chromatography (unpublished results).

These results indicate that the xylosylation occurs on nascent peptides. The synthesis of chondroitin sulfate chains has been studied extensively by the author, Lennart Roden, Nancy B. Schwartz and students and other colleagues (7). The pathway is illustrated in Fig. 2. This pathway has been largely established by the use of Smith degraded core protein and small molecular acceptors for the individual enzymes. Xylosylation of core protein occurs by transfer of xylose from uridine diphosphate (UDP)-xylose by a specific glycosyl transferase which has been isolated (10). The linkage region is then completed by the concerted action of three specific glycosyl transferases, namely: galactosyl transferase I, galactosyl transferase II, and glucuronosyl transferase I. Subsequently, the repeating disaccharides are added by the action of two other glycosyl transferases, N-acetyl-galactosamine transferase and glucuronosyl-transferase II. The molecules are completed by the transfer of sulfate

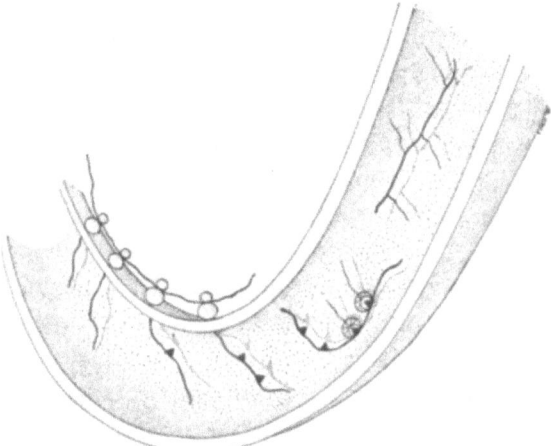

FIG. 3.   Localization of synthesis of cartilage chondroitin sulfate proteoglycan.

groups to the 4 or 6 hydroxyls of N-acetyl-galactosamine residues by the action of specific sulfate transferases. Although there is evidence that sulfation may occur concomitant with polysaccharide chain extension (11), sulfation is not requisite for the growth of polysaccharide chains (2). As indicated above and illustrated in Fig. 3, xylosylation probably occurs in the rough endoplasmic reticulum on nascent peptides chains. There is evidence that chain extension and sulfation occurs in the Golgi apparatus (12, 13), however the site of completion of the linkage region has not been firmly established.

Although much remains to be learned regarding the mechanism of coordination of the various reactions involved, there are reasons to believe that the various transferases are bound to membrane in a multienzyme complex. The specific interaction of xylosyl transferase with galactosyltransferase I has been demonstrated by Schwartz (14). The former enzyme appears to be considerably more soluble than the other transferases involved in the complex. It seems reasonable that the binding of xylosyltransferase to its substrate and to the next enzyme in the pathway, which is membrane bound, serves to coodinate protein and polysaccharides synthesis.

Details concerning the initiation of keratan sulfate and its synthesis remain unknown. The recent discovery of N-glycosyl bound oligosaccharides as part of the proteoglycan molecule further complicates the biosynthetic problem (4). It is to be expected that such oligosaccharides are added from dolichol-containing intermediates and are perhaps processed in a manner similar to that which occurs in the synthesis of glycoproteins (8).

The available information now makes possible an approach to a number of interesting problems. A key question of modern biology is the mechanism of cellular differentiation. This problem is being investigated in many biological systems. An attractive model for this process is the differentiation of mesenchyme to cartilage which may be investigated under well controlled tissue culture conditions. Stage

24 chick limb cells when cultured at high density differentiate to chondrocytes (15). The emergence of the biochemical phenotype is readily monitored by the rapid increase of synthesis of cartilage proteoglycan and type II collagen. It has been demonstrated in our laboratory that this change is accompanied by the increased synthesis of the core protein (16). Somewhat surprisingly, it has been found that predifferentiated cells as well as many other cell types possess the enzymic machinery for the synthesis of chondroitin sulfate (17). The problem of differentiation then becomes one of elucidation of the mechanisms which determine the onset of synthesis of the specific proteins, core protein and type II collagen. At present, we are attempting to clone the genes for these proteins in order to obtain probes for the onset of mRNA synthesis and determine whether any genomic rearrangement is involved in the differentiation events. It should be noted that although the synthesis of a large molecular weight protein appears to be reasonably specific for chondrocytes, smaller chondroitin sulfate proteoglycans are present in a variety of tissues (17). The structual and genetic relationship of such proteins to the core protein of cartilage proteoglycan remains unclear, although recent studies with Sugahara has demonstrated immunological crossreactivity of calvarial proteoglycan with cartilage proteoglycan (unpublished results).

The onset of synthesis of type II collagen represents an equally interesting problem, particularly since it appears to mirror a cessation of synthesis of type I collagen. Immunohistochemical studies with Vertel (18) have indicated the single chondrocyte contain proteoglycan and type II collagen and rarely proteoglycan and type I collagen.

Unanswered questions regarding the mechanism of biosynthesis of proteoglycan may be enumerated as follows:

1) Which end of the core protein represents the amino terminal?
2) Is the nascent peptide processed?
3) At what stage of synthesis are keratan sulfate chains added?
4) Where are the various polysaccharides completed?
5) What mechanisms are responsible for termination of chondroitin sulfate and keratan sulfate chains?
6) At what stage of synthesis are N-glycoside oligosaccharides added? Are these added and processed by the established mechanisms for glycoprotein synthesis?
7) How are the completed proteoglycans excreted from the cell?
8) What is the site of interaction of proteoglycan subunits with hyaluronic acid and link protein to form the typical aggregates of matrix?

The eventual aim of this work is answer all of the above questions, but most importantly to achieve an understanding of the mechanisms by which the cell coordinates this myriad of processes. The progress in recent years in understanding of the functions of the cell are a source of great satisfaction to biologists, but the unanswered questions represent a continuing, exciting challenge. The evolving pattern of coordinate, intricate mechanisms can only add to the appreciation of inherent beauty of living systems.

Acknowledgment

Studies described have been supported by Grants: AM05996, HD9402, and HD 4583 of the National Instuittes of Health.

REFERENCES

1. Markovitz, A., Cifonelli, J.A., and Dorfman, A.: 1959. *J. Biol. Chem.* 234, 2343–2350.
2. Perlman, R.L., Telser, A., and Dorfman, A.: 1964. *J. Biol. Chem.* 239, 3623–3629.
3. Hascall, V.C. and Heinegard, D.K.: 1979. *Glycoconjugates*, Vol. 1 (Jeanloz, R.W., ed.) pp. 341–364, Academic Press, New York.
4. DeLuca, S., Lohmander, L.S., Nilsson, B., Hascall, V.C., and Caplan, A.I.: 1980. *J. Biol. Chem.* 255 (13), 6077–6083.
5. Upholt, W.B., Vertel, B.M., and Dorfman, A.: 1979. *Proc. Natl. Acad. Sci. USA* 76, 4847–4851.
6. Dorfman, A., Hall, T., Ho, P-L., and Fitch, F.: 1980. *Proc. Natl. Acad. Sci. USA* 77, 3971–3973.
7. Rodén, L.: 1980. *The Biochemistry of Glycoproteins and Proteoglycans* (Lennarz, W.J., ed.) pp. 267–371, Plenum Press, New York.
8. Struck, D.K. and Lennarz, W.J.: 1980. *The Biochemistry of Glycoproteins and Proteoglycans* (Lennarz, W.J., ed.) pp. 35–83, Plenum Press, New York.
9. Strous, G.J.A.M.: 1979. *Proc. Natl. Acad. Sci. USA* 76, 2694–2698.
10. Schwartz, N.B. and Roden, L.: 1974. *Carbohydr. Res.* 37, 167–180.
11. DeLuca, S., Richmond, M.E., and Silbert, J.E.: 1973. *Biochemistry* 12, 3911–3915.
12. Godman, G.C. and Lane, N.: 1964. *J. Cell Biol.* 21, 353–366.
13. Freilich, L., Lewis, R.G., Reppuci, A.D., Jr. and Silbert, J.E.: 1975. *Biochem. Biophys. Res. Commun.* 63, 663–668.
14. Schwartz, N.B.: 1975. *FEBS Lett.* 49, 342–345.
15. Levitt, D. and Dorfman, A.: 1974. *Current Topics in Developmental Biology*, Vol. 8, (Moscona, A.A. and Monroy, A., eds.) pp. 103–149, Academic Press, New York.
16. Ho, P-L., Levitt, D., and Dorfman, A.: 1977. *Dev. Biol.* 55, 233–243.
17. Dorfman, A., Vertel, B.M., and Schwartz, N.B.: 1980. *Current Topics in Developmental Biology*, Vol. 14 (Friedlander, M., ed.) pp. 169–198, Academic Press, New York.
18. Vertel, B.M. and Dorfman, A.: 1979. *Proc. Natl. Acad. Sci. USA* 76, 1261–1264.

Enigma on the Function of Glycolipids and Glycoproteins

SEN-ITIROH HAKOMORI

S. HAKOMORI is currently the Head of the Division of the Biochemical Oncology, at the Fred Hutchinson Cancer Research Center, and Professor of Pathobiology, Microbiology, and Immunology at the University of Washington. He was educated in Sendai, Japan, undergraduate study at Tohoku University Medical School and graduate studies at the Department of Biochemistry of the same University (with late Professor Hajime Masamune) and earned Doctor of M. Sci. degree in 1957. He immigrated to USA in 1963 assuming positions at Massachusetts General Hospital (1963–1966), Brandeis University (1966–1967), and at University of Washington in Seattle (1967–present). He has been in the present position since 1975. His major work is the studies on structure and function of glycolipids of animal cells, and the transformation dependent changes of cell membranes. He is among the few who discovered a cell surface "fibronectin" in 1973.

Optimistic views on the possible function of carbohydrate moiety of glycoproteins and glycolipids have been discussed extensively based on a large variety of experimental data, some of them are listed in Table 1 for glycolipids and Table 2 for glycoproteins, respectively.

Defective cell surface function has been found as the common property of tumor cells, and many of them have been correlated with the anomalous structure of cell surface carbohydrates. Remarkable changes in glycolipid structure and metabolism (1, 2) and in certain carbohydrate chains of glycoproteins (3) due to an increased branching at the core domain (4) have been the major consistent changes found in the majority of oncogenic transformants. Thus, these carbohydrates at the cell surface have been implicated to have the essential function in cellular interaction

TABLE 1.   Function of Glycosphingolipids

---

1. Constitutive component of outer leaflet of lipid bilayer; conferring structural rigidity
2. Cell surface markers and antigens
    a. Blood group ABH, Le^{a,b}, Ii, P$_1$, P, P^k antigens
    b. Heterophile Forssman and H-D antigen
    c. Lymphoid subpopulation markers
    d. Differentiation antigens, *e.g.*, stage-specific embryonic antigen (SSEA-1), teratoma antigen (TerC)
    e. Glycolipid reacting to autoimmune antibodies
    f. Glycolipid tumor antigens
3. Cellular interaction and recognition
    a. Neuro-glial, retino-tectal, and neuro-muscular interactions
    b. Cell contact response of glycolipids and "contact inhibition"
4. Differentiation markers
    a. Early embryo and teratocarcinoma
    b. Crypt to villus cell differentiation
    c. Erythrocyte differentiation (glycolipid branching, and i to I conversion)
    d. Myogenic differentiation
    e. Neuroblastoma
    f. Lymphoid cell differentiation and mitogenesis
5. Cell growth regulation and oncogenesis
    a. Modified cell growth and glycolipid alteration induced by glycolipid addition, differentiation inducers (butyric acid, retinoids, and cyclic AMP), and by anti-glycolipid antibodies
    b. Defective glycolipid structures in oncogenic transformants
6. Interaction of bio-active factors and implication as receptors

---

TABLE 2.   Implication of Carbohydrate Function in Glycoproteins

---

1. Regulators of protein processing, mobilization, and degradation
    a. Mobilization of nascent polypeptide to Golgi and to plasma membrane (Blobel/Lodish)
    b. Protection from protein degradation by protease
    c. Regulation of protein release from cells (Ed Eylar hypothesis)
    d. Regulation of circulating blood protein *via* recognition of galactosyl/glucosaminyl/or mannosyl residue of glycoprotein by membrane proteins (Ashwell and Morrell phenomenon)
    e. Lysosomal hydrolase-uptake is based on the recognition of mannose phosphate residue of the enzyme by membrane proteins (Neufeld/Sly/Jourdian phenomenon)
2. Transmembrane control of cell proliferation
    a. Mitogenesis caused by lectin or by modification of cell surface carbohydrates
    b. Oncogenic transformation associates the alteration of glycoprotein carbohydrate chains (Warren-Glick glycopeptide, *etc.*)
3. Cellular interaction, differentiation
    a. Markers for the stage-specific early embryo cell surfaces (*e.g.*, F9 antigen) could be glycoprotein (Muramatsu)
    b. Tunicamycin inhibits development of sea-urchin and mouse embryo (Lennarz/Surani)
    c. Morphogenesis and histotypic aggregation can be inhibited by the specific glycoprotein (Moscona)
4. Antigens and regulation of antigenic expression
    a. Blood group ABH, Le^{a,b} antigens in secretions and ABH, Ii antigens carried by membrane protein
    b. Regulation of polypeptide determinants, *i.e.*, blood group MN, and some viral membrane antigens

---

and cell growth regulation. The possible regulation of cell growth through glycolipids has been suggested by the observation that cell-cell contact induces enhanced synthesis of certain glycolipid species (termed "glycolipid response on cell contact") and the response was lost associated with the loss of contact inhibition in oncogenically transformed cells (5, 6). The role of cell surface carbohydrates on neuro-glial, retino-tectal, neuro-muscular interaction has been studied (7–10) and the change of the cell surface carbohydrate during various phases of ontogenesis has been substantiated (11–20). Various carbohydrate determinants showed the clear stage-specific changes at the early stage of embryonic development as well as in the later stage of differentiation. The molecular changes during the development of early embryo have been detected by immunological procedures and have been described as the "developmentally regulated antigens." These include ABH (11), Forssman (12, 13), Ii (14–16), and the stage-specific embryonic antigen (17) as well as complex mixture of antigens detected by antisera raised against embryonal carcinoma cell F9 (18, 19), and teratoma cells TerC (20). All of these antigens showing the dramatic changes during ontogenesis were found to be glycolipids or glycoproteins. A sequential change in the branching process of a certain carbohydrate chain termed "polylactosaminoglycan" (21) showed a remarkable change in branching at the early embryo (14) and in the later stage of erythrocyte differentiation (15, 16, 21).

In contrast to the glycolipid functions which may primarily mediate the interaction, transduction, and mobility of plasma membranes, the carbohydrate chains linked to the proteins may control primarily the mobilization, metabolism and interaction of proteins. Transmembrane control of cellular function and proliferation (item #2, Table 2), and cellular interaction and differentiation (item #3, Table 2) could be regulated through cell surface carbohydrates linked to proteins. But, these effects could be the secondary ones. The recognition of the glycoprotein carbohydrate by hepatocyte membrane protein as described by Ashwell and Morrell (22), and the recognition of lysosomal enzymes containing phosphomannosyl residues by cell surface proteins which induces endocytosis as described by Neufeld, Sly, and Jourdian (23) are the two clearest demonstrations for glycoprotein recognition through cell surface membranes.

Despite all of these positive and optimistic views of carbohydrate functions in cellular interaction, differentiation, growth control, and oncogenesis, we have been puzzled by the enigma as described below.

One is the presence of "normal" individuals, who by some genetic alteration, lack the major glycosphingolipids or the major glycoproteins in specific cells, tissues or in organs. For example, the major neutral glycolipid globoside (24, 25) is completely absent in erythrocytes and probably in various other tissues of the individual with blood group p-status (26). The total absence of ABH-determinants in various tissues of individuals with blood group Bombay population (27), the absence of the branched polylactosaminoglycan with blood group I-structure in blood group i-individuals (28), and the absence of glycophorin A in En^{a-} erythrocytes (29) have been well documented. These individuals have no recorded symptoms and their life is normal, although their susceptibility to disease is not known. Obviously, those carbohydrate chains whose absence does not threaten the function of cells or individual's life could be regarded as non-essential and a luxury product of cell.

Another feature of some carbohydrate chains at the cell surface is their well documented differences and specificities; e.g., the major carbohydrate of glycolipids of erythrocytes is clearly distinctive for each species (30), although function of erythrocytes and their developmental feature between different species are essentially the same. This again suggests that the major glycolipid and possibly the major carbohydrate chain in glycoprotein of erythrocytes may have no essential function.

For the sake of simplicity, I would like to continuously discuss taking globoside and Forssman glycolipid as the example. Globoside is absent in erythrocytes of blood group p-individuals (26), the very rare genetic trait (1 out of 150,000 population). Whether globoside is absent in all tissues and organs of p-individuals is not known*; it may probably be absent because sera of p-individuals contain anti-P, P_1, and P^k antibodies. If globoside were present in tissue, anti-P (anti-globoside) response may not occur in p-individuals. The life of p-individuals is perfectly normal as in P-individuals except that p-women show much greater chance of miscarriage due to blood group P incompatibility (31). Nevertheless, no data is available whether globoside exists in the various stages of development of p-individuals. It is entirely possible that globoside may be present at the early stage of development, and it functions as the essential developmental marker, through which cells interact, migrate, and organize properly during ontogenesis and morphogenesis. In fact, recently, one of the monoclonal antibodies directed to the stage-specific antigens of the early human embryo and human teratocarcinoma was found to be reactive to globoside (personal communication; B.B. Knowles, Wistar Inst., Philadelphia). Expression of globoside, therefore, must be extremely stage-specific in various tissues during ontogenesis. Globoside must be expressed in the "embryo to be p-individual" and functioning as "area code molecules" (32). Globoside expression must be turned off after organogenesis and differentiation are completed, such individuals are known as "blood group p". However, suppression of globoside expression must be uneconomical, therefore globoside is continuously expressed in the majority of humans even after its missions are accomplished. This is seen as blood group P.

Humans have been regarded to be Forssman negative (F⁻) species (33), however, some of the population (20–30%) contain Forssman glycolipid in gastrointestinal tissue and probably in other tissues as well (34), because the majority of humans has a considerable level of anti-Forssman antibodies in their serum, whereas, the minority population has no detectable anti-Forssman antibodies (35). In this aspect, Forssman glycolipid could be a blood group antigen limited in tissue ("tissue group antigen"). There is a possibility, however, in the early stage of human embryo, Forssman antigen could be expressed normally and is suppressed after

---

* Since this manuscript was submitted, a paper was published describing that the stomach tissue of a patient with blood group P does not contain globoside (P) or globo-triosylceramide (P^k) (Breimer, M.E., Cedergen, B., Karlsson, K-A., Nilson, K., and Samuelsson, B.E., FEBS Letts 118, 209–211, 1980).

morphogenesis and differentiation are completed. In mice embryo, Forssman antigen showed a striking stage-specific expression, and its expression was suppressed after development (12, 13). Some mice tumors contain a high level of Forssman antigen (36), therefore, the expression of Forssman antigen must be "oncofetal." In a similar sense, expression of Forssman antigen in humans must be also "oncofetal" for some human tumors contain Forssman antigen (34, 37). Interestingly, one case of tumor derived from a patient with a rare blood group p-status showed a serological evidence that tumor tissue may contain globoside-like structure (the appearance of P-antigen) (38, 39). In an analogy, the phenomenon can be regarded as oncofetal expression of globoside.

The stage-specific expression of blood group ABH as well as Ii antigens in early embryo has been well described. The early human embryo, the 5th post-fertilization week, showed a clear expression of ABH antigens. At about the end of the 1st trimester of pregnancy, the antigens of many organs undergo a systematic, well ordered recession coincident with morphologic maturation. ABH antigen disappeared from various organs such as pituitary, thyroid, pancreas, gastrointestinal mucosa, etc. By contrast, some tissues such as squamatous membranes retain ABH antigens (11). The cell surface carbohydrate structures recognized by monoclonal anti-I (Ma), anti-I (Step), and anti-i (Dench) have been shown to be expressed on the early post-implantation mouse embryos and embryonal carcinoma cells. Undifferentiated teratocarcinoma stem cells were rich in surface-associated and cytoplasmic I-antigen; whereas i-antigen appeared when cells were differentiated into primary endoderm (15). Interestingly, the antigen I is converted to i at the early differentiation process of mouse embryo, whereas, the antigen i is converted to I in the later stage of erythrocyte differentiation (16). The well known developmental antigen of mouse embryo is the F9 which was detected by syngenic and heterogenic immunization with mouse embryos and with embryonal carcinoma cells. The antigen is absent at 4-cell stage, appeared at 8-cell stage, maximally expressed at morulae stage and greatly diminished or disappeared after blastocyte (8). Significant changes of glycopeptide containing polylactosaminoglycan type associated with differentiation was described by Muramatsu and his colleagues (19). Recently, a part of such stage-specific antigens was identified as glycolipid similar to Ii antigens (40). All these ABH and Ii determinants at the erythrocyte surfaces are regarded to be "luxury" components and have no apparent specific functions.

Thus, many of these "luxury" carbohydrates at the cell surface which have no apparent function after development could have important function during ontogenic development as have been described in this essay. Theoretically, their synthesis should have been turned off after cells or tissues were differentiated, and in fact, some population of individuals are capable to turn off their synthesis but some are not. There is a possibility that blood group may be related to the "turning off" mechanism of the apparently "luxury" but developmentally "essential" carbohydrate chain.

Many of the cell surface carbohydrates are ontogenically *paleo*-functional; they have lost their functions after development, or they may have different functions. The situation is essentially the same as the monarchy of our society, the Queen of England or the Emperor of Japan; they had essential functions in the past! A possible role of cell surface carbohydrates in phylogenesis can be analogous as I have discussed in this essay, thus a great many diversities of cell surface carbohydrates of microbial organism, plants, protozoan, and animals can be collectively understood.

REFERENCES

1. Hakomori, S.: 1973. *Adv. Cancer Res.* 18, 265–315.
2. Brady, R.O. and Fishman, P.: 1974. *Biochim. Biophys. Acta* 335, 121–148.
3. Buck, C.A., Glick, M.C., and Warren, L.: 1971. *Biochem. J.* 7, 4567–4576.
4. Takasaki, S., Ikehira, I., and Kobata, A.: 1980. *Biochem. Biophys. Res. Commun.* 92, 735.
5. Hakomori, S.: 1970. *Proc. Natl. Acad. Sci. USA* 67, 1741–1747.
6. Robbins, P.W. and Macpherson, I.: 1971. *Nature* 229, 569–570.
7. Mandel, P., Dreyfus, H., Yusufi, A.M.K., Sarlieve, L., Robert, J., Neskovic, N., Harth, S., and Rebel, G.: 1980. *Structure and Function of Gangliosides (Adv. Exp. Med. Biol.* 125) (Svennerholm, L., Mandel, P., Dreyfus, H., and Urban, P-F., eds.) pp. 515–531, Plenum Press, New York.
8. Marchase, R.B.: 1977. *J. Cell Biol.* 75, 237–257.
9. Piearce, M.: 1980. Thesis, University of Pennsylvania, PA (Mentor, S. Roth).
10. Obata, K., Oide, M., and Handa, S.: 1977. *Nature* 266, 369–371.
11. Szulman, A.E.: 1977. *Human Blood Groups, 5th International Convocation on Immunology* (Mohn, J.F., Plunkett, R.W., Cunningham, R.K., and Lambert, R.M., eds.) pp. 426–436, S. Karger, Basel.
12. Stern, P.L., Willison, K., Lennox, E., Galfre, G., Milstein, C., Secher, D., and Ziegler, A.: 1978. *Cell* 14, 775–783.
13. Willison, K.R. and Stern, P.L.: 1978. *Cell* 14, 785–793.
14. Marsh, W.L.: 1961. *Br. J. Hematol.* 7, 200–209.
15. Kapadia, A., Feizi, T., and Evans, M.J.: 1981. *Exp. Cell Res.* 131, 185–195.
16. Fukuda, M., Fukuda, M.N., and Hakomori, S.: 1979. *J. Biol. Chem.* 254, 3700–3703.
17. Solter, D. and Knowles, B.B.: 1978. *Proc. Natl. Acad. Sci. USA* 75, 5565–5569.
18. Artzt, K., Dubois, P., Bennett, D., Condamine, H., Babinet, C., and Jacob, F.: 1973. *Proc. Natl. Acad. Sci. USA* 70, 2988–2992.
19. Muramatsu, T., Gachelin, G., Damonneville, M., Delarbre, C., and Jacob, F.: 1979. *Cell* 18, 183–191.
20. Larraga, V. and Edidin, M.: 1979. *Proc. Natl. Acad. Sci. USA* 76, 2912–2916.
21. Fukuda, M., Fukuda, M.N., Papayannopoulou, T., and Hakomori, S.: 1980. *Proc. Natl. Acad. Sci. USA* 77, 3474–3478.
22. Ashwell, G. and Morrell, A.G.: 1977. *Trends Biochem. Sci.* 2, 76–79.

23. Sly, W.S. and Stahl, P.: 1978. *Transport of Macromolecules in Cellular Systems* (Silver-stein, S.C., ed.) pp. 239–244, Dahlen Konferenzen, Berlin.

24. Yamakawa, T. and Suzuki, S.: 1952. *J. Biochem.* 39, 393–402.

25. Hakomori, S., Siddiqui, B., Li, Y-T., Li, S-C., and Hellerqvist, C.G.: 1971. *J. Biol. Chem.* 246, 2271–2277.

26. Marcus, D.M., Naiki, M., and Kundu, S.K.: 1976. *Proc. Natl. Acad. Sci. USA* 73, 3262–3267.

27. Bhende, Y.M., Deshpanda, C.K., Bhatia, H.M., Sanger, R., Race, R.R., Morgan, W.T.J., and Watkins, W.M.: 1952. *Lancet* i, 903.

28. Hakomori, S. and Fukuda, M.: 1980. *27th International Congress of Pure and Applied Chemistry* (Varmavuori, A. ed.) pp. 161–170, Pergamon Press, Oxford and London.

29. Tanner, M.J.A. and Anstee, D.: 1976. *Biochem. J.* 153, 271–277.

30. Yamakawa, T.: 1966. *Lipoide 16 Colloquium Mosbach/Baden* (Schutte, E., ed.) pp. 87–111, Springer-Verlag, Berlin, New York.

31. Levine, P.: 1978. *Seminars Oncol.* 5, 25–34.

32. Hood, L., Huang, H.V., and Dreyer, W.J.: 1977. *J. Supramol. Struct.* 7, 531–559.

33. Buchbinder, L.: 1935. *Arch. Pathol.* 19, 841–880.

34. Hakomori, S., Wang, S-H., and Young, W.W., Jr.: 1977. *Proc. Natl. Acad. Sci. USA* 74, 3023–3027.

35. Young, W.W., Jr., Hakomori, S., and Levine, P.: 1979. *J. Immunol.* 123, 92–96.

36. Stern, K. and Davidsohn, I.: 1956. *J. Immunol.* 77, 305–312.

37. Kawanami, J.: 1972. *J. Biochem.* 72, 783–785.

38. Levine, P., Bobbitt, O.B., Waller, R.K., and Kihmichel, A.: 1951. *Proc. Soc. Exp. Biol. Med.* 77, 403–405.

39. Levine, P.: 1976. *Ann. N.Y. Acad. Sci.* 277, 428–435.

40. Nudelman, E., Hakomori, S., Knowles, B.B., Solter, D., Nowinski, R.C., Tam, M.R., and Young, W.W., Jr.: 1980. *Biochem. Biophys. Res. Commun.* 97, 443–451.

Molecular Exploration in Unknown Metabolic Worlds

A. A. BENSON

A.A. BENSON was born on September 24, 1917 in Modesto, California, and educated at University of California, Berkeley, 1935–1939, College of Chemistry, and California Institute of Technology, 1939–1942, Department of Chemistry. Ph.D. degree awarded May, 1942. During 1955–1961, he was associate professor of Agricultural and Biological Chemistry, Pennsylvania State University, where he discovered phosphatidyl glycerol and the plant sulfolipid. Since 1962, he has been Professor of Biology, Scripps Institution of Oceanography, University of California, San Diego. His research covers the following fields: bio-organic chemistry; nuclear applications: radioactive tracer methodology, neutron activation analysis, hot atom chemistry. Photosynthesis. Lipid metabolism and cell membrane biology. Marine biology: wax ester biochemistry and utilization in crustaceans and fishes. Salmon physiology. Deep ocean fish physiology, symbiotic biology of corals. Halophyte biology: salt tolerant bacterial, plant, and mangrove physiology. Arsenic metabolism, excretion.

Man has always sought new worlds. To see and understand the unknown fascinates and gives us a feeling of satisfaction, whether it be a novel ascent of Everest, a visit to the moon, or to the sea bottom. There are many such worlds to explore whether they be physical or conceptual. Each requires special training, special tools, and a special vision of the goal. I share with Professor Egami the fascination of Molecular Exploration and the satisfaction of discoveries in the metabolic world at the molecular level.

Molecular exploration and discovery has the same fascination and provides, it seems, the same satisfaction which explorers of other worlds enjoyed. It involves risks and often has pitfalls and disappointments like other adventurous explorations. It exposes the explorer to the dread and uncertainty of error or success. So often the explorer cannot bear to recall or to write of his countless steps in the wrong direction. In the light of the ultimate beauty and simplicity of Nature the explorer cannot bear to expose his human frailty and unsuccessful gropings toward a goal which lacks reality until it is passed. To look back at the path of discovery is too painful; it is never revealed.

I, like Professor Egami, have enjoyed molecular exploration. We have, from time to time, been successful, sometimes as the result of premonitions which are hard to define; sometimes as the result of a "burning bush" or revelation or previously unanticipated goals or phenomena. Such can only be described by example. The feelings of the explorers can only be left to the readers' imagination. I offer three examples. Each elicited similar doubts and fears of error, and astonishment at the simplicity and molecular elegance of Nature.

Discovery of Phosphatidylglycerol

A tantalizing unknown phosphorus compound appeared in our two-dimensional radiochromatograms (1) of the ^{32}P components of plants, first in algae, then in every plant we looked at. Phosphate esters had been known for years. Explorers like Neuberg, Fischer, Meyerhoff, Lohmann, and Leloir had discovered important phosphorylated mediators of metabolism. It seemed there could hardly be any more, especially one which was the major phosphorus compound of green plants. It had to be a well-known substance; but all attempts to show that failed. Futile search for the "probable" solutions to the problem of identity consumed an embarrassing year or more. Only by the good fortune of having Professor Bunji Maruo as collaborator did the real compound reveal itself. It was so simple that its identity was painful to accept. Maruo's identification of glycerophosphate as the hydrolysis product and the recognition of glycerol as another hydrolysis product indicated that the unknown contained glycerol and phosphate. What could be simpler, and at the same time more perplexing? Both were too obvious. With ^{32}P and ^{14}C radiochromatographic exploration it slowly dawned on us that the unknown was GPG or glycerophosphorylglycerol, a frighteningly simple but completely novel combination of glycerol and phosphate (2). Why should Nature construct anything so simple?

$$H_2CO-P-OCH_2$$
$$H-C-OH \quad HO-C-H$$
$$H_2COH \quad H_2COH$$

The glycerolphosphoryl diesters were well known as skeletons of the glycerolphosphatides. Could GPG be such a skeleton of a phospholipid of plants? We turned our attention to the ^{32}P-labeled lipids of algae. When the fatty acids were removed, GPG appeared. It sometimes even exceeded the amounts of the known esters of choline, ethanolamine, and inositol. How could such a simple lipid be such a major component of plants without it having been known long ago? It was more likely that Nature was playing a trick on us than that we could be looking at a new and important member of the membrane world. Phosphate, glycerol, and fatty acids; there must be another novel component! Even as we sent the manuscript off to "BBA" (3) it was hard to believe that Nature could have withheld such a simple sight for so long. Now we know that phosphatidylglycerol may be the major membrane lipid of living things. Not only plants but bacteria and fungi chose to use it in their membranes. Such simplicity was hard to imagine and even now is amazing to comprehend.

Could Life on Earth Have Succumbed to Arsenic Poisoning?

No one would have imagined that neither we nor the other inhabitants of the earth might be here today if it were not for the merciful metabolism of arsenic in the seas. Our inherent fear of arsenic and oftentimes fascination with its role in history and medicine places it in a special position among the elements. This tasteless poison is so remote from most of our lives that it hardly seems a threat. Yet, life could not exist were it not for the molecular ingenuity of the algae of the ocean. They solved the problem of arsenic poisoning for us long ago; only now do we realize our debt to their successful accomplishment and to their still-continuing efforts.

For years I had wondered how the algae of the sea survive in waters so depleted of their phosphate by algal growth and photophosphorylation that phosphate concentrations drop to those of seawater arsenate which comes from volcanism and submarine hot springs. Both concentrations are about 10^{-8} molar. And algae, like other cells, can hardly discriminate between the phosphate they must absorb and the arsenate which would kill them. If an alga had even one-tenth as much arsenic as phosphate, it could never survive. Yet at least that much arsenic is absorbed by algae of much of the ocean, especially in the tropics. Fortunately, of course, Japan and many of the fish harvesting countries are bathed by seas rich in phosphate from upwelling of the nutritious waters of the depths. Man did not concern himself with the plight of algae and animals in the less fortunate deserts of the sea.

Recently we explored the world of arsenic metabolism, with radioactive arsenic metabolism revealed themselves clearly in our radiograms. Our experiences with carbon, sulfur, and phosphorus metabolism prepared us for recognition of long-kept secrets of arsenic metabolism. The chromatograms revealed active metabolic products of arsenic interconversions and turnover of arsenic in algae. The excitment of the revelation was tempered by our recognition of the facts that arsenate absorption was obligatory under the circumstances and that detoxicative metabolism was necessary to explain the relatively low concentrations of arsenic in algae and most marine organisms.

In time the chemical nature of the unknown arsenical metabolites was revealed. Although the innocuous nature of the products had been determined forty years ago, we were unaware of the fact that such substances played a role in the salvation of the algae of the sea and consequently of those of us who live on the products of the sea. Now, we understand that the algae developed metabolic processes and products to avoid accumulation of the toxic intermediates. Further, they devised a mechanism for release of their arsenical products to the sea. Even this "depuration" mechanism is novel to contemporary biochemistry. We are exploring a new world, the molecular world of arsenic metabolism.

Nature's solution was, again, elegant, simple, and successful. The hazards of dealing with an element mimicking phosphorus but having dangerous reactivities were avoided by converting it quickly to trimethylarsonium derivatives where the arsenic mimics nitrogen (4).

$$H_2C-\overset{+}{As}(CH_3)_3$$
$$|$$
$$H-\overset{|}{C}-OH$$
$$|$$
$$COO^-$$

Such compounds like trimethylarsoniumlactate seem not to be toxic or metabolically hazardous or reactive. The really elegant metabolic maneuver was to convert the trimethylarsoniumlactate to its phosphatidyl ester, a membrane phospholipid, where the trimethylarsonium group could lie in the seawater surrounding the membrane of the algal cell. There the ubiquitous symbiotic bacteria could gain energy by oxidative cleavage of the arsenic-carbon bond and consequent release of the methylated arsenical to the sea. As a result the alga becomes free of arsenic and ready to deal again with its problems.

Even marine animals use such a process to free themselves of arsenic. The giant clams of the South Pacific divert their arsenic to the external membranes of their gills. There the arsenic of membrane-associated lipids is released to the sea. Nature could hardly have devised a more elegant solution to this frightening predicament.

The Sulfonic Sphinx of Green Plants

Like the Sphinx which silently withstands passing time, the sulfonic acid metabolism of plants has perplexed us all. For some yet unknown reason plants chose to use these unlikely compounds in their chloroplast membranes. Now, 26 years after their revelation we still fail to see Nature's reason for having designed them.

Years ago, when searching for lipoic acid in algae, I labeled them by illumination with radioactive sulfate in the medium. The radiograms revealed nearly complete conversion of the ^{35}S-sulfate to a lipid, easily resolved and unique. For a year or more it confounded and tantalized us. The lipid was clearly a simple glycolipid but contained its sulfur in an acidic group. Seeking to break its bonds to the sugar I searched the works and methods of Professor Egami whose skill had elucidated much of sulfate ester biochemistry and enzymology. His sulfatases failed to produce sulfate from our sulfolipid or from its derivatives. Nothing could cleave the sulfur from its molecular bonds.

The natural conclusion for an organic chemist like Dr. Helmut Daniel in our laboratory was that it must be a sulfonic acid. For a biochemist, that conclusion was unacceptable. No one had ever heard of a sugar sulfonic acid in Nature! It simply could not be! But Daniel persisted and the search for evidence of sugar sulfates was nearly exhausted. The manuscript was written, claiming identification of a "stable" sulfate ester of a glycolipid akin to cerebroside sulfate. Finally, Daniel's patient persistence convinced us that we were in a new world of sulfo-carbohydrate metabolism (5). The chemical skills of Drs. Masateru Miyano (6), Isao Shibuya (7), and Tatsuhiko Yagi (7) placed the new compounds and their metabolism in a logical metabolic framework. And, finally, Dr. Yoshi Okaya dispelled any remaining uncertainty by his determination of the molecular structure by X-ray diffraction (8). The sulfolipid was, indeed, α-sulfoquinovosyldiglyceride (9).

$$
\begin{array}{ll}
\text{H--C--O} & \text{CH}_2 \\
\text{H--C--OH} & \text{H--C--O--CO--C}_{15}\text{H}_{31} \\
\text{HO--C--H} & \text{H}_2\text{CO--CO--C}_{17}\text{H}_{29} \\
\text{H--C--OH} & \\
\text{H--C--O} & \\
\text{H}_2\text{C--SO}_3\text{H} &
\end{array}
$$

Molecular exploration must have many more worlds to reveal and many more goals to attain. The three we have just described were as invisible as those yet to be recognized. New explorations will tantalize and surprise us again and again. The uncertainty, skepticism, and apprehension of their reality will lead us ultimately to appreciation of the simplicity and elegance of Nature.

Having followed some of Professor Egami's interests and achievements for 25 years, it is clear that he too has enjoyed and transmitted to many others as well the excitement and delights of "Molecular Exploration."

REFERENCES

1. Benson, A.A.: 1979. *Curr. Contents* 10, 16.
2. Maruo, B. and Benson, A.A.: 1957. *J. Am. Chem. Soc.* 79, 45–64.
3. Benson, A.A. and Maruo, B.: 1958. *Biochim. Biophys. Acta* 27, 189–195.
4. Cooney, R.V., Mumma, R.O., and Benson, A.A.: 1978. *Proc. Natl. Acad. Sci. USA* 75, 4262–4264.

5. Benson, A.A., Daniel, H., and Wiser, R.: 1959. *Proc. Natl. Acad. Sci. USA* 45, 1582–1587.
6. Miyano, M. and Benson, A.A.: 1962. *J. Am. Chem. Soc.* 84, 59–62.
7. Shibuya, I., Yagi, T., and Benson, A.A.: 1963. *Studies on Microalgae and Photosynthetic Bacteria* (Japan Soc. Plant Physiologists, ed.) pp. 627–636, Univ. Tokyo Press, Tokyo.
8. Okaya, Y.: 1964. *Acta Crystallogr.* 17, 1276–1282.
9. Benson, A.A.: 1977. *Radioisotopes* 26, 348–356.

Amino Acids and Brain Function*

J. H. QUASTEL

J.H. QUASTEL was born in Sheffield in 1899 and graduated from Imperial College of Science, London, 1921 and Trinity College, Cambridge in 1924. After he served World War I, he worked as a fellow of Trinity College since 1924. He was Director of Research Cardiff City Mental Hospital from 1929 to 1941 and of Unit of Soil Metabolism, Agricultural Research Council from 1941 to 1947. From 1947 to 1966, he was Professor of Biochemistry McGill University, Montreal, Canada. During those times, he was concurrently Director of McGill, Montreal General Hospital Research Institute from 1947 to 1965 and Director of McGill Unit of Cell Metabolism from 1965 to 1966. Since 1966, he is Professor of Neurochemistry, University of British Columbia. He received Meldola Medal in 1927 and other medals, an Award from Canadian Microbiological Society in 1965 and the Gairdner International Award for Medical Research, 1974. He was made companion of the order of Canada, 1970. He was the President of Canadian Biochemical Society (1963) and Honorary President of the International Biochemical Congress in Canada, 1979. His research covers various fields of biochemistry and neurochemistry including metabolic inhibitors and chemistry of brain metabolism.

Amino acids of the brain have now attained a major importance in the study of brain function. Investigations of the cerebral metabolism, locations, movements and electrical activities of the amino acids and their metabolic products are now beginning to dominate the fields of neurochemistry and neurophysiology. The reason for this outburst of activity in the last two or three decades is the realization that the cerebral amino acids and their immediate derivatives influence and control mental behaviour and that, in some manner, mental abnormalities are a reflection of neurochemical and neurophysiological changes in which the amino acids are directly or indirectly involved.

* Based on the article "The Role of Amino Acids in the Brain" by J.H. Quastel (54).

Clinical Conditions Associated with Abnormal Cerebral Amino Acid Concentrations

1. Parkinsonism

About 20 years ago, it was shown (*1*) that dopamine (dihydroxyphenethyl-amine) is a normal constituent of the brain and, moreover, that it may be increased in quantity after the administration of dopa (3,4-dihydroxyphenyl-alanine) especially when given together with a monoamine oxidase inhibitor. A little later it was found that most of the dopamine is present in the striatum (caudate uncleus and putamen) and that this is due to the existence of a large dopamine-containing neuronal pathway originating in the substratia nigra and ending in the striatum. The distribution in the brain of the enzymes, tyrosine hydroxylase and dopa decarboxylase, responsible for the formation of dopamine, is similar to that of dopamine, a finding also true for human brain (*2*).

Parkinson's disease is one of the most common of the motor disturbances of extrapyramidal (striatal) origin and many investigations have been made concerning it. Studies, begun in 1959, of human post-mortem material, in which the cerebral distribution of dopamine was found to be similar to that in freshly obtained animal brains, showed that Parkinson's disease is characterized by a deficiency of dopamine in the nigro-striatal pathway. It is now thought that degeneration of the dopamine neurons, proceeding from the substantia nigra to the striatum, is the factor responsible for the dopamine deficiency (*3*). Within a year of the discovery, in 1961, of dopamine deficiency in Parkinsonism, it was found that L-dopa administration, given intravenously or orally, results in amelioration of the symptoms of this disease, such as akinesia and rigidity. The oral administration of this substance also increases brain levels of dopamine and its metabolites (*3*). Although side effects may occur, owing to the formation of O-methyldopa and various condensation products or to the competition between dopa and tryptophan for transport into the brain, there seems to be little doubt that the major effect of dopa administration, seen in the marked improvement found in many cases of Parkinsonism, is connected with increased cerebral level of dopamine.

2. Huntington's chorea

This severe degenerative disorder of the brain is characterized by progressive chorea and dementia, from early to middle adult life, terminating in death 12–15 years after the appearance of the early symptoms. The disease is said to account for at least 1% of all patients now chronically hospitalized in mental hospitals. It has been found (*4*) that γ-aminobutyrate and a related dipeptide (homocarnosine, γ-aminobutyrylhistidine) are diminished in quantity in the substantia nigra, caudate nucleus, and putamen-globus pallidus of choreic patients, as compared with normal subjects. There is also a fall in the glutamate decarboxylase and choline acetyltransferase activities of the same areas of the brain of choreic patients (*5*). It seems definite that in Huntington's chorea, there is a marked and specific loss of an amino acid, γ-aminobutyrate, in special areas of the brain, but whether this loss is solely responsible for the symptoms characteristic of the degenerative cerebral disorder or whether it is one of a number of biochemical changes that together create the neurological disturbance is at present unknown. It is possible that the deficiency of the amino acid reflects a decrease in the number of neurons that normally utilize γ-aminobutyrate as a transmitter (4).

3. Phenylketonuria

This disease, once called Folling's disease, which is associated with mental defect, is characterized by a congenital biochemical abnormality in which there is a disturbed phenylalanine metabolism with excretion of phenylpyruvic acid and phenylalanine in the urine (6). The disease is inherited as a simple recessive. The biochemical disturbance responsible for the disorder, is a deficiency of the liver enzyme, phenylalanine hydroxylase, that converts phenylalanine into tyrosine (7). As phenylalanine is a normal ingredient of most protein foodstuffs, it accumulates in the blood and its metabolism, in phenylketonuria, is diverted towards the production of phenylpyruvic acid and its metabolites. Phenylpyruvic acid is metabolized with difficulty by phenylketonuria patients (8) presumably because its oxidation, after transformation to phenylalanine, is diminished by the absence of phenylalanine hydroxylase. Tyrosine is normally metabolized in these patients (8). The profound mental retardation that is observed in phenylketonurics, some of whom show epileptic seizures and almost all of whom show abnormal electroencephalograms, is thought to be largely due to the abnormally high concentrations of phenylalanine, and perhaps of its metabolites, in the brain. This has led to the testing of diets, low in phenylalanine content, on phenylketonuric patients. It has been found that when dietary treatment commences in early infancy, the child develops normally, or almost normally, intellectually and exhibits a normal electroencephalogram and no behavioural disturbances (9). These results support the current view that the mental and neurological features of phenylketonuria are due to the abnormally high cerebral concentrations of phenylalanine or its metabolites. These, in turn, are due to the adsence or inactivity of phenylalanine hydroxylase and presumably therefore to the mutation of a single gene (9).

4. "Maple-syrup" -urine disease

Another, though rare, familial disease in which amino acids in the brain are associated with mental disturbance is "maple-syrup"-urine disease in which there is an abnormally large excretion of valine, leucine, and isoleucine together with their metabolites (10).

The disease first described as a clinical entity in 1954, occurs in very young children who succumb to this rapidly progressive neurological disorder in the first few weeks of life. The urine of the children has an odour similar to that of maple syrup. The first symptoms appear at about 3–5 days of age. Convulsions followed by coma and disturbances of respiration then occur and this is followed by a steady downhill fatal course unless therapy is instituted immediately (*11*).

Oxidative decarboxylation, in the normal pathway of amino acid degradation (*11, 12*), is impaired in these patients and an excess of branched-chain oxo acids appears in the urine (*13*). Deficiency of the enzyme, α-oxoacid decarboxylase, is thought to result in the accumulation of both the oxo acid derivatives and the branched-chain acids in the child having the classical form of the disease. The manner, however in which the metabolic block produces the gross neurological disorder is at present unknown. It has been shown (*14*) that the α-oxoacids may inhibit the activity of glutamate decarboxylase in rat brain, affecting in this manner certain aspects of brain glutamate metabolism.

A striking fact is that the cerebral concentrations of glutamate, glutamine, and γ-aminobutyrate may all be greatly diminished in maple-syrup urine disease (*15*), pointing to some effect other than, or in addition to, diminished glutamate decarboxylase activity in the brain. It is evident that abnormal biochemical changes occur in the brains of these victims of the disease. One of these apparently is a suppression of protein biosynthesis, as shown by the inhibitory effects of the branched-chain α-oxoacids on the cerebral protein incorporation of valine (*16*). New results (*17*) indicate that high concentrations of the branched-chain amino acids affect the kinetics of the glutamate-glutamine cycle in brain (*18*) and possibly, therefore, brain function.

Treatment of the disease consists, immediately after diagnosis, of providing a diet completely free, initially, of the branched-chain amino acids. As the plasma concentration of each of the branched-chain amino acids approaches normal, individual supplementation is begun (*11*). Considerable improvement occurs as a result of therapy and normalization of the electroencephalogram occurs within 3 months of dietary control. Late initiation of therapy is accompanied by mental retardation (*11*).

Administration of Tryptophan or Methionine

In addition to the facts such as those given above, it is now well known that the administration of some amino acids, *e.g.*, tryptophan and methionine, to patients undergoing treatment with certain drugs, such as monoamine oxidase inhibitors, leads to markedly enhanced mental disturbances.

Evidence is accumulating, in this manner, to indicate the profound role of amino acids and their metabolites in the central nervous system. We are, however, as yet only in the early stages of learning the manner in which these substances operate to influence animal, including human, behaviour.

Amino Acids and Brain Proteins

Amino acids are incorporated into brain proteins both *in vivo* and *in vitro*. Whereas this process must obviously occur in developing brain, there is also considerable incorporation into adult brain protein, the speed of synthesis in the adult brain balancing that of breakdown. Experiments using labelled amino acids, such as leucine or lysine, have shown that the half-life (that is the time to replace half) of cerebral proteins can be measured in days, the proteins having varying rates of turnover. There are regional differences between these turnover rates. The proteins of immature brain have a higher turnover rate than those in the adult, the rapid synthesis of newly formed brain being superimposed on the turnover rates of the existing proteins. A dynamic state of the cerebral proteins exists in both immature and adult animals, indicating that there exists in the adult brain at all times pools of amino acids that will vary in quantity and composition from region to region in the brain. Such pools are reinforced, of course, in the living animals by amino acids derived from the blood.

Many results illustrate the dynamic aspects of amino acid-protein interrelations in brain cells and the fact that these are closely involved in the functional activity of the central nervous system.

Neural Peptides

1. Endorphins and enkephalins

Among the various substances now known to be implicated in the functional machinery of the brain are the neural peptides. A group of substances, occurring in the brain and in the pituitary gland, have effects similar to those of opiates, such as morphine and heroin, and are termed endorphins (*19*). These substances, found in a search for endogenous compounds that might combine with specific opiate receptors (*20–22*), contain among them two related pentapeptides (methionine) enkephalin and (leucine) enkephalin, which have been isolated from pig brain (*23*). These peptides possess similar amino acid sequences, namely Tyr-Gly-Gly-Phe-Met for the methionine-containing peptide and Tyr-Gly-Gly-Phe-Leu for the leucine-containing peptide. They have a high affinity for opiate receptors and are active at morphine-sensitive neuro-effector junctions. This and related work also show that the enkephalins are stored in a stable form in the brain nerve terminals, but are rapidly inactivated in the presence of cerebral peptidases. It is thought possible that these readily inactivated enkephalins may act as neurotransmitters.

The same sequence of five amino acids is found in a polypeptide, β-lipotropin, which contains 91 amino acids. The entire peptide appears to have no opiate activity, but the moiety containing the sequence of amino acids 61–91 (β-endorphin) is active. Apparently various breakdown products of this fragment are potent, provided that tyrosine occupies the N-terminal position (24). The smallest active fragment proves to be (methionine) enkephalin, and all active fragments contain this pentapeptide sequence. Removal of methionine leaves a virtually inactive tetrapeptide. Direct application of the β-endorphin (β-lipotropin, 61–91) by the intracerebral route leads to persistent analgesia (25, 26), a result that points to this molecule being the normal mediator of effects at the opiate receptor *in vivo*. Recent results (27) have shown that extracellular metabolism of β-endorphin may occur to form the series of opiate-like peptides that have been isolated from brain. However, it has been found that enkephalin is present in rat brain soon after killing by microwave injection (28).

This work on the effects of neural peptides in cerebral activity is still in its initial stages. It is, however, clear that certain specific sequences of amino acids can bind, possibly, to synaptic membranes and so affect neurotransmission.

2.   *Amino acid entry into the brain*

The concentration of free amino acids in the brain is usually six- or seven-fold that in the blood plasma but the concentrations of some of the less quantitatively prominent amino acids, including those essential for growth such as valine, leucine, isoleucine, tryptophan, methionine, and phenylalanine approximate those in the plasma. The relatively low concentrations of the essential amino acids in the brain may increase with an increase in their blood concentration. This is not so with an amino acid such as glutamate, which normally attains a high concentration, although the amino acid can pass freely in and out of the brain cell. A group of amino acids, containing glutamate, glutamine, γ-aminobutyrate and aspartate (which will be referred to as the glutamate system), makes up about 70% of the total free amino acids in the adult animal brain. Although the total amount of free amino acids per g wet weight changes but little from infancy to maturity, the infant brains have a markedly high concentration of taurine and relatively low one of the glutamate system. As the brain develops to maturity the value for the taurine concentration falls and that for the glutamate system becomes correspondingly greater. The relatively high brain concentrations of taurine in infancy applies to animals as widely separate as mouse and man. The marked increase in brain glutamate concentrations, which occurs on maturity, takes place in most of the animals that have been investigated, usually about 14 days after birth, but not in the guinea pig which is well known to be born with many mature characteristics. γ-Aminobutyrate seems to be absent from very immature brain; it appears in the chick embryo during the fourth to the sixth day of incubation. This amino acid is confined largely to the nervous system. The concentration of glutamine in the brain increases a little during development.

3.   Cerebral movements of amino acids in vivo

Glutamate administration, for example by intravenous injection, does not result in an increased cerebral concentration of glutamate, but if radioactive glutamate is injected into a rat or mouse, the cerebral glutamate becomes radioactive. Glutamate evidently enters the brain in the living animal, its uptake being due to an exchange process, as there is no net uptake even when the plasma concentration is increased 50 times. Injection of glutamine, however, does result in an increased uptake of glutamine in the brain.

The process of cerebral uptake that occurs with glutamate may occur either by exchange diffusion, whereby one molecule of glutamate in the plasma replaces one molecule of glutamate (or another related amino acid) in the brain, or by active transport, due to a carrier-mediated energy-assisted process, leading to accumulation of the amino acid against a concentration gradient. The uptake process is accompanied by diffusion out of the brain cell, so that a steady state is eventually reached. The process of exchange diffusion, first shown with inorganic ions and now known to apply to amino acids and sugars, may occur more rapidly than active transfer. It seems to be energy independent and not even affected by ouabain which suppresses the energy-assisted transfer process.

The permeability of brain cells to amino acids seems to vary with age, for the cerebral uptake of certain amino acids (e.g. lysine) can occur to a greater extent in the immature animal than in the adult (29).

There is stereospecificity in cerebral amino acid uptake, the L-amino acid usually penetrating to a greater extent than the corresponding D-isomer.

4.   Carrier-mediated transport

Intraperitoneal injection of L-tyrosine results in an elevated concentration of this amino acid in the brain, which may be 5 times that of the normal. The uptake is inhibited by leucine, isoleucine, valine, histidine, tryptophan but not by serine, arginine, lysine, glutamine, or glutamate. Such results indicate that the cerebral transfer of L-tyrosine in the living animal occurs by a carrier-mediated mechanism. There is a marked lowering of the uptake of L-tyrosine in the brain in vivo, if the plasma levels of L-phenylalanine, or other aromatic amino acids, or long-chain aliphatic amino acids are elevated. L-Methionine, administered at a concentration of 60 mg/kg, reduces the transport of dopa into the brain, but not that of 5-hydroxytryptophan. L-Tryptophan, given at 80 mg/kg, has a similar effect on the transport of dopa in vivo. Such results as these indicate how the uptake of an amino acid into the brain from the blood may be affected by the presence of other amino acids, a fact of much clinical importance when considering therapy by amino acid administration or the consequences of high concentrations of amino acids in the blood as in amino acidurias.

The regional pattern of distribution of amino acids in the brain is not necessarily identical with that of the uptake rate (30). The uptake process is a function of both transport and rate of incorporation into tissue constituents and may also depend on the rate of blood flow which varies greatly in different parts of the brain.

5.   Cerebral transport of amino acids in vitro

Much has been learned of the nature of cerebral transport by studies of the brain in vitro, particularly by studies of brain slices incubated immediately after removal of the brain from the animal in an oxygenated glucose-containing physiological saline medium. Such experimental material has shown results similar, both quantitatively and qualitatively, to those observed in vivo. Through the use of drugs and metabolic inhibitors, by kinetic studies and by close examination of different types of brain preparation, studies of the brain in vitro can yield more information than can be obtained at present from studies only with living animals. It has even been shown that slices of cerebellum show electrical phenomena, such as spontaneous action potentials, that resemble in many aspects, including their reactions to possible amino acid transmitters, those found in studies of the brain in situ.

6.   Energy dependence of amino acid accumulation

Brain cortex slices are able to concentrate many amino acids against a concentration gradient, e.g. L- and D-glutamate, γ-aminobutyrate, glycine, serine, alanine, proline, histidine, tyrosine, α-methyltyrosine, 5-hydroxytryptophan. Experiments on the uptake of glycine in rat brain cortex slices show that the extent of accumulation of the amino acid under a wide variety of experimental conditions may be proportional to the ATP level. Such conditions include changes in the glucose concentration in the incubation medium, the presence of metabolic inhibitors, alteration of the ionic composition of the medium and anaerobiosis. Nevertheless, although the level of ATP has a controlling influence on amino acid transport, it is by no means the only factor involved.

7.   Na+ dependence of amino acid transport

The original finding, made in 1958 (31), that the active transport of glucose into the isolated surviving intestine is Na+ dependent was succeeded by the finding that the active transport of many amino acids into the brain is also Na+ dependent. The transport of glucose, however, into brain cells, though carrier-mediated, seems to be independent of Na+.

A decrease in the concentration of Na+ in the incubation medium leads to a decrease in the extent of cerebral glycine transport without an appreciable change in the concentration of ATP. It seems now that the concentration of Na+ in the incubation medium, taken in conjunction with that in the brain cell, plays a major role in the control of amino acid uptake (32). The Na+ gradient appears greatly to control the amino acid-transport system; this, in turn, depends partly on the

ATP concentration. Sucrose or choline, which is often used to maintain the iso-osmotic level in the incubation medium, cannot replace Na^+ for the maintenance of the Na^+ gradient.

8.   High- and low-affinity amino acid-uptake processes

Apart from passive diffusion of an amino acid into a brain cell, which may not be a carrier-mediated process, there are at least two other processes of uptake, one of which is marked by a high affinity to the transport carrier, i.e. one in which half the maximum rate of entry occurs at very low concentrations (e.g. 20 μM) of the amino acid in the medium bathing the cell, and another process which is char-acterized by a low affinity to the transport carrier. Specific high-affinity uptake mechanisms exist in brain slices, or in suspensions of nerve terminals, for γ-amino-butyrate, glutamate, aspartate, taurine, proline, and glycine. Such high-affinity processes appear to be dependent on external Na^+, and in this respect they differ from the low-affinity processes, where relatively high concentrations (e.g. 0.5 mM) of amino acids are required for optimal uptakes.

It has been demonstrated that both glia and neurons possess high-affinity uptake systems for possible amino acid transmitters (33, 34). It seems likely that such systems represent a means whereby transmitter function may be terminated (35). This inference, however, need not always be true, because some high-affinity systems may only represent exchange diffusion processes and may well have the characteristics of concentrative carrier-mediated uptakes at low concentrations. Simple exchanges need not reduce the amino acid concentration in the synaptic cleft and may not therefore terminate transmitter function.

Amino Acids as Neurotransmitters

Early studies (36, 37) have pointed to the association of changes in cerebral concentrations of certain amino acids, such as γ-aminobutyrate and glu-tamate, with neurological disturbances. For example, microinjections of glutamate into mammalian brain have a direct excitatory effect, and γ-aminobutyrate and related neutral amino acids are potent depressants. Investigations of the effects of the addition of amino acids to the medium bathing single cortical neurons, and other neurons, in the brain show that glutamate and γ-aminobutyrate have short-lived excitatory and inhibitory effects and it was suggested that these substances may be natural transmitters in the brain (38). Numerous investigations have in-dicated that L-glutamate and L-aspartate are excitatory amino acids in many parts of the brain, and glutamate has been considered to be the excitatory transmitter in the spinal cord. L-Glutamate has no effects at vertebrate neuromuscular junc-tions and its excitatory activities are apparently confined to the central nervous system. Aspartate has been proposed as an excitatory transmitter released from spinal interneurons because it is more concentrated in the ventral grey matter than in the dorsal. Moreover, following anoxic lesions, there is a correlation between the decrease in concentration of aspartic acid and loss of motoneurons (39).

Among the potent excitatory molecules that are related in chemical structure to glutamate are D-aspartate, L-cysteate, L-cysteinesulphinate, and threo-3-hydroxy-DL-aspartate. These amino acids excite neurons in the feline spinal cord when administered by micro-electrophoresis (40). Even more potent excitants are N-methyl-D-aspartate and D-homocysteate.

Amino acids that have large inhibitory effects on cerebral neurons are γ-aminobutyrate, taurine and glycine and these substances have been proposed as inhibitory transmitters in the vertebrate central nervous system. β-Alanine exerts potent depressant effects in many regions of the mammalian central nervous system by an action resembling that of taurine. In most areas of the brain, γ-aminobutyrate is an effective depressant of neuronal firing, but there are regional differences in sensitivity to glycine or taurine. For example, in the cortex, γ-aminobutyrate is more effective than glycine, but the opposite is the case with spinal motoneurons, and taurine resembles glycine in potency (41–43).

Iontophoretically administered glycine reduces, or abolishes, the firing of most spinal cells independently of the method by which these cells are excited (44). In investigations on the distribution of glycine in the neuraxis of a variety of animal species, cat, rat, pigeon, bullfrog, alligator, catfish, boa constrictor, it has been found (45) that much higher glycine contents are present in the medulla oblongata than in the cerebral hemispheres and cerebellum. Moreover, glycine is more plentiful in those areas of the spinal cord supplying the limbs, the cervical and lumbar enlargements, than in the thoracic region which is supplied with a rather scanty musculature. Neurophysiological evidence supports the conclusion that glycine may be a transmitter in the medulla but probably not in the cerebral cortex.

Studies with unanesthetized decerebrate cats have shown that DL-homocysteate is a more potent excitant than L-glutamate or L-aspartate on most bulbar reticular neurons tested, whereas glycine, β-alanine, and γ-aminobutyrate depress the activity of most neurons tested (46). The activities of brain-stem neurons are depressed by β-alanine and γ-aminobutyrate, but they are enhanced by L-glutamate (43).

In the cerebellum, γ-aminobutyrate is a potent depressor of the firing of Purkinje cells, and β-alanine and glycine have definite but weaker depressing effects. It is thought that the inhibitory transmitter released from Purkinje-cell axon terminals is likely to be γ-aminobutyrate. This amino acid, when administered electrophoretically, depresses the firing of neurons throughout the central nervous system.

L-Glutamate, the parent amino acid of γ-aminobutyrate, and the most abundant amino acid in the brain, is concentrated more in the nervous system than in any other vertebrate tissue. Its content is higher in the cerebral hemisphere and in the cerebellum than in the mid-brain and medulla. The high level of L-glutamate in dorsal roots, dorsal grey and dorsal white matter may indicate involvement of this amino acid in excitation by primary afferent fibres (47). Both the onset and offset of L-glutamate and L-aspartate excitations are rapid, and the threshold extracellular concentration for the firing of neurons in the brain cortex and spinal cord is about 0.1 mM (47).

1.   Mode of action of amino acid transmitters

Postsynaptic inhibition of mammalian motoneurons is associated with membrane hyperpolarization as a result of a brief increase in permeability to K^+ and Cl^- (48). Extracellularly administered γ-aminobutyrate increases neuronal membrane conductance (47). Neuroglial cells and spinal axons are not affected by γ-aminobutyrate, and it is thought that the hyperpolarizing action of this amino acid, which is rapidly reversible, is restricted to inhibitory synapses. Both γ-aminobutyrate and the naturally occurring inhibitory transmitter increase membrane conductance and hyperpolarize the neuronal membrane. These processes involve similar ionic changes. In particular, there is an increased permeability to Cl^-. The superior cervical ganglion also contains receptors for γ-aminobutyrate which increases permeability to Cl^- just as it does in the central nervous system and these receptors are blocked by picrotoxin (49).

The excitatory amino acids exert their effects by depolarization, involving, particularly, an increase in Na^+ permeability (38, 39). It has also been suggested (42) that the depolarizing action of glutamate involves displacement of Ca^{2+} from particular sites in the membrane and that this displacement initiates the permeability change as an excitant. It is to be noted that glutamate, like some other transmitters, is only effective when injected outside the cells (42).

It is found that excitation by L-glutamate, as shown by its acceleration of the frequency of these potentials, is considerably diminished by reducing the external Na^+ concentration from 150 to 75 mM (50). This result, taken in conjunction with other electrophysiological evidence, supports the conclusion that excitation by L-glutamate is associated with its property of increasing neuronal permeability to Na^+.

It has been found (50) that the response to L-glutamate is unaltered by replacement of the Cl^- in the external medium by another anion. On the other hand, replacement of the Cl^- by an impermeant anion, e.g. sulphate or isothionate, brings a profound change in the effects of γ-aminobutyrate, glycine, taurine, and β-alanine. Their normal inhibitory activities are converted into excitations (50). This effect of replacing Cl^- ions by impermeant ions may be simply explained by the increase in permeability to Cl^- brought about by the inhibitory amino acids. In the absence of extracellular Cl^-, the net anion flux will become outward from the cell instead of inward and this will result in depolarization and therefore excitation. It is now well known that removal of Cl^- abolishes inhibitory synaptic potentials without disturbing excitatory transmission (51–53).

Further results with cerebellar slices show that, whereas deprivation of Na^+ in a low chloride (sulphate) medium leads to a suppression of excitation by glycine, taurine and β-alanine, there is no suppression of the activity of γ-aminobutyrate (50). This leads to the inference that the Cl^- efflux mechanism for excitation, in media containing impermeant anion, may largely account for the behaviour of γ-aminobutyrate. All the inhibitory amino acids investigated appear to increase permeability of the neuronal membrane to K^+. Glycine, taurine, and β-alanine, in addition, affect permeability to Na^+ as well as to Cl^-. The actual receptor site for γ-aminobutyrate is presumably located on the outer surface of the neuronal membrane, because intracellular injections of this amino acid are without effect (53).

Thus the amino acids, whether excitatory or inhibitory, accomplish their activities by specific changes at the neuronal membrane, resulting in changes in inorganic ion permeabilities.

REFERENCES

1. Carlsson, A.: 1959. *Pharmacol. Rev.* 11, 490–493.
2. Lloyd, K. and Hornykiewiez, O.: 1972. *J. Neurochem.* 19, 1549–1559.
3. Hornykiewiez, O.: 1973. *Fed. Proc. Fed. Am. Soc. Exp. Biol.* 32, 183–190.
4. Perry, T.L., Hansen, S., and Kloster, M.: 1973. *N. Engl. J. Med.* 288, 337–342.
5. McGeer, P.L., McGeer, E.G., and Fibiger, H.C.: 1973. *Neurology* 23, 912–917.
6. Closs, K.: 1966. *Molecular Basis of Some Aspects of Mental Activity*, Vol. 1 (Walaas, O., ed.) pp. 231–247, Academic Press, New York.
7. Jervis, G.A.: 1947. *J. Biol. Chem.* 169, 651–656.
8. Penrose, L. and Quastel, J.H.: 1937. *Biochem. J.* 31, 266–274.
9. Woolf, L.I.: 1966. *Molecular Basis of Some Aspects of Mental Activity*, Vol. 1 (Walaas, O., ed.) pp. 249–264, Academic Press, New York.
10. Westfall, R.G., Davies, J., and Miller, G.: 1957. *Am. J. Dis. Child.* 94, 571–572.
11. Snyderman, S.E.: 1975. *Treatment of Inherited Metabolic Diseases* (Raine, D.N., ed.) pp. 71–90, Medical Technical Publishing Co., Lancaster.
12. MacKenzie, D.Y. and Woolf, L.I.: 1959. *Br. Med. J.* 1, 91–95.
13. Menkes, J.H.: 1959. *Pediatrics* 23, 348–357.
14. Tashian, R.E.: 1961. *Metab. Clin. Exp.* 10, 393–402.
15. Prensky, A.L. and Moser, H.W.: 1966. *J. Neurochem.* 13, 863–874.
16. Appel, S.H.: 1966. *Trans. N.Y. Acad. Sci.* 29, 63–70.
17. Benjamin, A.M., Verjee, Z.H., and Quastel, J.H.: 1980. *J. Neurochem.* 35, 78–87.
18. Benjamin, A.M. and Quastel, J.H.: 1972. *Biochem. J.* 128, 631–646.
19. Goldstein, A.: 1976. *Science* 193, 1081–1086.
20. Goldstein, A., Lowney, L.I., and Pal, B.K.: 1971. *Proc. Natl. Acad. Sci. USA* 68, 1742–1747.
21. Pert, C.B. and Snyder, S.H.: 1973. *Science* 179, 1011–1014.

22. Simon, E.J., Hiller, J.M., and Edelman, I.: 1973. *Science* 70, 1947–1949.
23. Hughes, J., Smith, T.W., Kosterlitz, H.W., Fothergill, L.A., Morgan, B.A., and Morris, H.R.: 1975. *Nature* 258, 577–579.
24. Cox, B.M., Goldstein, A., and Li, C.H.: 1976. *Proc. Natl. Acad. Sci. USA* 73, 1821–1823.
25. Loh, H.H., Tseng, L.F., Wei, E., and Li, C.H.: 1976. *Proc. Natl. Acad. Sci. USA* 73, 2895–2898.
26. Feldberg, W. and Smyth, D.G.: 1976. *J. Physiol.* 260, 30P–31P.
27. Austin, B.M., Smyth, D.G., and Snell, C.R.: 1977. *Nature* 269, 619–621.
28. Yang, J.S., Hong, L., and Costa, E.: 1977. *Neuropharmacology* 16, 303–307.
29. Waelsch, H.: 1962. *Neurochemistry*, 2nd ed. (Elliot, K.A.C., Page, I.H., and Quastel, J.H., eds.) pp. 288–320, C.C. Thomas, Springfield, Ill.
30. Levi, G., Cherayil, A., and Lajtha, A.: 1965. *J. Neurochem.* 12, 757–770.
31. Riklis, E. and Quastel, J.H.: 1958. *Can. J. Biochem. Physiol.* 36, 347–362.
32. Schultz, S.G. and Curran, P.F.: 1970. *Physiol. Rev.* 50, 637–717.
33. Logan, W.J. and Snyder, S.H.: 1972. *Brain Res.* 42, 413–431.
34. Henn, F.A., Goldstein, M.N., and Hamberger, H.: 1974. *Nature* 249, 663–666.
35. Iversen, L.L.: 1971. *Br. J. Pharmacol.* 41, 571–591.
36. Hayashi, T.: 1956. *Chemical Physiology of Excitation in Muscle and Nerve*, Nakayama-Shoten, Tokyo (in Japanese).
37. Eliott, K.A.C. and Jasper, H.H.: 1959. *Physiol. Rev.* 39, 383–406.
38. Krnjevic, K. and Phillis, J.W.: 1973. *J. Physiol.* 165, 274–304.
39. Straughan, D.W.: 1976. *Biochem. Soc. Spec. Publ.* 2, 213–231.
40. Curtis, D.R., Duggan, A.W., Felix, D., Johnston, G.A.R., Tebecis, A.K., and Watkins, J.C.: 1972. *Brain Res.* 41, 283–301.
41. Curtis, D.R. and Johnston, G.A.R.: 1974. *Ergeb. Physiol. Biol. Chem. Exp.* 69, 97–188.
42. Krnjevic, K.: 1974. *Physiol. Rev.* 54, 418–540.
43. DeFeudis, F.V.: 1975. *Annu. Rev. Pharmacol.* 15, 105–130.
44. Werman, R., Davidoff, R.A., and Aprison, M.H.: 1968. *J. Neurophysiol.* 31, 81–95.
45. Aprison, M.H., Shank, R.P., and Davidoff, R.A.: 1969. *Comp. Biochem. Physiol.* 28, 1345–1355.
46. Hosli, L. and Tebecis, A.K.: 1970. *Brain Res.* 11, 111–127.
47. Curtis, D.R. and Johnson, G.A.R.: 1970. *Handb. Neurochem.* 4, 115–134.
48. Eccles, J.C.: 1964. *The Physiology of Synapses*, Springer-Verlag, Berlin and Heidelberg.
49. Bowery, N.G. and Brown, D.A.: 1974. *Br. J. Pharmacol.* 50, 205–218.
50. Okamoto, K., Quastel, D.M.J., and Quastel, J.H.: 1976. *Brain Res.* 112, 147–158.
51. Takeuchi, A. and Takeuchi, N.: 1967. *J. Physiol.* 191, 575–590.
52. Takeuchi, A.: 1976. *GABA in Nervous System Function* (Roberts, E., Chase, T.N., and Tower, D.B., eds.) pp. 253–267, Raven Press, New York.
53. Krnjevic, K. and Schwartz, S.: 1967. *Exp. Brain Res.* 3, 320–336.
54. Quastel, J.H.: 1979. *Essays Med. Biochem.* 4, 1–48.

γ-Aminobutyric Acid (GABA): A Major Inhibitory Transmitter in the Vertebrate Nervous System

EUGENE ROBERTS

E. Roberts has been Chairman of the Division of Neurosciences (formerly Department of Biochemistry) at the City of Hope Research Institute, Duarte, California, since 1954 and adjunct professor of Biochemistry at the University of Southern California School of Medicine, Los Angeles, California since 1970. He has published some 275 scientific articles on his research in such fields as enzymology, metabolism, neurochemistry, cancer biochemistry, epilepsy, Huntington's disease, and schizophrenia. A major theme of his work for the last 30 years has been the exploration of the roles of γ-aminobutyric acid (GABA) in nervous system function. He currently serves as associate editor for "Brain Research," "Journal of Neurochemistry," "Biochemical Pharmacology," "Analytical Biochemistry," and "Molecular Pharmacology." He has served on many scientific advisory boards and national committees, and currently is a member of the Board of Scientific Counselors of the National Institute of Neurological and Communicative Disorders and Stroke, Scientific Advisory Committee of the Foundation for Research in Hereditary Disease, Board of Scientific Advisors of the La Jolla Cancer Research Foundation, professional advisory section of the Scottish Rite Schizophrenia Research Program, board of trustees of the California Foundation for Biochemical Research, international advisory board of the Israeli Center for Psychobiology, and Research Group on Huntington's Chorea of the World Federation of Neurology.

Recently, much attention has been devoted to the function of the γ-aminobutyric acid (GABA) system because GABAergic neurons are important for the control of activity in all parts of the vertebrate central nervous system (CNS) (1–3). Disarrangements in function of these neurons may be involved in various forms of epilepsy, in Huntington's and Parkinson's diseases, in schizophrenia, and possibly in the genesis of brainstem- and hypothalamus-related "psychosomatic disorders." Many GABA studies have dealt with a) its synthesis from glutamate, b) its release, c) its postsynaptic action, d) its synaptic inactivation by carrier-mediated transport, and e) its metabolic destruction by transamination and oxidation of the carbon chain. Of course, identification and characterization at the molecular level of the components of the receptor-ionophore-ion pump complexes and inactivation uptake systems are a *sine qua non* for our understanding of the underlying mechanisms involved in GABA action on membranes. The immunocytochemical localization of GABAergic neurons at the light and electron microscopic levels has allowed us to begin to study identified inhibitory local circuit and projection neurons in various parts of the normal vertebrate CNS and to apply our findings to the study of focal epilepsy and related problems (4).

In 1946, I was interested in the study of amino acid metabolism in normal and neoplastic tissues in experimental animals. The analytical procedures available at that time were laborious and not always specific microbiologic assays. With the introduction of two-dimensional paper chromatographic techniques, it became possible to examine the free or easily extractable ninhydrin-reactive constituents in animal tissues. Our earliest observations (5) showed that, in a given species at a particular stage of development, each normal tissue has a characteristic distribution of easily extractable ninhydrin-reactive constituents, whereas the patterns of free amino acids in many different types of transplanted and spontaneous tumors are quite similar.

In the course of this work, the free amino acid content of the C1300 transplantable neuroblastoma, then available only in solid form, was compared with several mouse brain extracts. An unidentified and previously undetected ninhydrin-reactive material was observed on chromatograms of the brain extracts. Only traces of this material had been noted in samples of urine and blood or in a large number of extracts of many other normal and neoplastic tissues previously examined. Further study of the substance revealed it to be GABA. The initial identification was based on the co-migration of the unknown with GABA on paper chromatography in three different solvent systems (6, 7), and an absolute identification of the GABA in our extracts was made by the isotope derivative method (8). Independently, GABA was found in rat brain in another laboratory (9). The first complete papers dealing with the occurrence of GABA in brain appeared in the same issue of the Journal of Biological Chemistry (7, 9).

The identification of GABA in brain extracts was facilitated by the previous finding that GABA was found to be prominent among the soluble nitrogenous components of the potato tuber (10). GABA had been found in nature long before. In 1910, Ackermann found it to be produced in putrefying mixtures as a result of bacterial action (11, 12), and, subsequently, many reports had been made about the occurrence of GABA and/or its formation in bacteria, fungi, and plants.

Even in those early days, I believed that GABA might serve an important function in brain. I remember vividly a conversation with a friend in which I jokingly said that GABA might possibly even play a role in epilepsy and schizophrenia. There now are good reasons to believe that abnormalities in the function of GABA neurons might be involved in various neurologic and psychiatric disorders.

For several years, the unique presence of relatively large amounts of GABA in the tissue of the CNS of various species remained a puzzle. The late great neurochemist, Heinrich Waelsch, once discouragingly remarked that GABA was probably a metabolic wastebasket. In the first review of the subject in 1956 (*13*), I concluded in desperation: "Perhaps the most difficult question to answer would be whether the presence in the gray matter of the central nervous system of uniquely high concentrations of γ-aminobutyric acid and the enzyme which forms it from glutamic acid has a direct or indirect connection to conduction of the nerve impulse in this tissue." However, later that year, the first suggestion that GABA might have an inhibitory function in the vertebrate nervous system came from studies in which it was found that topically applied solutions of GABA exerted inhibitory effects on electrical activity in the brain (*14, 15*). In 1957, from studies with convulsant hydrazides (*16, 17*), the suggestion was made that GABA might have an inhibitory function in the CNS. Also in 1957, definitive evidence for an inhibitory function for GABA at synapses was derived from studies that established GABA as the major factor in brain extracts responsible for the inhibitory action of these extracts on the crayfish stretch receptor system (*18*). In a short period of time, activity in this field increased greatly, and the research ranged from a study of the effects of GABA on ionic movements in single neurons to the clinical evaluation of the role of the GABA system in epilepsy, schizophrenia, and some types of mental retardation.

A memorable interdisciplinary conference was held in 1959 (*19*) and was attended by most of the individuals who had a role in opening up this field. Excitement was pervasive because everyone sensed the beginning of a new era in neuroscience. The subject of neural inhibition finally had returned to the forefront (for background and history (*19–23*)). It was obvious that future progress in the field would greatly depend on interdisciplinary efforts and that we would have to begin to learn each other's terminologies and ways of thinking. We were all optimistic that we would help each other learn enough so that effective communication soon would take place. It was a particularly heartening occasion because scientists from all over the world met in enthusiastic amity and forged long-lasting scientific and personal links.

In the years that followed, there were good times and fallow periods. For some years many doubted that GABA was a true neurotransmitter. The status of this substance now is that of a major inhibitory transmitter in the vertebrate CNS and in invertebrate central and peripheral nervous systems. Great advances in our understanding of neural circuitry based on neuroanatomic and physiologic observations gave us the opportunities and incentives to attempt to devise more specific ways of localizing GABA neurons *in situ*, eventually leading to the development of immunocytochemical tools for visualizing them at both the light and electron microscopic levels. Advances in techniques of cell fractionation, autoradiography, nerve tracing with isotopes and peroxidase, lesioning procedures, and clinical neurology led to the convocation of the second GABA meeting (*1*), seventeen years after the first one. Rapid developments in understanding general membrane and specific receptor properties, the effects of drugs on these properties and on ionic movements, and the role of the GABA system in disease states led to the organization of a third meeting devoted solely to GABA (*2*). Since that time, a number of other such meetings have been held in rapid succession.

I believe that we have more than scratched the surface of knowledge of the function of GABA in nervous system function but that many further advances remaining to be made in this field will have to take place hand in hand in a contingent fashion with other types of progress. For example, important controls in regulation of the GABA system may be exerted at points related to the availability of L-glutamic acid, the substrate for GABA synthesis in nerve endings by L-glutamic acid decarboxylase (GAD), which is the terminal enzyme in GABA biosynthesis. Experience with immunocytochemistry of the GABA system suggests that the most likely hope for a definitive resolution of the GABA-precursor question would be to achieve direct visualization, at light and electron microscopic levels, of pertinent substances or enzymes in glutamate biosynthesis that would show specific localization in neurons previously shown to be GABAergic (*e.g.*, cerebellar Purkinje cells). An urgent need also is to be able to trace the connectivities of the mainline and local circuit excitatory neurons that impinge on inhibitory local circuit and projection neurons and, in turn, whose activities the latter serve to regulate. Glutamate and aspartate, or still undefined derivatives thereof, may be major excitatory transmitters in vertebrate and invertebrate nervous systems. If this is so, eventually it will be necessary to identify the relationships of the neurons liberating these substances by techniques similar to those developed for GABAergic neurons. Currently, knowledge is lacking as to what enzymes may be rate-limiting in the biosynthesis of these amino acids in the presynaptic endings of the neurons that may liberate them or, what may be less likely, in glial cells that may "feed" precursors (*e.g.*, glutamine) to the presynaptic endings of excitatory neurons. If these enzymes were identified with certainty, their purification, development of antisera to them, and application of immunocytochemical procedures to CNS tissue for their visualization probably could be achieved rapidly. The major difficulty for the biochemist lies in the multiplicity of metabolic pathways in which these ubiquitously occurring amino acids participate.

It seems to me that in the case of glutamate, for example, either as precursor for GABA synthesis or possibly as an excitatory transmitter, itself, or as excitatory transmitter precursor, that quantitatively less important metabolic pathways, such as arginine→ornithine→glutamate and proline→Δ^1-pyrroline-4-carboxylate→

glutamate more likely would be used for transmitter or transmitter-precursor synthesis than would the more common aminotransferases (aspartic or alanine), glutaminase, or glutamic dehydrogenase. This would be in keeping with the need for specificity of control and sequestration that generally seem to be part of neuro-transmitter logistics. Of course, a variant of one or more of the generally occurring enzymes with unique control properties also might possess suitable characteristics. At this point, it seems that a systematic search for such enzymes might yield rich scientific dividends.

It would be important to have additional biochemical markers for GABAergic neurons, the only current one being GAD, which would be useful in studies of development of the GABA system and in disease states such as Huntington's disease and seizure disorders. In addition, knowledge of specific properties of the enzymes involved might furnish new pharmacologically active agents for manipulation of the GABA system or help explain the action of some old ones. Specific localization of a particular glutamate-forming enzyme (or enzymes) in the presynaptic endings of known excitatory neurons, e.g., primary afferents in spinal cord, would strengthen the possibility that glutamate is, indeed, an excitatory transmitter and also might give possibilities of developing specific tools for pharmacologic manipulation of such glutamatergic neurons. It would then be possible to determine whether or not the same pathways furnish substrate glutamate for GABAergic nerve endings and transmitter glutamate for the putative glutamatergic ones or whether different pathways are employed for them, e.g., proline as precursor for the former and arginine as precursor for the latter. If the latter were the case, immunocytochemical tools would be at hand for distinguishing the two types of endings and for the development of agents for the specific manipulation of one system or the other.

Even though it is not feasible at present to work with ultra-thin sections to visualize two antigens in the same synaptic terminal or to routinely perform experiments in which two or more antigens can be visualized on the same section, a judicious blending of approaches visualizing one antigen at a time should make the proposed tracing techniques yield valuable information. For example, the somata of neurons in the nucleus interpositus in the rat are literally studded with GABAergic terminals coming from the Purkinje cells of the cerebellar cortex; the terminals, as well as somata and dendrites of Purkinje, stellate, basket, and Golgi II cells in the cerebellar cortex, are GAD-positive after colchicine injection (4). The visualization of other antigens by this technique can be extended to the several neural regions in which GAD already has been studied in order to increase the confidence that the other antigens either are or are not found at the same sites as GAD. Work is in progress in which techniques are being sought for the visualization of more than one antigen in a single section. Visualizing different combinations of two enzymes of a series at one time would, of course, make it much easier to determine whether or not the enzymes of a particular metabolic sequence coexist in a particular neural structure.

There is much work currently being done in which potential clinical implications related to function of GABA neurons are being explored. For example, recent experiments have begun to delineate the roles of GABA neurons in control of both central sympathetic and parasympathetic outflows to the heart and vasculature. It is possible that incoordinations in specific cardioregulatory neural regions resulting from inadequate function of GABA neurons could lead to cardiac arrhythmias and large fluctuations in blood pressure that would predispose to cardiac and vascular damage (24–26). A role for GABA in cerebral vascular function is supported by the observation that specific GABA receptors and GABA related enzymes are associated with cerebral blood vessels (27–30). Our current work suggests that a non-neural, indigenous GABA system in cerebral blood vessels may be involved in the control of cerebral vascular tone. Thus, GABA probably plays an important role in the complex processes involved in normal regulation of cerebral circulation and its dysfunction. It is anticipated that work going on in many laboratories will show important roles for GABA neurons in central control of other basic bodily functions.

It is my hope that the fundamental advances being made in this field eventually will lead not only to a better understanding of nervous system function but also to the development of substances and procedures that are useful in the alleviation of some neurologic and psychologic disorders.

Acknowledgment

This work was supported in part by USPHS Grants NS 1615 and NS 12116 from the National Institute of Neurological and Communicative Disorders and Stroke.

REFERENCES

1. Roberts, E., Chase, T.N., and Tower, D.B. (eds.): 1976. *GABA in Nervous System Function*, Raven Press, New York.
2. Krogsgaard-Larsen, P., Scheel-Krüger, J., and Kofod, H. (eds.): 1979. *GABA-Neurotransmitters*, Munksgaard, Copenhagen.
3. Roberts, E.: 1980. *Nerve Cells, Transmitters, and Behavior* (Levi-Montalcini, R., ed.) pp. 163–213, Pontifical Academy of Sciences, Rome.
4. Roberts, E.: 1980. *Antiepileptic Drugs: Mechanisms of Action* (Glaser, G.H., Penry, J.K., and Woodbury, D.M., eds.) pp. 667–713, Raven Press, New York.
5. Roberts, E. and Simonsen, D.G.: 1962. *Amino Acid Pools* (Holden, J.T., ed.) pp. 284–349, Elsevier, Amsterdam.
6. Roberts, E. and Frankel, S.: 1950. *Fed. Proc.* 9, 219.
7. Roberts, E. and Frankel, S.: 1950. *J. Biol. Chem.* 187, 55–63.
8. Udenfriend, S.: 1950. *J. Biol. Chem.* 187, 65–69.
9. Awapara, J., Landua, A.J., Fuerst, R., and Seale, B.: 1950. *J. Biol. Chem.* 187, 35–39.
10. Steward, F.C., Thompson, J.F., and Dent, C.E.: 1949. *Science* 110, 439–440.

11. Ackermann, D.: 1910. *Z. Physiol. Chem.* 69, 273–281.

12. Ackermann, D.: 1910. *Z. Physiol. Chem.* 69, 1265–1272.

13. Roberts, E.: 1956. *Progress in Neurobiology*, Vol. 1 (Korey, S.R. and Nurnberger, J.I., eds.) pp. 11–25, Hoeber-Harper, New York.

14. Hayashi, T. and Suhara, R.: 1956. Abstr. the 20th Int. Physiol. Congr., Brussels, p. 410.

15. Hayashi, T. and Nagai, K.: 1956. Abstr. the 20th Int. Physiol. Congr., Brussels, p. 410.

16. Killam, K.F. and Bain, J.A.: 1957. *J. Pharmacol. Exp. Ther.* 119, 255–262.

17. Killam, K.F.: 1957. *J. Pharmacol. Exp. Ther.* 119, 263–271.

18. Bazemore, A.W., Elliott, K.A.C., and Florey, E.: 1957. *J. Neurochem.* 1, 334–339.

19. Roberts, E., Baxter, C.F., Van Harreveld, A., Wiersman, C.A.G., Adey, W.R., and Killam, K.F. (eds.): 1960. *Inhibition in the Nervous System and Gamma-Aminobutyric Acid*, Pergamon Press, Oxford.

20. Diamond, S., Balvin, R.S., and Diamond, F.R.: 1963. *Inhibition and Choice*, Harper and Row, New York.

21. Eccles, J.C.: 1969. *The Inhibitory Pathways of the Central Nervous System*, C.C. Thomas, Springfield.

22. von Euler, C., Skoglund, S., and Söderberg, U.: 1968. *Structure and Function of Inhibitory Neuronal Mechanism*, Pergamon Press, Oxford.

23. von Békésy, G.: 1967. *Sensory Inhibition*, Princeton University Press, Princeton.

24. DiMicco, J.A., Gale, K., Hamilton, B., and Gillis, R.A.: 1979. *Science* 204, 1106–1109.

25. Williford, D.J., DiMicco, J.A., and Gillis, R.A.: 1979. *Neuropharmacology* 19, 245–250.

26. Gillis, R.A., DiMicco, J.A., Williford, D.J., Hamilton, B.L., and Gale, K.N.: *Brain Res. Bull.* 5 (Suppl. 2), 303.

27. Edvinsson, L., Larsson, B., and Skärby, T.: 1980. *Brain Res.* 185, 445–448.

28. Edvinsson, L. and Krause, D.N.: 1979. *Brain Res.* 173, 89–97.

29. Fujiwara, M., Muramatsu, I., and Shibata, S.: 1975. *Br. J. Pharmacol.* 55, 561–562.

30. Krause, D.N., Wong, E., Degener, P., and Roberts, E.: 1980. *Brain Res.* 185, 51–57.

Ion Channels of Excitable Membranes

Yu. A. OVCHINNIKOV

Yu.A. Ovchinnikov was born in 1934 and graduated from Moscow State University in 1957. Since 1970, he has been the Director of Shemyakin Institute of Bioorganic Chemistry and the Head of Department of Protein and Peptide Chemistry of the Institute, Moscow, and also a member of the Academy of Sciences of the USSR In 1978, he received the Lenin Prize for studies of the molecular basis of transmembrane ion transport. His research covers various fields of biochemistry and molecular biology including the chemistry and biochemistry of peptides and proteins (structural elucidation of several biologically active proteins such as aspartate amino transferase, neurotoxins, ribosome proteins, amino acid-binding proteins, RNA-polymerase, bacteriorhodopsin, etc.); application of mass spectrometry to sequential studies of proteins; application of physico-chemical methods to study of the stereochemistry of peptides—gramicidin S, serratamolide, sporidesmolide (NMR, ORD, etc.); the physico-chemical basis of functioning of biological membranes; structure and function of ionophores such as valinomycin, etc.; membrane channels (bacteriorhodopsin, channels of exitable membranes, etc.).

M ass transport across biological membranes is a key problem of the modern physicochemical biology. Nowadays one can assume it generally accepted that among the two alternative ways of transmembrane transfer of substances and ions by mobile carrier—ionophores or by channels the latter is, as a rule, more preferable in Nature.

The existence of ion-conducting membrane channels was postulated several decades ago, in particular when studying the ways of the nerve impulse transmission in animals as well as the possible mechanism of the active transport of ions and metabolites across plasmic cell membranes ("biological pumps", etc.). The channels most efficiently complied with various theoretical concepts, especially in neurophysiology, and the channel transfer mechanism fit best the kinetic data.

The progress achieved in the membrane biology of the latter years, namely the new possibilities for the isolation and structural elucidation of the integral membrane proteins made the direct study of the structure and function of channels of various biological membranes quite possible and stimulated its impetious development.

The most promising approach to the identification and isolation of the nerve membrane channels is the usage of various neurotoxins—extremely selective inhibitors of different stages of nerve impulse transmission. Along with the derivatives of tetrodotoxin, batrachotoxin, histrionicotoxin, and other compounds of polyfunctional acyclic origin, peptide toxins from snakes, scorpions, spiders, bees, and other animals are successfully applied for these purposes. In particular in our laboratory many neurotoxins of a Middle Asian scorpion were isolated and their structures were studied; on the basis of these neurotoxins the radioactive photosensitive inhibitors of fast sodium channels of axonal membranes were obtained. By means of these neurotoxin derivatives we succeeded in the identification of the channel protein components.

One system should be mentioned whose functioning leads to changes in membrane ionic permeability and, apparently, involves specific ionic channels—namely, nerve impulse transmission through a neuromuscular junction mediated by the acetylcholine interaction with the acetylcholine receptor of the postsynaptic membrane. The most extensively studied is the acetylcholine receptor protein, isolated in a purified form from the electric organs of some rays (see Refs. *1*, *2*). Still the crucial problem remains unsolved: what is the channel moiety of the receptor. The existence of the two functional units is assumed: the receptor component which binds the mediator, and the moiety playing a role of the ionic channel, so-called ionophoric component. Using cobra neurotoxins it is possible to block the acetylcholine-binding function of the receptor, whereas some other compounds, like histrionicotoxin from Colombian frog, can selectively inhibit the ion-translocating component. However, it is not clear whether an ion-translocating function can be ascribed to any of the receptor subunits (attribution of this function excusively to the 43 KD subunit (*3*), removable on alkaline treatment, was later disproved (*4*)). Spatial relationships of the receptor and ion-translocating components are not known; even about binding of the neurotransmitter and its antagonists (*e.g.*, cobra neurotoxins) within the receptor component are available only fragmentary and indirect data.

In this respect, new possibilities open the spectral studies of acetylcholine receptor interaction with the neurotoxins bearing "spectroscopic" labels (spin, fluorescence labels, *etc.*). At our Institute a series of derivatives of neurotoxin II from the venom of Middle Asian cobra *Naja naja oxiana* has been prepared, each containing one spin or fluorescence label in the identified position of the molecule. EPR and fluorescence spectroscopy were used to monitor the interaction of labeled neurotoxins with the purified acetylcholine receptor from the *Torpedo marmorata*

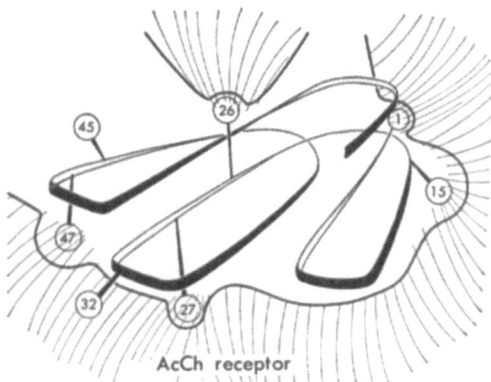

FIG. 1. A scheme for neurotoxin binding of ACh as inferred from the EPR and fluorescence studies.

electric organs (5) (this part of the study was done in cooperation with Dr. E. Karlsson from Uppsala). Thus direct experimental data were obtained about the mode of neurotoxin binding (*i.e.* about the "active site" of neurotoxins), and also shed some light on the topography of the respective binding region of the receptor.

Fluorescence measurements showed that dansyl groupings attached at positions Lys 27 and Lys 47 of neurotoxin are transferred into a hydrophobic environment upon binding with the acetylcholine receptor. The EPR data (using also the paramagnetic probe technique) revealed that the spin label at position Lys 45 is exposed into aqueous media and takes no part in the formation of the toxin-receptor complex, whereas the labels attached at N-terminal amino group of Leu 1 and to the side chains of Lys 15, Lys 26, Lys 27, His 32, and Lys 47 to larger or lesser degree are involved in binding the acetylcholine receptor (the interaction is most pronounced for Lys 27 spin label). This means that considerable area of the neurotoxin surface comes into contact with the receptor (Fig. 1). A comparison of binding constants for native neurotoxin and its derivatives suggests the role of charged and hydrophobic residues of the neurotoxin in association with the receptor.

A characteristic feature of the postsynaptically acting neurotoxins isolated from snake venoms is the rigidity of their spatial structure. Their conformation is essentially the same for various homologous neurotoxins both in solution and in crystalline state, and is preserved at those modifications which were utilized for incorporating the spin and fluorescence labels. Therefore it might be concluded that the distances between the labels found experimentally by EPR for di-labeled derivatives of neurotoxin II in solution (6) or deduced from the X-ray structure of the related neurotoxin (7), should not greatly change in the neurotoxin complex with the acetylcholine receptor and thus it is possible to get a rough estimation of the dimensions of the acetylcholine receptor binding site for neurotoxins.

The distances between various labels range from 12 to 35 Å. Since the diameter of one acetylcholine receptor molecule, which binds two neurotoxin molecules, is about 80 Å (2), it becomes obvious that neurotoxins cover a considerable portion of the receptor surface and have contacts not only with two α-subunits, the sites for acetylcholine binding, but with the neighbouring subunits. Further studies, utilizing also photoactivable neurotoxin derivatives (8, 9), may help to determine the subunit topography and possibly to identify the ion-translocating structure.

As one of the convenient tools for the study of sodium channels of electrically excitable membranes the neurotoxin M_{10} from the venom of a Middle Asian scorpion *Buthus eupeus* was used. M_{10} consists of 65 amino acid residues and comprises 4 intramolecular disulfide bridges and the high quantity of lysine and aspartic acid residues, being characterized by the absence of threonine, methionine, and histidine residues (10).

According to voltage clamp data the neurotoxin M_{10} interacts with the nerve membrane (equilibrium constant 10–100 nm); its affinity of the receptor decreases by the membrane depolarization just as for other toxins of scorpion venom (11).

The usage of the radioactive photosensitive derivatives of scorpion toxin could be effective for the identification of its receptor molecular organization. 2,4–Dinitro-5-fluorophenylazide was used for the preparation of such derivative of the toxin M_{10}. The separation of the reaction mixture by means of ion-exchange chromatography resulted in the isolation of the monosubstituted derivative of M_{10}, (DNPA-M_{10}), with the high yield. The toxin derivative possesses the characteristic absorptions at 355 and 415 nm, and it might be photoactivated at long wavelengths thereby avoiding the photodestruction of common aromatic protein groups (12).

After the iodination by means of lactoperoxidase method the obtained photosensitive analog of the neurotoxin M_{10} possessed the high level of biological activity

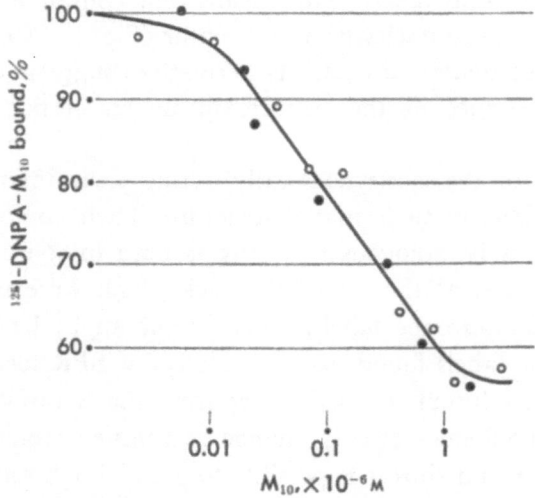

FIG. 2. Displacement of ^{125}I-DNPA-M_{10} by native toxin M_{10} upon photoaffinity modification of rat brain synaptosomes.

that was shown by the voltage clamp tests on neuroblastoma cell membranes. Thus it was found that the scorpion toxin derivative unreversibly modifies sodium channels, *i.e.* its radicals generated at UV-light irradiation interact with the regulatory component of the channel (*12*).

Upon the competition study of the modified neurotoxin for the membrane receptors of rat brain synaptosomes with the native toxin it was shown that the specific binding of ^{125}I-DNPA-M_{10} does not exceed more than 50% and the apparent equilibrium dissociation constant of the toxin-receptor complex is about 10^{-7} M (Fig. 2). The number of binding sites is within the range from 80 to 120 fmol/mg of synaptosomal protein (*13*) assuming the equimolar binding of the toxin with the receptor.

The sea anemone *Anemonia sulcata* toxin II also inhibits the reception of the photolabeled derivative of the M_{10} toxin with the apparent dissociation constant 200 nM. On the other hand calcium ions within the range of concentrations 0–10 mM completely suppress the reception of the toxin derivative and simultaneously increase the level of non-specific absorption.

The nature of the receptors of scorpion neurotoxins was established by polyacrylamide gel electrophoresis of photoaffinity labeled membranes in the presence of sodium dodecylsulfate. The identification of the recepting membrane components was performed by the difference analysis of radioactivity profiles of electrophoregrammes that were obtained in the absence or in the presence of the excess of the native toxin M_{10} in the control run. In case of rat brain synaptosomes the presence of two specifically binding zones was detected. Both of them could be specifically suppressed by the excess of the native toxin, molecular weights of these

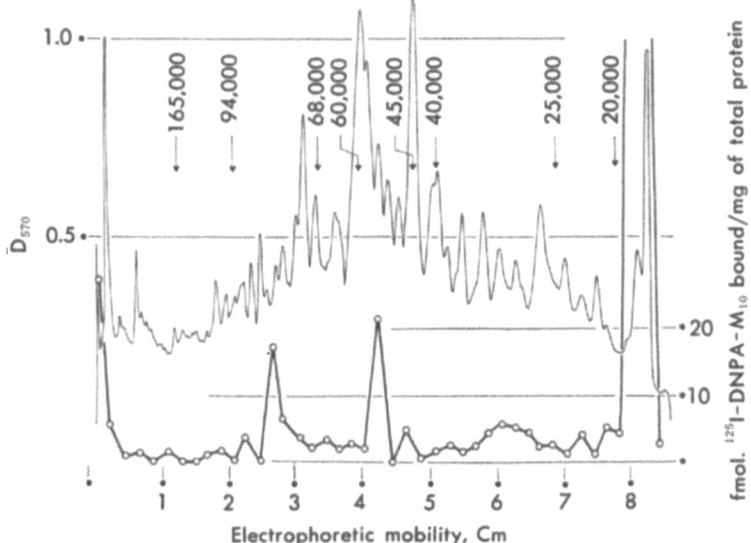

FIG. 3. The electrophoretical analysis of rat brain synaptosomes photomodified by ^{125}I-DNPA-M_{10}.

zones being equal to 84 and 58 KD (Fig. 3). The true molecular weights of the receptor components are equal to 77 and 51 KD in assumption that the binding of the photoaffinity toxin with the receptor is equimolar and its own molecular weight is about 7 KD.

It is interesting to note that the photomodification of the neuroblastoma cells resulted in the binding of the photosensitive analog of the M_{10} toxin by two receptor components of the molecular weights 73 and 52 KD.

We believe that the modification of two membrane components could be accounted for their spatial proximity relative to the point of fixation of the scorpion toxin in the membrane. In addition one of the components could belong also to the tetrodotoxin receptor, *i.e.* to be a component of the ion-conducting part of the sodium channel.

To elucidate this question the walking leg nerve plasma membrane from crabs *Paralithodes camchatica* were used. The preparative method was developed to isolate the membranes from crab nerves that contain more than 40 protein components of the molecular weights from 12 to 260 KD and possess the tetrodotoxin binding capacity about 4 pmol of tetrodotoxin/mg protein (*14*). Upon the photomodification of crab nerve plasma membranes by the photosensitive analog of the M_{10} toxin the label was specifically incorporated in two proteins of the molecular weights equal to 78 and 51 KD. The modified membrane preparation was solubilized by Lubrol PX and then subjected to sucrose gradient density sedimentation. Here it was found that the covalent toxin-receptor complex is present in two fractions with sedimentation coefficients 9, 5, and 16 S. Both fractions also contained the receptor components of tetrodotoxin.

The further purification of tetrodotoxin receptor by means of ion-exchange chromatography led to the isolation of the membrane receptor complex, consisting of 5 protein subunits with molecular weight $\sim$230, 85, 77, 70, and 69 KD (*15*). The analysis of the radioactivity distribution evidences that only the component of the molecular weight 77 KD is specifically bound with the photoactivated derivative of the scorpion toxin. Apparently the true receptor component for the scorpion neurotoxin is presented by the protein of the molecular weight 77 KD (*16*).

We hope that the forthcoming investigation of the membrane receptor complex and its components will shed light upon the molecular organization and functioning of the channels of the electroexcitable membrane.

REFERENCES

1. Heidmann, T. and Changeux, J-P.: 1978. *Annu. Biochem.* 47, 317–357.
2. Barnard, E.A., Lamprecht, J., Lo, M., Nockes, E., Simikawa, K., Cavanagh, J., and Dolly, J.O.: 1980. *Synaptic Constituents in Health and Disease* (Brzin, M., Sket, D., and Bachelard, H., eds.) pp. 224–240, Mladinska Knjiga-Pergamon Press, Ljubljana-Oxford.

3. Sobel, A., Heidmann, T., Hofler, J., and Changeux, J-P.: 1978. *Proc. Natl. Acad. Sci. USA* 75, 510–514.

4. Neubig, R.R., Krodel, E.K., Boyd, N.D., and Cohen, J.B.: 1979. *Proc. Natl. Acad. Sci. USA* 76, 690–694.

5. Tsetlin, V.I., Karlsson, E., Arseniev, A.S., Utkin, Yu.N., Surin, A.M., Pashkov, V.S., Pluzhnikov, K.A., Ivanov, V.T., Bystrov, V.F., and Ovchinnikov, Yu.A.: 1979. *FEBS Lett.* 106, 47–52.

6. Ivanov, V.T., Tsetlin, V.I., Karlsson, E., Arseniev, A.S., Utkin, Yu.N., Pashkov, V.S., Surin, A.M., Pluzhnikov, K.A., and Bystrov, V.F.: 1980. *Natural Toxins* (Eaker, D. and Wadström, T., eds.) pp. 523–530, Pergamon Press, Oxford.

7. Kimball, M.R., Sato, A., Richardson, J.S., Rosen, L.S., and Low, B.W.: 1979. *Biochem. Biophys. Res. Commun.* 88, 950–959.

8. Witzemann, V. and Raftery, M.A.: 1978. *Biochem. Biophys. Res. Commun.* 85, 623–631.

9. Hucho, F.: 1979. *FEBS Lett.* 103, 27–32.

10. Grishin, E.V., Soldatova, L.N., Shakhparonov, M.I., and Kazakov, V.K.: 1980. *Bioorg. Khim.* 6, 714–723.

11. Mozhayeva, G.N., Naumov, A.P., Nosyzeva, E.D., and Grishin, E.V.: 1980. *Biochim. Biophys. Acta* 597, 587–602.

12. Grishin, E.V., Soldatov, N.M., Ovchinnikov, Yu.A., Mozhayeva, G.N., Naumov, A.P., Zubov, A.N., and Nisman, B.Ch.: 1980. *Bioorg. Khim.* 6, 724–730.

13. Grishin, E.V., Soldatov, N.M., and Ovchinnikov, Yu.A.: 1980. *Bioorg. Khim.* 6, 914–922.

14. Grishin, E.V., Kovalenko, V.A., Pashkov, V.N., and Litvinov, I.S.: 1979. Abstr. the 1st Soviet-Swiss Symposium "Biological Membranes: Structure and Function," Tbilisi, p. 67.

15. Soldatov, N.M., Kovalenko, V.A., and Grishin, E.V.: 1980. Abstr. the III USSR-FRG Symposium of Chemistry of Peptides and Proteins, Makhachkala, p. 28.

16. Kovalenko, V.A., Pashkov, V.N., and Grishin, E.V.: 1980. Abstr. the V All Union Symposium on Chemistry and Physics of Peptides and Proteins, Baku, p. 14.

Precursors and Intermediates in Enkephalin Biosynthesis

SIDNEY UDENFRIEND

S. UDENFRIEND was born in New York City on April 5, 1918. He studied biochemistry at New York University where he received the Ph.D. degree in 1948. He was a instructor at Washington University Medical School (1948) and then moved to the NIH in 1950 as a biochemist in the Laboratory of Chemical Pharmacology (National Heart Institute), where he became Head of the Section on Cellular Pharmacology in 1953, and Chief of the Laboratory of Chemical Pharmacology in 1956. Since 1968, he has been Director of the Roche Institute of Molecular Biology. He currently serves on the editorial boards of many scientific journals and on the advisory boards of academic institutes. He is a member of the National Academy of Sciences (1971). He was elected to Honorary Membership in the Japanese Biochemical Society (1973) and the Japanese Pharmacological Society (1977). He is also a member of the American Academy of Arts and Sciences (1974), a Fellow of the New York Academy of Sciences (1977) and an Honorary Member of the Congress of International Neuropsychopharmacologists (1980). In 1979 he became a member of the board of governors of the Weizmann Institute of Sciences. To date he has received many awards: Arthur S. Fleming Award (1958), Hillebrand Award of the American Chemical Society (1962), Superior Service Award, Department of Health, Education and Welfare (1965), Van Slyke Award (1967), Gairdner Award (1967), Ames Award (American Association of Clinical Chemists (1969)), City of Hope Research Award (1975), Townsend Harris Medal (1979), and Rudolf Virchow Gold Medal Award (1979).

M any biologically active polypeptides are present at concentrations below a microgram per gram of tissue and biological assays that are used to study them are orders of magnitude more sensitive than the more commonly utilized chemical procedures. Hence, it is often not possible to purify sufficient quantities of polypeptides for adequate chemical characterization, including evaluation of purity, molecular weight determination, amino acid analysis, preparation of peptide fragments, and sequencing. We have introduced fluorometric and chromatographic procedures for isolation and characterization of polypeptides, applicable at the picomole level, and have applied them to precursors and intermediates in the biosynthesis of β-endorphin and the enkephalins.

Fluorescamine, the reagent we have used for monitoring elutions from chromatographic columns is ideally suited for this purpose since, on mixing with a solution containing a primary amine, a stable fluorophor is produced within milliseconds, and excess reagent is concomitantly hydrolyzed to nonreactive, water-soluble products within minutes. Both the reagent and its hydrolyzed products are nonfluorescent and the reaction goes to near completion, producing a monosubstituted derivative with each primary amino group. This senstiive fluorometric detection system for amino acids, peptides, and proteins has been coupled to high performance liquid chromatography (HPLC), along with bioassay, radioimmunoassay and/or radioreceptor assay, for the identification and quantitation of specific peptides, as well as for their isolation in pure form (1). More recently, this combination of methodologies has been applied to the purification of proteins as large as 300K daltons (2).

Before enkephalin was discovered, our laboratory had developed specific fluorometric assays for peptides, with a sensitivity in the picomole range, and applied them to various aspects of peptide chemistry, including the assay of oxytocin and vasopressin in the pituitary gland (3) and of carnosine (β-alanyl histidine) in the olfactory bulb (4).

Following the discovery of the enkephalins (5), we applied our methods to the opioid peptides and initiated studies on their biosynthesis. Studies in other laboratories had already shown that the pentapeptide sequence of Met-enkephalin was present in the 3,000 molecular weight pituitary polypeptide, β-endorphin. β-Endorphin was itself known to be present as a sequence in a still larger peptide, β-lipotropin (β-LPH). Other congeners of enkephalin and endorphin had also been reported in pituitary extracts of large animals (cow, sheep, pig). At that point in time, it appeared that the pentapeptide sequence was derived from a larger precursor molecule in the pituitary that led first to β-lipotropin, then to β-endorphin, and then to enkephalin.

To investigate such a pathway and its physiological significance our studies were directed to the rat, since this animal has been used traditionally for experiments involving opiate drugs. While the use of the rat was ideal from a biological standpoint, the amounts of material available were comparatively small. However, only in such small animals can carefully controlled biochemical and pharmacological studies be carried out with precision. Tissues can be removed and extracted so rapidly that little or no autolysis takes place. In contrast, when accumulating large amounts of tissue from the abattoir, autolysis is inevitable. The quantitative and even qualitative significance of the residual peptides isolated from such autolyzed material is questionable. The combination of microfluorometric assay and HPLC coupled with a receptor-binding assay utilizing neuro-blastoma-glioma hybrid cells (6) provided the sensitivity necessary for the isolation and characterization of the small amounts of opioid peptides present in rat pituitary gland.

The initial chromatographic separation of pituitary extracts was by molecular size on a Sephadex G-75 column, with the appearance of several distinct areas of opioid activity. The largest of the intermediate size molecular weight opioid sub-stances in the pituitary gland was eluted at the position of the β-endorphin marker. To obtain enough of this material for isolation, 200 rat pituitaries were employed. Following chromatography on Sephadex G-75, the area corresponding to β-endor-phin was extracted and subjected to two steps of HPLC, yielding a single sharp peak of activity. Amino acid assay and comparison of tryptic fragments showed that rat β-endorphin was identical to β-endorphin from large mammals (7). Suffi-cient amounts of pure β-lipotropin (11K daltons) from forty rat pituitaries were collected for characterization. Most interestingly, however, a fifth area of activity was revealed that was eluted just after the void volume. From its elution on gels the large opioid protein was estimated to have a molecular weight of approximately 30,000. When an extract containing this protein was digested with trypsin and the fragments resolved by HPLC, the only biologically active substance generated was the nonapeptide β-LPH$_{61-69}$, the same peptide that is formed when β-LPH or β-endorphin is treated with trypsin. At about the same time Mains et al. (8), utiliz-ing cultured pituitary cells and radiolabeled amino acids, obtained evidence of a protein of about 31K daltons that reacted with antibodies to both β-endorphin and ACTH. We subsequently showed that the partially purified rat protein also interacts with antibody to ACTH. To indicate the precursor relationship to both ACTH and opioid peptides, the name pro-opiocortin was proposed for this sub-stance (9). We have purified pro-opiocortin from camel pituitaries to homogeneity and characterized it (7). Rat proopiocortin has now been cloned and its sequence determined by DNA sequencing (10). Since no evidence for a physiologic mech-anism for converting β-endorphin to Met-enkephalin has been obtained, it is generally believed that in the pituitary gland, β-endorphin is the final opioid peptide in the synthetic pathway from the 30K daltons precursor. β-Endorphin and ACTH appear to be released simultaneously from the pituitary gland in re-sponse to stress; both may be considered to be hormones. Having been unable to demonstrate the conversion of β-endorphin to enkephalin in the pituitary gland, we turned to other tissues. Since the enkephalins were originally isolated from the brain, we explored this area first. Utilizing the same type of methodology, we demonstrated large amounts of the enkephalins as well as two large enkephalin-containing-proteins—one of about 40K daltons and another of the order of 100K daltons (11). Significantly, none of the polypeptides of the series pro-opiocortin-β-lipotropin-β-endorphin could be detected; furthermore, the residual peptides with opioid activity that resulted from trypsin digestion of the two large

Peptide F

$\overset{1}{\text{Tyr}}$- Gly-Gly-Phe-$\overset{5}{\text{Met}}$-Lys-Lys-Met-Asp-$\overset{10}{\text{Glu}}$-Leu-Tyr-Pro-Leu-$\overset{15}{\text{Glu}}$-

Val-Glu-Glu-Glu-$\overset{20}{\text{Ala}}$-Asn-Gly-Gly-Glu-$\overset{25}{\text{Val}}$-Leu-Gly-Lys-Arg-$\overset{30}{\text{Tyr}}$-

$\overset{34}{\text{Gly}}$-Gly-Phe-Met

Peptide I

$\overset{1}{\text{Ser}}$-Pro-Thr-Leu-$\overset{5}{\text{Glu}}$-Asp-Glu-His-Lys-$\overset{10}{\text{Glu}}$-Leu-Gln-Lys-Arg-$\overset{15}{\text{Tyr}}$-

Gly-Gly-Phe-Met-$\overset{20}{\text{Arg}}$-Arg-Val-Gly-Arg-$\overset{25}{\text{Pro}}$-Glu-Trp-Trp-Met-$\overset{30}{\text{Asp}}$-

Tyr-Gln-Lys-Arg-$\overset{35}{\text{Tyr}}$-Gly-Gly-Phe-$\overset{39}{\text{Leu}}$

FIG. 1.   Complete amino acid sequence of adrenal peptide F and adrenal peptide I.

proteins were not the characteristic $\beta\text{-LPH}_{61-69}$ of the endorphin pathway of biosynthesis in the pituitary gland.

While working with extracts of the brain, we heard of reports that adrenal medulla contains large amounts of material that crossreacts immunologically with enkephalin (12). On investigation we found a great number of large and intermediate sized polypeptides that contained the enkephalin sequence; presumably enkephalin precursors. All these polypeptides were in the chromaffin granules of the adrenal medulla. We used beef adrenal medulla to obtain sufficient material for isolation and purification. When acetic acid extracts of adrenal chromaffin granules were subjected to gel filtration a large number of opioid-containing polypeptides were found. These were arbitrarily divided into five classes according to size (range: ca. 20,000 to 1,000) (13). Each class was further purified. Among the smallest substances, we found in addition to enkephalin, the hexapeptide Met-enkephalin-Arg (6), Met-enkephalin-Lys (6), and Leu-enkephalin-Arg (6), and the two heptapeptides Met-enkephalin-Arg (6) -Phe (7) and Met-enkephalin-Arg (6) -Arg (7). There were a number of polypeptides in the range of 2.2K to 5.0K daltons. Two of these were purified to homogeneity and characterized (14): one 3.8K daltons contains two Met-enkephalin sequences; the other, 4,900 daltons contains one Met-enkephalin sequence as well as a second Leu-enkephalin sequence. The complete sequences of the 3.8K dalton and 4.9K dalton peptides are shown in Fig. 1.

A series of larger and larger polypeptides, summarized in Fig. 2, were isolated, all containing one or more enkephalin sequences (15). Eight K, 14 K, and 22 K dalton polypeptides have been purified to homogeneity and characterized with respect to amino acid analysis and end-group analysis. It should be noted that the 14 K polypeptide contains three Met-enkephalin sequences, two of them internalized, and the 22K polypeptide contains five Met-enkephalin sequences, four of

Enkephalin Precursors

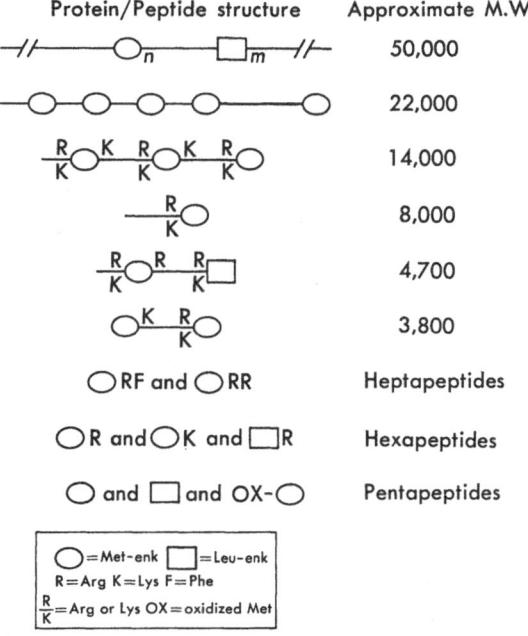

FIG. 2.   Structures of enkephalin-containing polypeptides: n is 6 to 7, m is 1, as determined by their ratio. (From Lewis *et al.* (*15*))

which are internalized. Polypeptides larger than 50 K have been partially purified and shown to contain several Met-enkephalin sequences and one Leu-enkephalin sequence (*15*). Interestingly, Hughes *et al.* (*16*) have described approximately 50 K and 80 K enkephalin-containing polypeptides in extracts of guinea pig brain. These may well be the same as those in the adrenal medulla. In our original report on enkephalin-containing polypeptides in brain (*11*), the molecular weights (100 K and 40 K) were merely estimates based on gel filtration. More recently we have reexamined guinea big brain extracts and have found that the large enkephalin containing polypeptides are closer to 80 K and 50 K as reported by Hughes *et al.* (*16*). In more recent studies we have found that the larger protein is an appregate of the ∼50 K protein.

The presence of enkephalin sequences in a series of polypeptides of decreasing size, ending with the pentapeptide enkephalins, suggests that all of these compounds represent a biosynthetic pathway starting from the largest (∼50 K) down to the smallest. Aside from such circumstantial evidence, we have gathered some factual support for such a pathway from pulse-chase experiments (*17*). More definitive experiments are underway to establish such a biosynthetic route.

An unexpected finding of our studies was that these putative precursors contain multiple copies of enkephalin per molecule. To my knowledge, this is a most unique occurrence in the biosynthesis of physiologically active peptides. It represents a heretofore unrecognized means of amplification of a functional peptide. This type of amplification is inherent in the DNA but only becomes apparent at the post-translational level. How such multiple sequences are introduced into a larger polypeptide sequence is an interesting challenge in genetics and development. We are in the process of isolating a messenger for this biosynthetic pathway and through recombinant DNA methodology, hope to establish the structures of the large precursor by DNA sequencing. Another surprising finding was the discovery of both Leu- and Met-enkephalin sequences in the same polypeptide. The ratio of approximately 7/1 (Met-/Leu-enkephalin) is in accord with the ratio of the two peptides found in tissue extracts. It would seem, therefore, logical to theorize that the ratio is set in the chromosome. The presence of both Met- and Leu-enkephalin in a common precursor also makes it unlikely that the two enkephalins are formed and act in separate neurons as has been suggested (18). It may well be that the Leu-enkephalin arose by a one-base mutation from Met-enkephalin. If so, then Leu-enkephalin may merely represent another active form of enkephalin or else that, having been formed by such a mutation, the Leu-enkephalin sequence then goes on to assume a distinct role of its own. The dynorphin reported by Goldstein (19) and the α-neoendorphin reported by Matsuo (20) may be indicative of unique activities for Leu-enkephalin polypeptides.

It is now apparent that there are two distinct genetic mechanisms for biosynthesis of the Met-enkephalin sequence—one in the pituitary, beginning with pro-opiocortin, followed by β-lipotropin and then β-endorphin. This pathway, although it leads to a Met-enkephalin sequence, does not appear to produce free enkephalin under physiologic conditions. The second pathway for enkephalin synthesis is in brain and peripheral tissues (i.e., adrenal gland and intestine). This pathway appears to start with an ∼50 Kdalton precursor, continuing on through a series of intermediates containing varying numbers of enkephalin sequences and ultimately to the free enkephalins. What is the significance of these findings? With respect to the pituitary, it would appear that either β-endorphin or a related larger enkephalin-containing polypeptide is a hormone secreted from the pituitary gland along with ACTH, since one may speculate that, like ACTH, β-endorphin is secreted in stress situations and stimulates peripheral end organs. The central and peripheral pathway may represent a mechanism designed for the production of free enkephalins. This possibility is quite likely for central neurons where a peptide with a limited life, such as enkephalin, may act as a neuromodulator. However, the adrenal gland, like the pituitary, is also an endocrine gland. Endocrine glands supposedly produce hormones for export to other target organs. Conceivably, the enkephalins could play such a role. On the other hand, it is also possible that some of the larger enkephalin-containing polypeptides represent the true adrenal opioid hormone. It should be noted that even the heptapeptide Met-enkephalin-Arg (6) -Phe (7) is more active and apparently more stable than free enkephalin, lending credence to the theory that some of the larger intermediates possess even more activity and therefore serve as the adrenal hormone (*21*).

To summarize, the enkephalin sequence Tyr-Gly-Gly-Phe-Met, when it occurs at the amino terminus of a peptide, is a signal for an opiate receptor. However, this signal sequence apparently occurs in peptides of varying size for different purposes—some centrally, some peripherally—some as hormones, and some as neuromodulators.

REFERENCES

1. Stein, S.: 1978. *Peptides in Neurobiology* (Gainer, H., ed.) pp. 9–37, Plenum Press, New York.
2. Lewis, R.V., Fallon, A., Stein, S., Gibson, K.D., and Udenfriend, S.: 1980. *Anal. Biochem.* 104, 153–159.
3. Gruber, K.A., Stein, S., Brink, L., Radhakrishnan, A., and Udenfriend, S.: 1976. *Proc. Natl. Acad. Sci. USA* 73, 1314–1318.
4. Harding, J., Graziadei, P.P.C., Monti Graziadei, G.A., and Margolis, F.L.: 1977. *Brain Res.* 132, 11–28.
5. Hughes, J., Smith, T.W., Kosterlitz, H.W., Fothergill, L., Morgan, B.A., and Morris, H.R.: 1975. *Nature* 258, 577–579.
6. Gerber, L.D., Stein, S., Rubinstein, M., Wideman, J., and Udenfriend, S.: 1978. *Brain Res.* 151, 117–126.
7. Rubinstein, M., Stein, S., and Udenfriend, S.: 1977. *Proc. Natl. Acad. Sci. USA* 74, 4969–4972.
8. Mains, R.E., Eipper, B.A., and Ling, N.: 1977. *Proc. Natl. Acad. Sci. USA* 74, 3014–3018.
9. Rubinstein, M., Stein, S., and Udenfriend, S.: 1978. *Proc. Natl. Acad. Sci. USA* 75, 669–671.
10. Nakanishi, S., Inoue, A., Kita, T., Nakamura, M., Chang, A.C.Y., Cohen, S.N., and Numa, S.: 1979. *Nature* 278, 423–427.
11. Lewis, R.V., Stein, S., Gerber, L.D., Rubinstein, M., and Udenfriend, S.: 1978. *Proc. Natl. Acad. Sci. USA* 75, 4021–4023.
12. Schultzberg, M., Hökfeld, T., Lundberg, J., Terenius, L., Elfirm, L-G., and Elde, R.: 1978. *Acta Physiol. Scand.* 103, 475–477.
13. Lewis, R.V., Stern, A.S., Rossier, J., Stein, S., and Udenfriend, S.: 1979. *Biochem. Biophys. Res. Commun.* 89, 822–829.
14. Jones, B.N., Stern, A.S., Lewis, R.V., Kimura, S., Stein, S., Udenfriend, S., and Shively, J.E.: 1980. *Arch. Biochem. Biophys.* 204, 392–395.
15. Lewis, R.V., Stern, A.S., Kimura, S., Rossier, J., Stein, S., and Udenfriend, S.: 1980. *Science* 208, 1459–1461.
16. Dell, A., Etienne, T., Morris, H.R., Beaumont, A., Burrell, R., and Hughes, J.: 1979. *Molecular Endocrinology* (MacIntyre, I. and Szelke, M., eds.) pp. 91–97, Elsevier/North-Holland, Amsterdam.

17. Rossier, L., Trifaro, J., Lewis, R.V., Lee, R.W.H., Stern, A., Kimura, S., Stein, S., and Udenfriend, S.: *Proc. Natl. Acad. Sci. USA*, in press.
18. Larsson, L.I., Childers, S., and Snyder, S.H.: 1979. *Nature* 282, 407.
19. Goldstein, A., Tachibana, S., Lowney, L.I., Hunkapiller, M.W., and Hood, L.E.: 1979. *Proc. Natl. Acad. Sci. USA* 76, 6666–6670.
20. Kangawa, K., Matsuo, H., and Igarashi, M.: 1979. *Biochem. Biophys. Res. Commun.* 86, 153–160.
21. Inturrisi, C.E., Umans, J.G., Wolff, D., Stern, A.S., Lewis, R.V., Stein, S., and Udenfriend, S.: 1980. *Proc. Natl. Acad. Sci. USA* 77, 5512–5514.

Endocrine Control of the Prostate

BERTIL HÖGBERG

K.B. Högberg was born on September 7, 1918, and educated at Stockholm Institute of Technology and University of Stockholm. During 1939–1946 he was Head of the Physical Chemistry Department of the Institute of Organic Chemistry and Biochemistry, Stockholm University, 1946~ AB Leo Helsingborg, Sweden, 1946–1973 Head of the Department of Biochemistry and Director of Research. Since 1973, he has been Vice President R&D and adjunct professor of Medical Biochemistry at the Karolinska Institutet, Stockholm. Fields of his present major scientific interest are steroid hormone receptors in human malignancy, the mode of action of TCDD and related compounds in chemical carcinogenesis, and inhibitors of chemical carcinogenesis. He received the following honors and awards: Honorary Doctor of Medicine, Karolinska Institutet, Stockholm, 1960. Regnells Prize, The Swedish Medical Association, 1958 and 1962. Elected member of the Swedish Medical Association (1961), Swedish Academy of Pharmaceutical Sciences (1963), New York Academy of Sciences (1971), and European Academy of Allergology and Clinical Immunology (1971).

I selected the prostate as my subject since we have focused a great deal of our interest to this gland over the last thirty years. The normal prostate gland is very responsive to androgen stimulation. The gland atrophies after castration but regains its normal size and characteristics after androgen administration. Cancer of the prostate is the most common malignant disease of the urinary tract, with a reported incidence of 50% in males over 80 years of age. Racial and geographical differences in the distribution are noticed but they remain unexplained. Androgenic secretions are necessary for the development of prostatic carcinoma but the endocrinological studies do not stress any statistical differences in the hormonal status of prostatic cancer men. Since the pioneering work of Huggins and Hodges (1) various programs of endocrine therapy have been employed for the treatment of patients with cancer of the prostate. These programs have included castration, estrogen administration, hypophysectomy, or adrenalectomy, and treatment with corticosteroids or antiandrogens. Adrenalectomy and hypophysectomy, or their combination were tried in the 1950's and the 1960's and although a small number of patients responded, the magnitude of the procedure relative to its minimal response rate does not presently seem to warrant its general use.

Also non-endocrine methods such as radiotherapy as well as non-hormonal chemotherapy have been found to be superior to adrenalectomy and hypophysectomy.

Even less is known concerning the pathogenesis of tumor development in the gland. For no evident reason one prostatic tumor will remain localized for decades while another spreads rapidly in spite of intensive treatment. Prominent urologists disagree on what constitutes adequate treatment. Nevertheless anti-androgenic therapy continues to be the mainstay for prostatic carcinoma.

While hormonal control was a real advance, the type of estrogen used and the mode of administration did not change appreciably until Flocks reported results with the intravenous administration of diethylstilbestrol diphosphate (2). Efforts to maintain adequate, constant therapeutic estrogen levels have not been achieved successfully over long periods on oral medication because of the unreliability of patients in self-medication and because of rapid excretion. Intravenous administration is expensive and is effective in maintaining levels only as long as administration continues. Pellets implantation has not proved to be generally satisfactory.

Delivery of active drug substances often can be improved substantially through application of the pro-drug concept either by altering their physical properties and/or by modification of their chemistry. Any systemic approach in considering possible application of the pro-drug concept to any active agent must include a review of various mechanisms which can be employed in retrieving the agent from its inactive form. A wide variety of pro-drug chemical linkages are known to be enzymatically hydrolyzed. If the bioactivation is inhibited a slow released drug effect can be obtained.

A large number of polymeric phosphoric acid esters and amides of polyphenols have been synthesized in our laboratories as inhibitors of several enzymes, e.g. hyaluronidase and phosphatases (3–6). One of the compounds, polyestradiol phosphate (Estradurin[R]) has found an application in the treatment of prostatic cancer (7). This compound is a water soluble polymerization product, in which several estradiol molecules are firmly united with phosphate ester bonds. It has an average molecular weight of about 4,400 (ultracentrifugation). As a strong inhibitor of phosphatases the polymer is very slowly hydrolysed resulting in a sustained release of the natural estrogen (estradiol-17β) with stable blood and tissue levels over a long period of time.

Endocrine therapy is effective in 60–70% of all patients with prostatic carcinoma, and in those responding the duration of response varies from individual to individual. Poorly differentiated tumors are often resistant to endocrine manipulations. Moreover, relapses occur sooner or later despite continued hormonal treatment. It is not known why certain cases do not respond to anti-androgenic treatment or why they first respond but later become resistant.

Lack of selective drug action against tumor cells, development of drug resistance in treated neoplasms and lack of weakness of host defence, are perhaps the most formidable difficulties facing tumor chemotherapy.

It has been proposed by several authors that the tumor inhibiting agents might be more effective, if the functional groups were present in a natural carrier such as an amino acid, sugar, or a nucleic acid base. Transport might then be favored, and the action of such compounds might gain in selectivity provided that the tumor cells are more permeable to the chosen "conductor" than the host cells.

From relatively recent studies on the molecular biology, we have arrived at a number of general concepts for the mechanism of hormone action. The most important factor for the specific retention of steroid hormones in their target cells is the presence of a cytoplasmic receptor protein in these cells, that binds the hormone with high affinity. The steroid-receptor complex then undergoes a temperature-dependent physical change and is transported into the nucleus. There the activated complex is bound to acceptor substances on chromatin thereby controlling the formation of specific mRNA and thus the synthesis of specific proteins (for review see Ref. 8).

The discovery of selective accumulation of steroids in special "target" tissues offered a rational approach for utilizing estrogens and other steroid hormones as carriers for cytotoxic agents. In 1961 a cancer chemotherapy program that included syntheses of several types of complexes between steroids and cytotoxic agents was started at our laboratoties.

The cytotoxic and antitumor effectiveness of alkylating agents attracted great interest, and much work was devoted to synthesizing steroid-nitrogen mustard derivatives. Initially, before the recognition of the structural requirements for receptor binding, many steroid derivatives were synthesized which later were shown not to bind specifically to "their" receptors, or to be taken up by the intended target organ.

The following aspects need to be considered for the mechanism of action for cytotoxicity and specificity. The antitumor effect could be exclusively dependent upon binding of the drug to the hormone receptor. Inactivation of the receptors following binding could lead to cell dysfunction; in addition, binding to the receptor could be the basis for tissue or tumor specificity of the drug. In this sense the drug would be acting as a potent antihormone.

One of the compounds synthesized was estramustine phosphate, a nornitrogen mustard carbamate derivative of estradiol-17β esterified with phosphoric acid to achieve water solubility. Estramustine phosphate (Estracyt[R]) was introduced as a therapeutic agent in the treatment of prostatic carcinoma in 1966 and appears to be a safe and effective treatment for patients with prostatic carcinoma who have failed to respond to conventional therapy. On the basis of long-term responses, it may also have a role in the treatment of earlier stages of prostatic cancer (for review see Ref. *9*).

In the search for the mechanism of action of estramustine phosphate it was found that its dephosphorylated metabolite estramustine is bound to a specific protein found in its highest concentration in rat ventral prostate. The protein has a molecular weight of about 50,000, sedimented at 3.7 S on a sucrose gradient and focused around pH 5. Estramustine-binding protein (EMBP) that composes 18–30% of the proteins in prostatic cytosol from adult rats is also present in prostatic fluid. EMBP was purified to homogeneity using chromatography on DEAE-cellulose, Sephadex G-100 and Octyl-Sepharose and polyacrylamide gel electrophoresis. The total amino acid composition has been determined and the N-terminal amino acid was found to be serine. Antibodies were raised against the protein in rabbits and a radioimmunoassay was developed for quantitation of the protein (*10*). A protein immunologically similar to EMBP has also been detected in prostate from other species including man (*11*). EMBP is an androgen-sensitive protein; its quantity in the prostate is reduced after castration and can be restored to normal levels after administration of androgens (*12*).

Purified EMBP was found to inhibit the nuclear uptake of the methyltrienolone-androgen receptor complex *in vitro*. Furthermore, purified EMBP inhibited the binding of the androgen-receptor complex to DNA-cellulose. These effects of EMBP may represent an important intracellular control of its synthesis. Administration of estramustine to rats was found to decrease the weight of the prostate gland to keep the concentration of EMBP at a constant level. In contrast, castration or administration of estrogens decreased the weight of the prostate gland as well as the concentration of EMBP (*13*).

These results indicate that even if estramustine seems to have similar effects to estrogens with regard to certain parameters (*e.g.* reduction of prostatic weight) it affects the prostate in a distinctly different way from that of estrogens or castration. What still remains to be investigated is the question if the different concentrations of EMBP in prostates from estramustine-treated as compared to estrogen-treated animals reflect differences in synthesis, storage, and/or secretion of EMBP.

The presence of receptor proteins in target cells is largely accepted as an important indicator of their steroid hormone "receptivity." From a clinical point of view, the basic information obtained from the study of estrogen receptors in experimental rat mammary tumors as well as in human breast cancer has been shown to be of value in the selection of patients with metastatic breast cancer for endocrine therapy (*14*).

The presence of receptor proteins has been demonstrated also in prostatic cancer. Some tumors contain androgen receptors, the amount of receptor protein varying greatly between different tumors. Available data indicate that patients with prostatic cancer containing androgen receptors have a greater chance to benefit from endocrine therapy than patients with a receptor-negative tumor (15). Improved assay techniques, including nuclear receptor studies, may increase the predictive value of such analyses.

If it will be possible to produce monoclonal antibodies against pure receptor preparations very sensitive and accurate immuno-assay techniques may be developed. Such techniques will permit scientists to entertain questions that earlier seemed beyond attack and to do experiments that provide interesting results leading to other fields of research activity.

REFERENCES

1. Huggins, C. and Hodges, C.V.: 1941. *Cancer Res.* 1, 293–297.
2. Flocks, R.H.: 1955. *J. Urol.* 74, 549–551.
3. Diczfalusy, E., Fernö, O., Fex, H., Högberg, B., Linderot, T., and Rosenberg, T.: 1953. *Acta Chem. Scand.* 7, 913–920.
4. Fernö, O., Fex, H., Högberg, B., Linderot, T., and Rosenberg, T.: 1953. *Acta Chem. Scand.* 7, 921–927.
5. Fernö, O., Fex, H., Högberg, B., Linderot, T., Veige, S., and Diczfalusy, E.: 1958. *Acta Chem. Scand.* 8, 1675–1689.
6. Diczfalusy, E., Fernö, O., Fex, H., Högberg, B., and Kneip, P.: 1959. *Acta Chem. Scand.* 13, 1011–1018.
7. Goodhope, C.D.: 1957. *J. Urol.* 77, 312–314.
8. Tymoczko, J.L., Liang, T., and Liao, S.: 1978. *Receptors and Hormone Action*, Vol. 11 (Birnbaumer, L. and O'Malley, B.W., eds.) pp. 122–156; Academic Press, New York.
9. Jönsson, G., Högberg, B., and Nilsson, T.: 1977. *Scand. J. Urol. Nephrol.* 11, 231–238.
10. Forsgren, B., Björk, P., Carlström, K., Gustafsson, J-Å., Pousette, Å., and Högberg, B.: 1979. *Proc. Natl. Acad. Sci. USA* 76, 3149–3153.
11. Högberg, B., Björk, P., Carlström, K., Forsgren, B., Gustafsson, J-Å., Hökfelt, T., and Pousette, Å.: 1979. *Prostate Cancer and Hormone Receptors*, pp. 181–199, Alan R. Liss, Inc., New York.
12. Pousette, Å., Björk, P., Carlström, K., Forsgren, B., Gustafsson, J-Å., and Högberg, B.: *Cancer Res.*, in press.
13. Pousette, Å., Björk, P., Carlstrom, K., Forsgren, B., Högberg, B., and Gustafsson, J-Å.: *Prostate*, in press.
14. Jensen, E.V., Block, B.E., Smith, S. Kyser, K., and De Sombre E.R.: 1971. *NCI Monogr. (Predict. responses cancer ther.)* 34, 55–61.
15. Gustafsson, J-Å., Ekman, P., Snochowski, M., Zetterberg, A., Pousette, Å., and Högberg, B.: 1978. *Cancer Res.* 38, 4345–4348.

Siderophores

J. B. NEILANDS

J.B. Neilands received a Ph.D. degree from the University of Wisconsin in 1949 and spent two years at the Nobel Medical Institute, Stockholm. He returned to Wisconsin as an instructor for the 1951–1952 year and then joined the University of California, where he is now Professor of Biochemistry. He is coauthor or editor of "Outlines of Enzyme Chemistry" (J. Wiley, 1955, 1958), "Harvest of Death—Chemical Warfare in Vietnam and Cambodia" (Free Press, 1972), and "Microbial Iron Metabolism" (Academic Press, 1974). He has written several hundred papers, mostly on microbial iron metabolism. Siderophores isolated and characterized include ferrichrome, ferrichrome A, 2,3-dihydroxybenzoylglycine, rhodotorulic acid, schizokinen, enterobactin, agrobactin, and parabactin. The iron repression of siderophore biosynthesis was reported in 1955 and in 1975 the tonA protein in the outer membrane of *Escherichia coli* was identified as the ferrichrome receptor. A founder of the Bertrand Russell Society, he claims to be "interested in everything except sports."

S iderophores (Gr.: *iron bearers*) are relatively low molecular weight, virtually ferric specific substances produced by aerobic and facultative anaerobic microbes. These novel chelating agents serve to solubilize and transport Fe(III) as part of a *high affinity* iron assimilation system in microorganisms.

It is supposed that siderophores were not required in the earliest stages of the development of life since iron was at that time in the reduced state and hence relatively soluble. It is further speculated that iron was then, as it is today, an element important in the metabolism of strictly anaerobic bacteria. With the advent

On the left, J.B. Neilands; on the right, M.A. McIntosh; in the flask, enterobactin from *E. coli ent D-* bearing lambda transducing phage 712. (Photo by John Heckendorn)

TABLE 1.   Presence of Siderophores in Microbial Species

Siderophore	Source
Ferrichrome-type	*Penicillia, Aspergilli, Neurospora, Ustilago*
Rhodotorulic acid	Yeasts
Schizokinen-type	*Bacillus megaterium, Anabaena* sp., *Aerobacter aerogenes, Arthrobacter* sp.
Mycobactins	Mycobacteria
Ferrioxamines	*Actinomyces* sp.
Enterobactin	Enteric bacteria
Agrobactin	*Agrobacterium tumefaciens*
Parabactin	*Paracoccus denitrificans*

of "plant-type," O_2-evolving photosynthesis the surface iron of planet Earth was switched from Fe(II) to Fe(III), and thereby became profoundly insoluble. In aqueous media the ferric ion forms oxyhydroxide polymers of over-all composition FeOOH; the solubility product constant of ferric hydroxide may be less than 10^{-38} M. The microbiological response to loss of dissolved iron must have been the elaboration of siderophores. The affinity of the siderophore ligand for Fe(III) is very great, the actual formation constants lying in the range 10^{30}–10^{50} M. As iron ranks just behind aluminum in *abundance* the real biological problem here is in how to make the former *available* to living cells. Obviously, as a necessary constituent of hemoproteins, iron-sulfur proteins and ribotide reductase—to mention a few important biocatalysts—iron is a truly "precious metal" for life.

Distribution of Siderophores

Table 1 gives a partial list of siderophores which have been characterized from bacterial and fungal species. It is obvious that the power to form siderophores may be found in gram-positive and gram-negative bacteria, spore formers, cyanobacteria (blue-green algae), soil bacteria, plant and animal pathogens and, among the fungi, in both molds and yeast. It is probably significant that siderophores have been characterized only in aerobic and facultative anaerobic species. They have not been found in *Clostridia* and lactic acid bacteria. In the former the reducing environment may solubilize the iron and in this way meet cellular demands which are known to be substantial. In the case of lactic acid bacteria it has not yet been proven that these unusual species require iron for growth. To sum up, the data in Table 1 compel the conclusion that siderophore synthesis is a common trait in the aerobic and facultative anaerobic microbial world.

Genetics and Biosynthesis

The most important point about the biosynthesis of siderophores is that the process is tightly regulated by iron. The quantity of iron required to shut down excretion of siderophores varies from species to species and with the composition of the medium. It is generally in the range 0.1 to 10 μM. In the systems which have been investigated all enzymes required for synthesis of siderophores and their transport are regulated, apparently coordinately, by iron. There is always a constitutive level of synthesis of all components of the high affinity pathway even at very high levels of iron. Enterobactin, the siderophore of enteric bacteria, is chromosomally programmed as is also agrobactin, the siderophore of *Agrobacterium tumefaciens*. Both of these siderophores belong to the catechol family. Hydroxamate type siderophores are formed generally among the fungi and by a few bacterial species. In the latter instance the genes for siderophore synthesis may be plasmid borne.

The genetic information for enterobactin synthesis is clustered at minute 13 on the *Escherichia coli* chromosome and situated in one or more iron operons. Genes *entA*, *B*, and *C* are located between chorismate and 2,3-dihydroxybenzoic acid while genes *D-G* are between dihydroxybenzoate and enterobactin. Additional genes, designated *fep* (permease) and *fes* (esterase-reductase) are part of the operon(s). The biosynthesis of the hydroxamate type siderophores proceeds in sequence through the amine and hydroxylamino functions. Much work remains to be done on the biosynthesis of both catechol and hydroxamate type siderophores and, especially, on the mechanism of iron as a regulator. In the author's laboratory the iron operon of *E. coli* has been cloned on phage λ (M. McIntosh and S. McDougal) as a prelude to a more incisive probe of the molecular mechanism of iron regulation.

Siderophore Structures

As chemical entities, siderophores may be generally classified as either hydroxamates or catechols. The coordinating atoms are usually all oxygen since, as a "hard acid," Fe(III) prefers to associate with a "hard base," such as oxygen. In order to retain iron in dilute solution, effective chelating agents should contain at least two, and preferably three, sets of bidentate ligands. Herewith the formulae of five siderophores, the structures of which have been determined in the author's laboratory.

Ferrichrome (Fig. 1) was isolated in 1952 from the basidiomycete *Ustilago sphaerogena* and was the first ferric trihydroxamate discovered in nature. It has since been detected in *Penicillia*, *Aspergilli*, *Neurospora*, and other fungi. It is considered to be the prototypical siderophore of fungal origin. The structure shown in Fig. 1 is a composite of X-ray diffraction and high resolution NMR work. A large number of ferrichrome type siderophores occur carrying variations in the neutral tripeptide moiety (glycine, alanine, serine) and/or the acyl substituent of the hydroxamate link. Ferrichrome itself is the ferric complex of cyclo-triglycyl-tri-N$^\delta$-acetyl-N$^\delta$-hydroxyornithyl. The antibiotic albomycin is closely related to ferrichrome and penetrates the cells of sensitive bacteria on a transport pathway designed for ferrichrome, *i.e.*, an example of "illicit transport."

Figure 2 gives the structure of rhodotorulic acid, produced in giant quantities by *Rhodotorula pilimanae* and similar yeasts. It is cyclic dipeptide of N^δ-acetyl-N^δ-hydroxyornithine and forms a 3:2 complex with Fe(III).

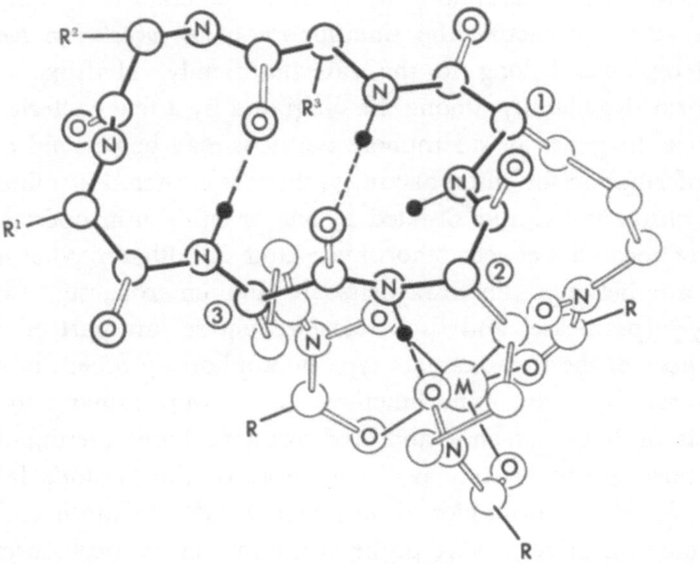

FIG. 1.   Basic structure of the ferrichrome type cyclohexapeptide ferric trihydroxamate siderophores. For ferrichrome itself, $R^1=R^2=R^3=H$; $R=CH_3$.

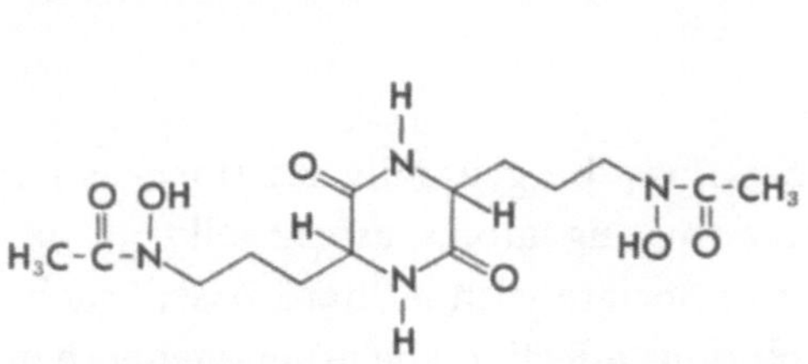

FIG. 2.   Rhodotorulic acid, a cyclic dipeptide of N^δ-acetyl-N^δ-hydroxy-ornithine, the siderophore of *Rhodotorula* and related yeasts.

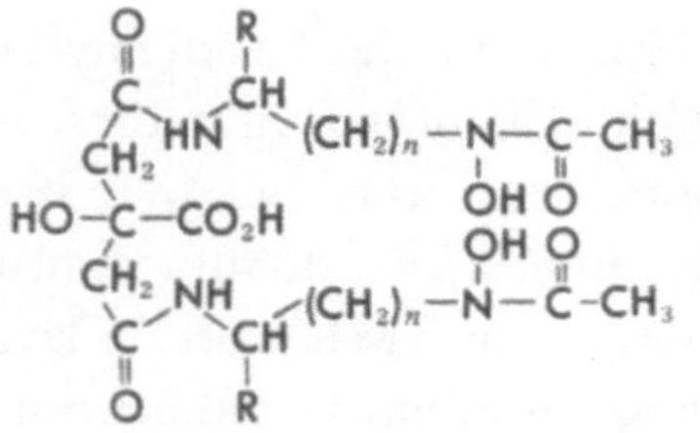

	R	n
Schizokinen	H	2
Aerobactin	COOH	4
Arthrobactin	H	4

FIG. 3.   Structures of three members of the schizokinen family of siderophores of bacterial origin, all of which are hydroxamic acid derivatives of citrate.

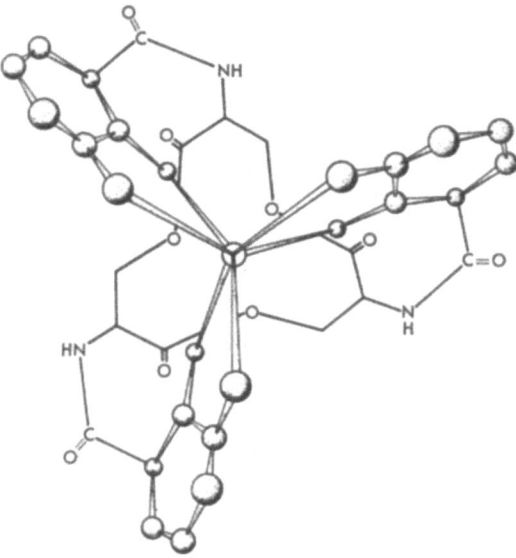

FIG. 4. Ferric enterobactin, the Fe(III) complex of cyclo-(2,3-dihydroxy-N-benzoyl-seryl)₃, a siderophore common to many enteric bacteria.

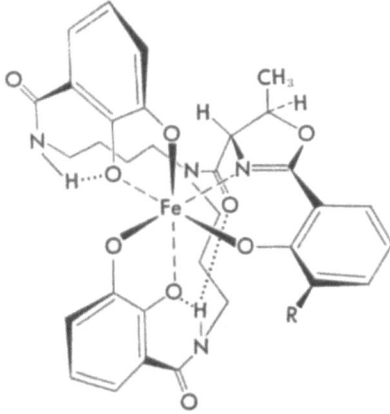

FIG. 5. Ferric complex is catechol type siderophores which are derivatives of spermidine. The known members include agrobactin, R=OH, from *Agrobacterium tumefaciens*, and parabactin, R=H, from *Paracoccus denitrificand*. The structure shown is a speculation based on NMR studies of the ligand, the constitution of which has been confirmed by X-ray diffraction.

In Fig. 3 a slightly different form of hydroxamate siderophore is depicted. The substance shown, schizokinen, was described first by Lankford and Byers in Texas as a growth factor for *Bacillus megaterium*. Several years later it was isolated in pure form in the author's laboratory and its structure determined. As may be seen, schizokinen is a derivative of citric acid in which the distal carboxyl groups bear side chains of 1-amino-3-acethydroxamidopropane. Analogous compounds have been obtained from *Aerobacter aerogenes* and *Arthrobacter* sp and named aerobactin and arthrobactin, respectively. As schizokinen has been isolated from the cyanobacterium *Anabaena*, it is possible that this type of siderophore is native to fungi and is programmed only by plasmids in bacteria.

The compound shown in Fig. 4 was first described in 1970. In that year groups working independently characterized it from *E. coli* and *Salmonella typhimurium* and gave it the trivial name enterochelin and enterobactin, respectively. It can be described chemically as cyclo-tri-2,3-dihydroxybenzoylserine. Enterobactin is formed by a large number of enteric bacteria.

The siderophore shown in Fig. 5 is agrobactin (R=OH) or parabactin (R= H) from *Agrobacterium tumefaciens* or *Paracoccus denitrificans*, respectively. In both of these siderophores coordination to the Fe(III) is through the oxazoline N and the steriochemistry is that of a left-hand propeller.

Mechanism of Iron Transport

The mode of iron transport promoted by some of the siderophores just described has been studied to some extent, most thoroughly in the case of ferrichrome and enterobactin in *E. coli*. Although ferrichrome is not synthesized by *E. coli* K-12 the organism maintains an efficient transport apparatus for this fungal siderophore. *Escherichia coli* and closely related species are the organisms of choice since the use of mutants is essential for elucidation of the most intimate details of transport. Thus one can work with mutants blocked at defined stages in siderophore mediated iron assimilation and in this way collect information on the mode of siderophore transport and iron utilization.

Ferric enterobactin first contacts an outer membrane receptor, the product of the *fep* gene. The SDS-gel molecular weight of this protein is 81,000 daltons. The receptor complex also serves as the binding site for colicin B; a phage has not yet been reported for this receptor but it is highly likely that one exists in Nature. It is assumed that the receptor forms a specific channel or pore and that it functions in cooperation with an inner membrane component such as a permease. Experiments using ^{14}C labelled ferric enterobactin indicate that the intact complex enters the cell. Mutants designated *fes* are unable to release enterobactin-complexed iron, possibly because such cell lack their esterase or a functional reductase. In the outer membrane of *E. coli* two proteins in addition to *fep* are strongly induced by growth of the organism in low iron media. These have SDS-gel molecular weights of 83,000 and 74,000 daltons. The former is of unknown activity and the latter is the colicin Ia receptor, the biochemical function of which is not understood.

The ferrichrome receptor was known as the *tonA* "T-one" protein long before its true biological purpose was defined. This receptor is probably not derepressed at low iron or at least not to the extent of *fep*, although efficient ferrichrome transport requires low iron grown cells. The *tonA* receptor acts as a common binding site for phages T1, T5, and φ80, colicin M, albomycin, and siderophores closely related to ferrichrome. Although ferrichrome is not degraded in the process of iron delivery it may be modified by acylation of one of the hydroxylamino oxygens following release of the complexed Fe(III). Ferrichrome is transported by inner membrane vesicles by a mechanism requiring the energized membrane. The finding that ferrichrome binds alkaline earth cations supports the view that the siderophore is transported by a non-proton symport process.

Several proteins appear in the outer membrane of *Pseudomonas aeruginosa* when the organism is grown at low iron. As one of these is absent in strains resistant to pyocin S2 it may serve as the receptor for this bacteriocin. *Pseudomonas* spp. have been reported to form siderophores such as ferribactin, pyoverdine, pyochelin, and pseudobactin but none of these has been characterized chemically.

Why are receptors required for siderophore transport? In order to make a really powerful chelator for Fe(III), it is necessary to design into one molecule 2 or 3 sets of conveniently disposed coordinating atoms. This requires that the molecular weight be in the range 500–1,000 daltons, a figure which exceeds the free diffusion limit of the outer membrane of gram-negaitve bacteria. Nature's response was to form a specific receptor which somehow facilitates the transport of these larger, water soluble molecules. In the final stages of molecular parasitism, phages and bacteriocins have adapted to use the siderophore receptors as attachment sites on the outer membrane.

Applied Aspects

Thus far the only practical application of the knowledge gleaned from studies of high affinity microbial iron assimilation is the inverse of the biological function, namely, deferration of patients suffering from transfusion induced siderosis. Ciba-Geigy markets Desferal, a siderophore from *Streptomyces pilosus*, for chelation therapy in Cooley's anemia (thalassemia). Desferal is the mesylate salt of deferriferrioxamine B and is a trihydroxamate comprised of repeating units of 1-amino-5-N-hydroxyaminopentane, succinic acid, and acetic acid. The drug is not absorbed and must be injected. Hence there is need for a compound more effective than Desferal.

The proposition that iron is a virulence factor in animal infections can be traced to the observation some 35 years ago by Schade and Caroline that the bacteriostatic activity of egg white and serum could be reversed with iron. Proteins such as ovotransferrin and transferrin bind iron very tightly and reduce its availability to microorganisms. In serum the free iron concentration may be as low as 10^{-12} μM and unless a pathogenic organism can transport heme it may very well face a severe iron deficit in certain environmental situations. Infection in which iron is thought to serve as a virulence factor include some of the most important human diseases.

It has been shown that mutants of *S. typhimurium* blocked in enterobactin biosynthesis possess greatly diminished virulence in the mouse. Chromatographic evidence was produced for the presence of enterobactin in peritoneal washings of guinea pigs lethally infected with 10^8 cells of *E. coli*. Recently the presence of an antibody to enterobactin in normal human serum has been reported.

Invasive strains of *E. coli*, clinical isolates, have been found to harbor the colicin V plasmid. A non-enterobactin, possibly hydroxamate type siderophore is associated with the plasmid, although colicin V itself is not a part of this unique iron assimilation systen.

While the studies just described would tend to confirm siderophores and iron assimilation as a virulence factor, a number of workers have failed to make this correlation. Infection must depend on a multiplicity of factors, of which iron is merely one.

Efforts are now being made to design drugs which will be recognized and transported by cell surface iron receptors.

This brief account can serve merely to sketch the general contours of high affinity, *i.e.*, siderophore-mediated, iron assimilation in microorganisms, which has been the subject of my research for almost three decades. References to the literature cited here may be found in one of two recent reviews (*1, 2*). The review by Lankford (*3*) on bacterial iron assimilation and by Emery (*4*) on aspects of fungal iron transport are recommended.

Ferrichrome (*5*) and rhodotorulic acid (*6, 7*) have been synthesized in Japanese laboratories. Uemura and Mizushima (*8*) were the first to note the effect of iron nutrition on the outer membrane proteins of *E. coli*. The low-iron induction of membrane proteins in *P. aeruginosa* and their possible correlation with the receptor for pyocin S2 was reported by Ohkawa *et al.* (*9*). Novel methods have been devised for isolation of ferrichrome-type siderophores from sake (*10, 11*). Thus Japanese scientists have made significant contributions to our knowledge of the chemistry and biology of siderophores.

Siderophores provide a novel opportunity for study of the molecular mechanics of transport and for the mode of regulation of this process at the membrane level. One can easily envisage practical developments in this field ranging from drug development, improved human nutrition, crop protection, and enhanced world food supply. These would seem to be a more worthy objective for the scientific community than the further design of weapons of mass destruction.

REFERENCES

1. Neilands, J.B.: 1981. *Annu. Rev. Biochem.* 50, 715–731.
2. Neilands, J.B.: 1980. *Annu. Rev. Nutr.* 1, 27–46.
3. Lankford, C.E.: 1973. *Crit. Rev. Microbiol.* 2, 273–331.
4. Emery, T.: 1978. *Metal Ions in Biological Systems*, Vol. 7 (Sigel, H., ed.) pp. 77–126, Marcel Dekker, New York.

5. Isowa, Y., Ohmori, M., and Kurita, H.: 1974. *Bull. Chem. Soc. Japan* 47, 215–220.
6. Isowa, Y., Takashima, T., Ohmori, M., Kurita, H., Sato, M., and Mori, K.: 1972. *Bull. Chem. Soc. Japan* 45, 1467–1471.
7. Fujii, T. and Hatanaka, Y.: 1973. *Tetrahedron* 29, 3825–3831.
8. Uemura, J. and Mizushima, S.: 1975. *Biochim. Biophys. Acta* 413, 163–176.
9. Ohkawa, I., Shiga, S., and Kageyama, M.: 1980. *J. Biochem.* 87, 323–331.
10. Tadenuma, M. and Sato, S.: 1967. *Agric. Biol. Chem.* 31, 1482–1489.
11. Narahara, H.: 1970. *Nippon Jozo Kyokai Zasshi* 65, 340–343 (in Japanese).

Studies with Nitrifying Bacteria *Nitrosomonas* and *Nitrobacter* Using Electrode and Fluorescence Techniques

D. J. D. NICHOLAS

D.J.D. NICHOLAS was born in Britain in 1924, and he was educated at the Universities of London, Cambridge and Bristol, England. He holds the M.A., Ph.D., and D.Sc. degrees in biochemistry. In 1946 he was appointed lecturer in Biochemistry at the University of Bristol, where he commenced studies in trace metal requirements of microorganisms, including their functions in enzyme systems. From 1952–1954 he held a Rockefeller fellowship at the McCollum-Pratt Institute, Johns Hopkins University, USA, where he identified molybdenum as a constituent of the enzyme nitrate reductase in microorganisms and plants. In 1961–1962 he was a Visiting Professor in Biochemistry at the University of Wisconsin, USA, where he worked on biochemical aspects of nitrogen fixation, especially the role of non-haem iron compounds. In 1964 he left Bristol to take up the Chair of Agricultural Biochemistry at the Waite Agricultural Research Institute, University of Adelaide, South Australia, where he has greatly expanded his work on the metabolism of inorganic nitrogen and sulphur compounds in microorganisms and plants. In 1969 he was a visiting research fellow in the Biochemistry Department, University of Cambridge and in 1975 held a similar position at the University of Oxford.

Nitrifying bacteria consist of two main genera, namely *Nitrosomonas* and *Nitrobacter*. Thus *Nitrosomonas* oxidises ammonia to nitrite, presumably *via* hydroxylamine according to the following equations:

$NH_4^+ + \frac{1}{2}O_2 \longrightarrow NH_2OH + H^+$	$\Delta F + 4$ kcal
$NH_2OH + O_2 \longrightarrow NO_2^- + H_2O + H^+$	$\Delta F - 69$ kcal
Sum $NH_4^+ + 1\frac{1}{2}O_2 \longrightarrow NO_2^- + H_2O + 2H^+$	$\Delta F - 65$ kcal

Nitrobacter oxidises nitrite to nitrate as follows:

$$NO_2^- + \tfrac{1}{2}O_2 \longrightarrow NO_3^- \qquad\qquad \mathit{\Delta F} - 17.8 \text{ kcal.}$$

These chemolithotrophic bacteria derive their energy (ATP) for CO_2 fixation and growth from the oxidation of these reduced inorganic nitrogen compounds.

Over the last few years several timely reviews have covered various aspects of the metabolism of the nitrifying bacteria (1–5).

In this essay it is my intention to consider briefly the use of electrode techniques for simultaneously and continuously monitoring (a) the uptake of NH_4^+ and O_2 by cells, spheroplasts, and vesicles of *Nitrosomonas*, (b) the uptake of O_2 and the extrusion of NO_3^- by cells of *Nitrobacter* in relation to the utilization of nitrite. In addition, the results of a fluorescence technique for measuring the extrusion of protons from cells of *Nitrosomonas* and *Nitrobacter* will be considered in relation to the electrode data.

One of the reasons for developing these techniques for use with cells of *Nitrosomonas* was the lack of success in isolating the ammonia oxygenase from this bacterium. Thus it was of interest to study the uptake of NH_4^+ by electrode techniques in washed cells, spheroplasts, and membrane vesicles respectively and then relate the data to the extrusion of protons (6–8). In *Nitrobacter* the relation between the uptake of O_2 and nitrite and the extrusion of nitrate also readily determined by these and other methods.

Electrode Systems

1. Nitrosomonas europaea

The uptake of NH_4^+ and O_2 in cell preparations were simultaneously and continuously measured in a magnetically stirred reaction mixture contained in a perspex vessel (9 ml or 5 ml capacity) maintained at 30°C by circulating water through an outer jacket. The ports in the vessel lid accommodated an O_2 electrode and an NH_4^+ electrode (Orion model 95.10). The NH_4^+ electrode was connected to a Beckman expanded scale meter (Model 76). The responses to the NH_4^+ and O_2 electrodes were recorded simultaneously using a two-channel Rikandenki potentiometric recorder (Model B181-H). A Unicam SP45 concentration readout was also used to convert the readings from a log to a linear scale (6–8).

Additions to the reaction mixture were made with an airtight microsyringe *via* a port in the lid of the vessel. O_2 was regenerated by adding 7 μl 2% v/v H_2O_2 into the reaction mixture which contained 5 μl catalase (2 mg/ml). The concentration of O_2 in air-saturated water at 30°C was taken to be 235 μmol/litre. The NH_4^+ electrode was calibrated for each experiment with NH_4Cl in 0.05 M Tris-HCl buffer (pH 7.8). Details of the reaction mixtures used are given in the legends to the figures.

2. Nitrobacter agilis

A nitrate electrode was substituted for the NH_4^+ electrode in a perspex vessel similar to the apparatus used for the *Nitrosomonas* experiments. In this way, O_2 uptake and the extrusion of nitrate from washed cells could be followed simultaneously and continuously. The utilization of nitrite was determined colorimetrically in aliquots of the reaction mixture sampled at intervals during the experiment (*9*).

Fluorescence Measurements for Protons

The fluorescence emission of the cells was followed as described previously (*6*). A fluorispec Model SF-1 fluorimeter (Baird Atomic, Cambridge, Mass., USA) at 420 nm excitation and 485 nm emission was linked to a potentiometric recorder (Servoscribe 1S). The reaction mixture in a 1 cm cuvette contained 25 μmol Tris-HCl buffer (pH 7.6) and washed cells 0.1 ml (20 mg wet wt.) in a total volume of 2.6 ml. The cells plus quinacrine dihydrochloride with or without an inhibitor

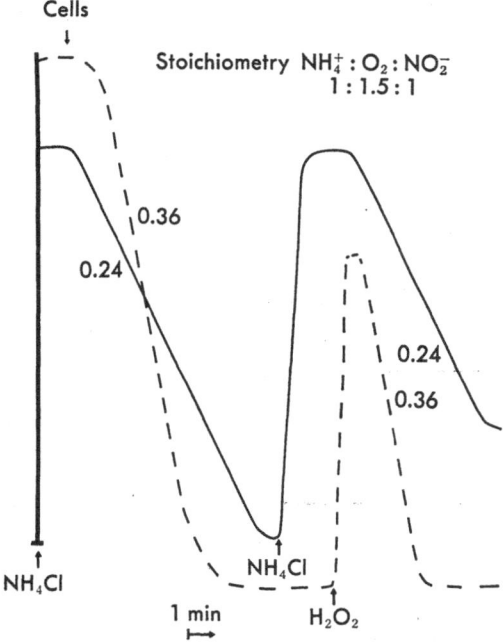

FIG. 1. Uptake of NH_4^+ and O_2 by *Nitrosomonas*. 0.1 ml of washed cells (40 mg, wet wt.) was added to a perspex reaction vessel containing 5 μl catalase (2 mg/ml), 1 μmol NH_4Cl and 0.42 mmol Tris-HCl buffer (pH 7.8) maintained at 30°C in a final volume of 8.4 ml. When the NH_4^+ had been ulitized within about 10 min, a further 1 μmol NH_4Cl was injected into the reaction mixture *via* a Hamilton microsyringe through a port in the lid of the vessel, as indicated. To regenerate O_2, 7 μl, 2% *v/v* H_2O_2 was injected into the reaction mixture, as indicated. The reaction mixture was continuously mixed with a magnetic flea. The maximum rates alongside the traces are expressed in μmol/min, NH_4^+ (—) and O_2 (---).

were preincubated in a 1 cm cuvette for 3 min before starting the reaction with 20 μmol NH₄Cl for *Nitrosomonas* or else 20 μmol NaNO₂ for *Nitrobacter*.

Results with Nitrosomonas

1. Washed cells

The results for the uptake of NH_4^+ and O_2 by washed cells determined simultaneously and continuously are presented in Fig. 1. In addition, nitrite production was also determined at intervals in aliquots of the reaction mixture. The stoichiometry of NH_4^+, O_2 uptake and NO_2^- production over a 10 min period was 1:1.5:1.0 respectively. When hydroxylamine was used instead of ammonia then the stoichiometry for O_2:NO_2^- was 1:1.

The uptake of NH_4^+ resulted in an outward extrusion of protons as shown by an increase in fluorescence emission activity (Fig. 2).

The electrode and fluorescence techniques were then used in a series of experiments designed to test the effects of various inhibitors on washed cells (6). The results indicate that the uptake of NH_4^+ by washed cells of *Nitrosomonas* is mediated by a Cu-dependent translocase in cell-membranes, which is markedly affected by inhibitors of the respiratory chain as well as uncouplers. This energy-

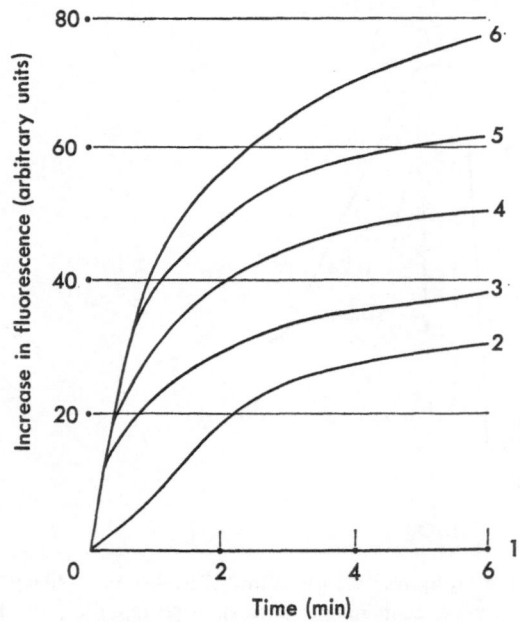

Fig. 2. Effect of NH_4^+ concentration on the quenching of the quinacrine fluorescence in washed cell suspensions of *Nitrosomonas*. 0.1 ml cell suspension (20 mg wet wt.) was preincubated for 3 min with 0.1 μmol quinacrine and 25 μmol Tris-HCl buffer (pH 7.6) in a final volume of 2.6 ml in separate 1 cm cuvettes before adding NH₄Cl at following concentrations: (1) 0, (2) 0.36 μM, (3) 0.71 μM, (4) 0.36 mM, (5) 3.6 mM, (6) 71 mM, respectively. The increase in fluorescence emission indicates an outward extrusion of protons.

dependent uptake of NH_4^+ results in the extrusion of protons from the cells and the latter process is affected similarly by the same inhibitors.

2. Spheroplasts and membrane vesicles

Spheroplasts which were prepared from 1 g washed cells suspended in 50 mM Tris-HCl buffer (pH 7.5) containing 200 mM sucrose, 2 mM Na$_2$EDTA, and 100 mg lysosyme gave similar results to washed cells (6, 7).

Membrane vesicles were then prepared from these spheroplasts, but it was found that sucrose interfered with the chemical determination of hydroxylamine, one of the oxidation products of NH$_4$Cl. This difficulty was overcome by substituting lithium chloride for sucrose (8).

Spheroplasts were then prepared as follows: 1 g washed cells were suspended in 50 mM Tris-HCl buffer containing 0.1 M lithium chloride (pH 7.3), 1 mM sodium EDTA and 100 mg lysosyme in a final volume of 40 ml in a 100 ml Erlenmeyer flask. The cell suspension was then incubated for 2 hr at 30°C with constant shaking. Then 3 mM magnesium acetate containing 1 μg DNAase was added to the reaction mixture and incubation continued for a further 2 hr. The suspension was then centrifuged at 15,000 g for 20 min at 4°C yielding a pellet of spheroplasts. Unless otherwise stated, these spheroplasts prepared in lithium chloride were used for the preparation of membrane vesicles.

Membrane vesicles, which were stable for 5 hr, were produced from these spheroplasts as follows: to the pellet of spheroplasts was added 10 vol of 10 mM

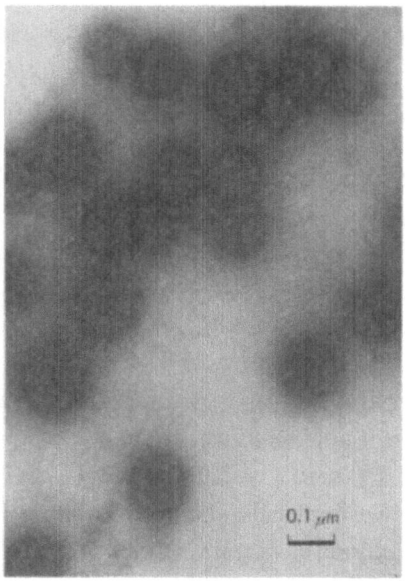

FIG. 3. Electron micrograph of membrane vesicles prepared from *Nitrosomonas europaea*. These were negatively stained with 3% v/v phosphotungstic acid (pH 6.6) and examined in an electron microscope JEOL (Model JEM-100 cx) at an accelerating voltage of 60 kV.

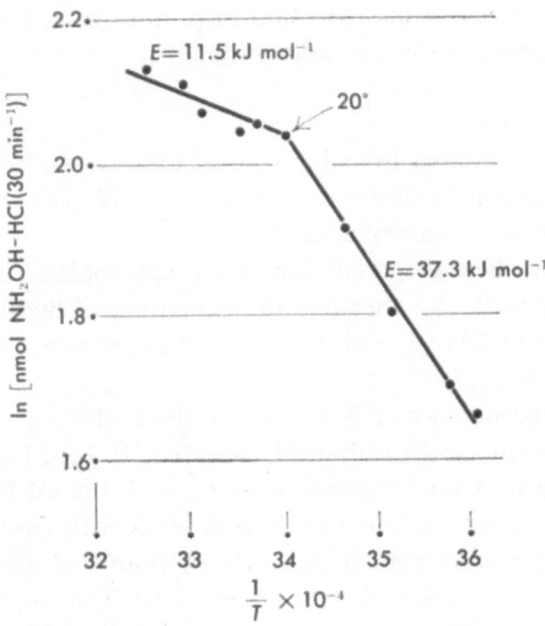

FIG. 4. Arrhenius plot for NH_2OH production from NH_4Cl by membrane vesicles. The vesicles (10 mg protein ml^{-1}) were prepared in 10 mM phosphate buffer (pH 7.3) instead of Tris-HCl buffer. The reaction was started by adding 20 μmol NH_4Cl to 1 ml suspensions of vesicles at the temperatures indicated. The production of NH_2OH was determined after a 30 min incubation.

Tris-HCl buffer (pH 7.3) containing 0.16 M LiCl, 30 mM $NH_2NH_2.H_2SO_4$ (neutralised), 3 mM Mg acetate and DNAase (2 μg ml^{-1}) and after centrifuging at 12,000 g for 10 min the pellet was resuspended in the same buffer. The spheroplast suspension (1 g wet. wt in 20 ml^{-1}) was then treated with a high frequency probe at 110 watts (Model B-12, Branson Sonic Power Company, Danbury, Connecticut) with short bursts of 15 sec over a 4 min period at 5°C in an ice bath. The suspension was then centrifuged at 500 g for 10 min to discard white heavy particles which included intact spheroplasts. The supernatant fraction, a turbid suspension which contained membrane vesicles, is illustrated in Fig. 3. Vesicles were also prepared from spheroplasts in the absence of Mg acetate. These were used to study the effects of metals on (a) the production of NH_2OH from NH_4Cl in the presence of hydrazine and (b) ATPase activity.

When the membrane vesicles were required for determining either the uptake of NH_4^+ and O_2 or ATPase activity, then hydrazine was omitted.

The stoichiometry for $NH_4^+:O_2:NO_2^-$ in the vesicles was 1:1.2:0.7 compared with 1:1.5:1.0 for either spheroplasts or washed cells. This probably resulted from a deactivation of the NH_2OH oxido-reductase system which. converts hydroxylamine to nitrite, presumably because cytochrome c of the respiratory chain is lost during the preparation of vesicles. The inhibitor effects were similar to those reported for cells and spheroplasts. An Arrhenius plot of the effect of temperature

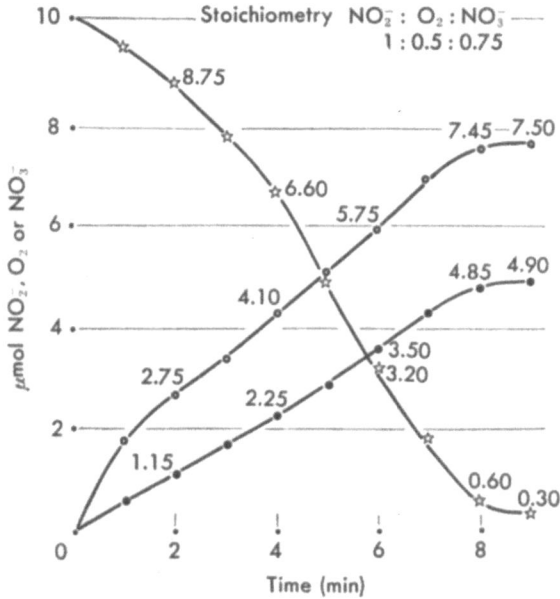

FIG. 5. NO_2^- and O_2 uptake and NO_3^- extrusion in wahsed cells of *N. agilis*. One hundred μl of washed cells (40 mg wet wt.) was added to a perspex vessel containing 10 μmol $NaNO_2$, 5 μl catalase (2 mg/ml), 0.2 mmol Tris-HCl buffer (pH 8.0) in a final volume of 5 ml. To regenerate O_2, about 5 μl of 2% v/v H_2O_2 was injected into the reaction mixture *via* a Hamilton Syringe through a port in the lid of the vessel. O_2 uptake and NO_3^- extruded from the cells were measured by electrode, while the utilization of NO_2^- was followed by a colorimetric method (6). The reaction mixture which was continuously stirred with a magnetic flea was maintained at 25°C. The values alongside the traces are concentrations of NO_2^-, O_2, and NO_3^- in μmol at the specified times. ☆ NO_2^- uptake; ● O_2 uptake; ○ NO_3^- production.

on the oxidation of NH_4Cl by membrane vesicles shown in Fig. 4 indicates a biphasic system with a break near 20°C. This reflects the importance of the lipid membrane fluidity in the functioning of the NH_4^+ translocase system. The activation energy between 33 and 20°C was 11.5 kJ mol⁻¹, whereas below 20°C there was a three-fold increase (37.3 kJ mol⁻¹). These membrane vesicles had ATPase activity, which was stimulated by Mg^{2+}.

These results confirm the importance of maintaining the integrity of membrane structures for the translocation of NH_4^+ into the cells. This is achieved in membrane vesicles prepared in lithium chloride since under these conditions the NH_4^+ translocase system with its attendant proton extrusion, as well as the monooxygenase enzyme, retain their activities.

Result with Nitrobacter

Washed cells

The results for the uptake of O_2 and the extrusion of nitrate, determined by electrode methods, in relation to nitrite uptake determined by a colorimetric method (6) are presented in Fig. 5. From these data the stoichiometry for the uptake of NO_2^- and O_2 and the extrusion of NO_3^- was calculated to be 1:0.5: 0.75. Thus, in washed cells after a 10 min incubation period 10 μmol NO_2^- was taken up and 7.5 μmol extruded, leaving 2.5 μmol to be assimilated into cell-nitrogen *via* the NADH-linked nitrate and nitrite reductase enzyme systems (10, 11).

Conclusions

I t is shown that electrode and fluorescence methods are useful in studying the uptake of O_2 and NH_4^+ and the extrusion of protons from washed cells, sphero-plasts and membrane vesicles of *Nitrosomonas*. Not only was the stoichiometry of these reactions determined by these methods, but it was also established that a Cu and energy-dependent NH_4^+ translocase was present even in membrane vesicles prepared from washed cells.

Preliminary data for *Nitrobacter* indicate that electrode methods can be used to continuously monitor the uptake of O_2 and the extrusion of NO_3^- when washed cells utilize nitrite.

REFERENCES

1. Wallace, W. and Nicholas, D.J.D.: 1969. *Biol. Rev.* 44, 359–392.
2. Aleem, M.I.H.: 1970. *Annu. Rev. Plant Physiol.* 21, 67–90.
3. Aleem, M.I.H.: 1977. *Microbial Energetics* (Haddock, B.A. and Hamilton, W.A., eds.) (27th Symposium Soc. Gen. Microbiol.) pp. 351–381, Cambridge University Press, London.
4. Hooper, A.B.: 1978. *Microbiology* (Schlessinger, D., ed.) pp. 299–304, American Society of Microbiology, Washington.
5. Nicholas, D.J.D.: 1978. *Microbiology* (Schlessinger, D., ed.) pp. 305–309, American Society of Microbiology, Washington.
6. Bhandari, B. and Nicholas, D.J.D.: 1979. *Arch. Microbiol.* 122, 249–255.
7. Bhandari, B. and Nicholas, D.J.D.: 1979. *FEMS Microbiol. Lett.* 6, 297–300.
8. Bhandari, B. and Nicholas, D.J.D.: 1980. *Anal. Biochem.* 109, 330–337.
9. Hewitt, E.J. and Nicholas, D.J.D.: 1964. *Modern Methods of Plant Analysis*, Vol. VII (Linskens, H.F., Sanwal, B.D., and Tracey, M.V., eds.) pp. 67–172, Springer-Verlag, Berlin.
10. Faull, K.F., Wallace, W., and Nicholas, D.J.D.: 1969. *Biochem. J.* 113, 449–455.
11. Wallace, W. and Nicholas, D.J.D.: 1968. *Biochem. J.* 109, 763–773.

Where There's Life There's Phosphorus

LORD TODD

A.R. Todd was born in 1907. He was graduated University of Glasgow and engaged in education in various universities such as Edinburgh, London, Manchester in England and Chicago, MIT in USA In 1948, he became the President of Chemical Society. He is the member council, Royal Society and Honorable Member of French, German, Spanish, Belgium, and Swiss Chemical Society. He received Lavoisier Medal from French Chemical Society in 1948, Pary Medal of Royal Society in 1949, Royal Medal of Royal Society in 1955. He is a Nobel Prize winner for chemistry. He was elevated to the peerage in 1962 as Lord Todd.

The past quarter of a century has seen a series of staggering advances in our understanding of living organisms and the ways in which they carry out their manifold activities. We know, for example, not only what the nucleic acids are but how the sequence of residues in DNA provides a genetic code which ensures accurate transmission of hereditary characteristics from one generation to the next; we know too the complex series of reactions leading to the synthesis of steroids, terpenoids, and alkaloids. What we do not know, however, is the "why" of all these things. Why does Nature use the intermediates and reactions which she does? In this essay I want to speculate a bit on this topic of "why". I may say at the outset that I believe that Nature does not necessarily use in every case the best conceivable reagent but she uses the best which, given the materials readily accessible to her, could be developed during the long process of evolution. Let us take her addiction to derivatives of phosphoric acid. When one looks at the living world it is at first sight astonishing to see how derivatives of phosphoric acid seem to turn up all over the place—in the nucleic acids, in the coenzymes, phospholipids, teichoic acids, and in many important intermediates in metabolism *e.g.* sugar phosphates, as well as being involved in the formation of carbon-carbon bonds. Some of them, especially polyphosphate esters, perform an important function not only in transferring phosphate but as energy stores, liberating energy in the course of phosphate transfer and facilitating endothermic reactions which would not otherwise be possible. Outstanding among these compounds is adenosine triphosphate (ATP). The most widely used energy-providing reaction in nature is that by which ATP transfers a phosphate residue to some substrate with liberation of energy: this enables organisms to use up energy for short periods at a rate much faster than they can generate it from food.

To understand the peculiar properties of ATP and other naturally occurring polyphosphates we much look for a little at the laboratory phosphorylation of organic compounds which was studied in depth by my colleagues and me in Cambridge a good many years ago. Phosphoric acid is a strong tribasic acid and its diesters are also very strong monobasic acids. The diesters readily form mixed anhydrides with other monobasic acids and the dialkyl phosphorochloridates, for example, are effective phosphorylating agents. They yield, of course, triesters of phosphoric acid whereas the naturally occurring phosphate esters are usually mono- or diesters so that using this method of phosphorylation it is necessary to have methods available for removing one or more of the original ester groups in the reagent. Fortunately a variety of methods for doing this have been discovered. One which is particularly significant from the standpoint of my talk today is fission by nucleophilic attack. If one of the ester groups is strongly electrophilic *e.g.* benzyl or allyl then treatment of the triester with a suitable nucleophile will remove that group cleanly and quantitatively. The more strongly acidic the diester produced (*i.e.* the more stable its anion) the more readily will the reaction proceed so that not only strong nucleophiles such as anions will react but even olefins. This is, of course, the reaction which is used by nature to make carbon-carbon bonds in *e.g.* the synthesis of terpenes and steroids.

$$\begin{array}{c} CH_3 \\ {}^{\textstyle >}C=CH-CH_2OPOP \\ CH_3 \end{array} \quad + \quad \begin{array}{c} CH_2 \\ {}^{\textstyle >}C-CH_2-CH_2OPOP \\ CH_3 \end{array}$$

$$\downarrow$$

$$\begin{array}{cc} CH_3 & CH_3 \\ | & | \\ CH_3-C=CH-CH_2-CH_2-C=CH-CH_2OPOP \end{array}$$

There is an interesting way of using a monoester of phosphoric acid of phosphorylate an alcohol yielding a diester of phosphoric acid. If we treat a monoester with an alkyl or arylsulphonyl halide there is first formed a mixed anhydride which then breaks up to produce the true phosphorylating agent—a monomeric ester of metaphosphoric acid ($ROPO_2$) the true internal anhydride of the phosphomonoester. Monomeric metaphosphoric acid (HPO_3), the internal anhydride of phosphoric acid itself, is an extremely powerful non-specific phosphorylating agent; although never isolated in the free state it has been shown to exist as a transient intermediate in a variety of reactions involving the transfer of phosphate residues.

Now let us return to ATP and its reactions and see whether our knowledge of phosphate chemistry helps to explain not only how it functions but why it is so important. As we have already noted, it functions as nature's agent for the transfer of energy and of phosphate. It also functions with the other nucleoside triphosphates in the synthesis of nucleic acids. Why the nucleoside adenosine should be the favoured material to carry the triphosphate residue I do not know but in the light of our knowledge of phosphate chemistry it is easy to see why nature uses the triphosphate part. First of all, contrary to what was often alleged by biochemists in the past there is nothing special about the energy content of the phosphorus-oxygen bond in the anhydrides of phosphoric acid; we know lots of acid anhydrides which, on fission, yield just as much or even more energy. So this is not the reason; the decisive feature is the reactivity and the great flexibility of the triphosphate system. To understand this fully let's look for a moment at the behaviour of esters of the simpler pyrophosphoric acid.

$$
\begin{array}{cc}
\underset{\text{I}}{\overset{RO}{\underset{RO}{>}}P\overset{\displaystyle O}{\overset{\|}{-}}O-\overset{\displaystyle O}{\overset{\|}{P}}\overset{OR}{\underset{OR}{<}}} &
\underset{\text{II}}{\overset{RO}{\underset{RO}{>}}P\overset{\displaystyle O}{\overset{\|}{-}}O-\overset{\displaystyle O}{\overset{\|}{P}}\overset{OR}{\underset{OH}{<}}}
\end{array}
$$

$$
\begin{array}{cc}
\underset{\text{III}}{\overset{RO}{\underset{HO}{>}}P\overset{\displaystyle O}{\overset{\|}{-}}O-\overset{\displaystyle O}{\overset{\|}{P}}\overset{OR}{\underset{OH}{<}}} &
\underset{\text{IV}}{\overset{RO}{\underset{RO}{>}}P\overset{\displaystyle O}{\overset{\|}{-}}O-\overset{\displaystyle O}{\overset{\|}{P}}\overset{OH}{\underset{OH}{<}}}
\end{array}
$$

$$
\underset{\text{V}}{\overset{RO}{\underset{HO}{>}}P\overset{\displaystyle O}{\overset{\|}{-}}O-\overset{\displaystyle O}{\overset{\|}{P}}\overset{OH}{\underset{OH}{<}}}
$$

Monoesters of pyrophosphoric acid (V) are rather stable unreactive compounds as are the symmetrical diesters (III). But as soon as more than two of the acidic hydroxyls are blocked (as in I–II) and in (IV) then the esters are active phosphorylating agents. Thus (I) and (II) phosphorylate by the normal acylation procedure of nucleophilic attack on phosphorus and expulsion of a stable anion but (IV) is a very unstable substance breaking up to give a phosphodiester anion and monomeric metaphosphoric acid, a powerful phosphorylating agent. At first sight this might suggest that adenine diphosphate would be quite adequate for nature's purposes. But a little reflection shows us just how superior ATP is. In the first place, of course, the more stable the anion liberated in generating metaphosphate the better will be the reagent and since pyrophosphate anion is more stable than phosphate, ATP will be a better metaphosphate producer (*i.e.* phosphorylating agent) than ADP. The beauty of ATP is that provided the acidic hydroxyls are free *i.e.* the substance can ionise, it is a stable relatively unreactive compound which can be transported and stored by organisms without difficulty.

$$A\alpha O-\overset{\overset{O}{\|}}{\underset{\underset{OH}{|}}{P_1}}-O-\overset{\overset{O}{\|}}{\underset{\underset{OH}{|}}{P_2}}-O-\overset{\overset{O}{\|}}{\underset{\underset{OH}{|}}{P_3}}-OH$$

If, however, we block the hydroxyls on P_1 and P_2—for example by complexing with a metal then we have a powerful reagent for the transfer of the P_3 phosphate as metaphosphate with concomitant formation of ADP; this is the normal behaviour of ATP as a phosphate transfer agent and energy liberator. Alternatively, if we block the hydroxyls on P_2 and P_3 in a similar fashion we will transfer nucleotide *via* the reactive intermediate adenosine metaphosphate. This explains the use of the other nucleoside triphosphates for synthesis both in the nucleic acid and coenzyme fields. I have already mentioned the formation of carbon-carbon bonds in nature by means of the enzyme controlled reaction of dimethyl allyl pyrophosphate and *iso*-pentenyl pyrophosphate; here too the process is easily understood and the beauty and economy of nature's methods stand out very clearly. These pyrophosphate esters are perfectly stable as long as the ionisable groups are free. When blocked the dimethyl allyl pyrophosphate will be the more ready for attack by the weakly uncleophilic isopentenyl pyrophosphate because the pyrophosphate anion to be set free will be more stable than a simple phosphate anion.

Now one begins to see why we encounter phosphates everywhere in living organisms. We do so, I believe, because of their flexibility which makes them very economical in use and capable of performing a great variety of functions which are essential if life is to proceed. Let us try to summarise what nature needs to facilitate all she has to do with carbon-based building bricks.

1) She needs a strong acid capable of forming anhydrides which can be used for energy storage and transport. Presumably other acids than phosphoric would do for that but phosphoric can fulfil many other purposes in addition.

2) The acid must be tribasic so that it may act as a link between two molecules or groups and still have one free acidic hydroxyl for further reaction.

3) The strength of the acid is important since it permits use in carbon-carbon bond synthesis.

4) As a tribasic acid both it and its monoesters should be capable of forming internal anhydrides which can be powerful acylating agents.

When one looks at the periodic table there is only one element which can give such an acid—phosphorus. The only other element of comparable character when it comes to forming oxy-acids is sulphur but it is quite useless for most of the above purposes because it gives only dibasic acids. It is my belief that the requirements listed above are so vital to the development of organised life that I would guess that if life exists anywhere else in the universe it will do so only on a planet on which phosphorus is readily available.

Thermodynamics and
Bioenergetics

E. C. SLATER

E.C. SLATER was born on January 16, 1917, in Melbourne, Australia. He was educated at Gee-long College and Melbourne University where he graduated in 1939. In the same year he was appointed Biochemist at the Australian Institute of Anatomy, Canberra where he remained until 1946 when he went to the Molteno Institute, University of Cambridge, England to work under D. Keilin. In 1949 he worked with S. Ochoa in New York returning to Cambridge in 1950. He received the Ph.D. degree in 1948 and the Sc.D. degree in 1960. In 1955, he was appointed Professor of Physiological Chemistry at the University of Amsterdam, The Netherlands, a position which he still holds. Since 1957 he has been Managing Editor of Biochimica et Biophysica Acta. From 1971 to 1979 he was Treasurer of the International Union of Biochemistry. He is a fellow of the Royal Society, member of the Royal Netherlands Academy of Sciences, foreign member of the Royal Swedish Academy of Sciences, foreign corresponding member of the Royal Belgium Academy of Medicine and of the Argentinian Academy of Sciences, and Honorary Member of the Society of Biological Chemists and the Japanese Biochemical Society.

Not so long ago, the question whether the second law of thermodynamics is applicable to living organisms was very controversial. In the non-animal world spontaneously occurring processes lead to an increase in disorder or, in the language of thermodynamics, to a state of thermodynamic equilibrium, or of "maximum entropy." The living organism, on the other hand, builds order out of disorder. For example, bacteria, starting from simple sugars and a source of nitrogen (in some bacteria atmospheric nitrogen serves), are able to build up the highly organized molecules of nucleic acid and protein and use them to construct new cells. Does this mean that bacteria do not obey the second law of thermodynamics? Or, to take another example, growing cattle synthesize the highly organized proteins, that we later enjoy as beef steak, from much simpler constituents present in the grass. Do cattle evade the second law of thermodynamics?

The famous physicist Max Planck was inclined to think so. He was reluctant to apply the principles of thermodynamics to biological phenomena since, in his view, biological systems lead to an up-grading ("veredelung") whereas the physical laws teach us that levelling is more probable ("das Gewöhnliche und Gemeine von vorherein wahrscheinlicher sind als das Geordnete, Vorzügliche, Hervorragende" cited in (*1*)).

Schrödinger in his book "What is life" (*2*), however, realized that living matter only seemingly evades the decay to equilibrium, since its organization is maintained by extracting order from the environment.

"Every process, event, happening—call it what you will; in a word, everything that is going on in Nature means an increase of the entropy of the part of the world where it is going on. Thus a living organism continually increases its entropy—or, as you may say, produces positive entropy—and thus tends to approach the dangerous state of maximum entropy, which is death. It can only keep aloof from it, *i.e.* alive, by continually drawing from its environment negative entropy."

The synthesis of protein in the bacterial cell or in the tissues of the cattle is not a thermodynamically closed system.

We now know that the chemical system responsible for the synthesis of protein is coupled to the hydrolysis of ATP to ADP and inorganic phosphate, a reaction that is characterized by a large increase in entropy. In the cattle, the ATP is resynthesized, with consequent loss of entropy, by coupling the reaction with the oxidation of foodstuffs such as carbohydrate by oxygen to carbon dioxide and water, a reaction with a large increase in entropy. The sum total of these processes taking place in the cells of the tissues of the animal is an increase in entropy:

	ΔS
Amino acids $\rightleftharpoons$ protein $+ H_2O$	—
$ATP + H_2O \rightleftharpoons ADP + P_i$	+
Foodstuff $+ O_2 \rightleftharpoons CO_2 + H_2O$	+
$ADP + P_i \rightleftharpoons ATP + H_2O$	—

sum: foodstuff $+ O_2 +$ amino acids $\rightleftharpoons CO_2 + H_2O +$ protein $+$.

Thus, we need to break down the thermodynamics of growing cattle into two parts linked by ATP—intracellular respiration with a positive ΔS and protein synthesis with a negative ΔS. To put it another way, the cattle feeding on grass become more ordered, because the grass becomes more disordered when it is oxidized to the simple molecules CO_2 and H_2O.

However, this is not the whole story. If more grass becomes disordered than beef steak becomes ordered, the total life on earth would decline; in fact, it could never have started in the first place. Thus, the existence of intracellular respiration and ATP as a means of feeding the organism with negative entropy is insufficient to still any doubts whether the second law of thermodynamics is obeyed by living things.

Two things have been left out of the picture, first the flow of radiant energy from the sun to the earth, which is a spontaneous process coupled with an increase in entropy, and, secondly, the presence in the green plant of a catalytic system capable of utilizing this energy for the synthesis of foodstuff from carbon dioxide and water.

$$CO_2 + H_2O \xrightarrow{h\nu} \text{foodstuff} + O_2$$

If we include this reaction in the sum reaction described above, we obtain

$$\text{Amino acids} \xrightarrow{h\nu} \text{protein.}$$

Thus, by the intervention of the plant, radiant energy has been utilized to make beef steak. Life on earth is not a closed system, since the earth receives radiant energy from the sun. Indeed, the phrase "life on earth" is incomplete, it should be "life on earth in the sunshine." Living things *appear* to evade the second law of thermodynamics by being fed with negative entropy from the sun.

As de Groot (1) has pointed out, they are not unique in this respect. In other thermodynamically open systems, it is also possible to maintain a loss of entropy. He gives as an example a metal rod that is heated on one side and cooled on the other. By virtue of exchange of energy with its environment, the temperature in the rod, which was originally uniform, becomes non-uniform, being higher on the side where there is a source of heat. Thus the entropy of the rod decreases—it has been fed with negative entropy from the environment.

The sub-discipline of biochemistry or molecular biology called bioenergetics is largely devoted to the study of the way reactions leading to greater disorder are coupled to those leading to greater order. The second law of thermodynamics in its statement that spontaneous chemical reactions proceed with a loss of Gibbs free energy (G), which is equal to $H + TS$, is rigorously obeyed, and imposes boundary limitations to any valid description of the process under investigation.

For example, it has been shown that when mitochondria catalyse the oxidation by molecular oxygen of succinate to fumarate [+malate], ATP is synthesized from ADP and P_i, until the quotient $[ATP]/([ADP] \cdot [P_i])$, measured with components in the suspension medium, reaches 2.5×10^5 M^{-1}. Expressing this in terms of phosphorylation potential.

$$\Delta G_p = \Delta G^0 + RT \ln \frac{[ATP]}{[ADP] \cdot [P_i]}$$

$$=30.9+36.4$$

$$=67.3 \text{ kJ/mol}$$

When the reaction reaches a steady state, with $p_{o_2}=0.2$, the ratio [NADH]/[NAD$^+$] in the mitochondria was found to equal 1.6 in a particular experiment. From this, it is possible to calculate the ΔG for the reaction

$$\text{NADH}+\tfrac{1}{2}\text{O}_2 \rightleftharpoons \text{NAD}^+ + \text{H}_2\text{O}$$

$$\Delta G_o = \Delta G^0 + RT \ln \frac{[\text{NAD}^+]}{[\text{NADH}]} \cdot (p_{o_2})^{\frac{1}{2}} = -212.3 \text{ kJ}.$$

If we write the overall phosphorylation reaction as

$$\text{NADH}+\tfrac{1}{2}\text{O}_2 + n\text{ADP} + n\text{P}_i \rightleftharpoons \text{NAD}^+ + \text{H}_2\text{O} + n\text{ATP} + n\text{H}_2\text{O}$$

then

$$\Delta G = \Delta G_o + n\Delta G_p$$

and since $\Delta G \leq 0$, n must be $\leq 212.3/67.3 = 3.16$.

Within the limits of the experimental accuracy of the measurements on which this value of n is calculated, the conclusion that it cannot exceed 3.16 is quite rigorous. In fact, since respiration still proceeds when the steady state (State 4) is reached, n must be less than 3.16.

This conclusion, which is consistent with more direct measurements of the P:O ratio in oxidative phosphorylation, and which is completely independent of the mechanism of coupling between respiration and phosphorylation, is a boundary condition that any description of the process must fulfill.

It is widely believed that coupling mechanism is provided by protons that are electrogenically pumped across the mitochondrial membrane from the matrix to the intermembrane phase, in a process coupled with electron transfer, and which are drawn in again through an ATP-synthesizing enzyme (3). It has been established (4) that for each molecule of ATP synthesized within the matrix and translocated to the outside, an extra H$^+$ atom must be pumped out by the respiratory chain. The uptake of phosphate is non-electrogenically accompanied by influx of a proton (3). The reactions may be written

$$\text{NADH}+\tfrac{1}{2}\text{O}_2 + (m+n)\ \text{H}_i^+ \rightleftharpoons \text{NAD}^+ + \text{H}_2\text{O} + (m+n)\ \text{H}_o^+ \tag{1}$$

$$m\text{H}_o + n\text{ADP}_i + n\text{P}_{ii} \rightleftharpoons n\text{ATP}_i + m\text{H}_i^+ \tag{2}$$

$$n\text{ATP}_i + n\text{ADP}_o \rightleftharpoons n\text{ATP}_o + n\text{ADP}_i \tag{3}$$

$$n\text{P}_{io} + n\text{H}_o^+ \rightleftharpoons n\text{P}_{ii} + n\text{H}_i^+ \tag{4}$$

Sum: $\text{NADH}+\tfrac{1}{2}\text{O}_2 + n\text{ADP}_o + n\text{P}_{io} + n\text{H}_o^+ \rightleftharpoons \text{NAD}^+ + \text{H}_2\text{O} + n\text{ATP}_o$

where the suffixes i and o refer to inside and outside the mitochondrial matrix, respectively.

Since the translocation of protons across the mitochondrial membrane is an electrogenic process, the difference in electrochemical activity of the protons across the membrane is given by the expression

$$\Delta\tilde{\mu}_{H^+} = \Delta\Psi + \frac{RT}{F} \ln \frac{[H_o^+]}{[H_i^+]} \, .$$

The second law of thermodynamics requires that each of these partial reactions proceeds in such a way that $\Delta G < O$. Applying this to Eq. 1,

$$\Delta G = -212.3 + (m+n)\,\Delta\tilde{\mu}_{H^+} \leq O$$

from which it follows that $(m+n)\,\Delta\tilde{\mu}_{H^+} \leq 212.3$ kJ. $\qquad(5)$

Similarly from Eqs. 2–4,

$$\Delta G = n \times 67.3 - (m+n)\,\Delta\tilde{\mu}_{H^+} \leq O \qquad(6)$$

from which it follows that $(m+n)\,\Delta\tilde{\mu}_{H^+} \gtrsim n \times 67.3$ kJ.

These boundary conditions, which are quite rigorous, indicate that if $n=3$, as is generally believed, and $m+n=6$ (5), $\Delta\tilde{\mu}_{H^+}$ is between 35.4 and 33.6 kJ, *i.e.* about 360 mV, which is greater than the value of 225 mV that has been reported (6). Either this value for $\Delta\tilde{\mu}_{H^+}$ is incorrect or $(m+n)$ is not equal to 6 or n is equal to 3, or a combination of these possibilities. For example, if it is accepted that $\Delta\tilde{\mu}_{H^+} = 225$ mV (6) $= 21.7$ kJ/mol, an upper limit of $(m+n)$ is 212.3/21.7 = 9.8. The second law of thermodynamics does not help in making a choice. Additional experimental information is required.

One of the approaches towards obtaining this information has been to make use of the principles of what has variously been termed non-equilibrium (7), ir- reversible (8), and mechanistic (9) thermodynamics. The description that follows is undoubtedly too simplistic, but is, I believe, sufficient for the purpose and will, perhaps, lift some of the fog that exists for many in the application of nonequili- brium thermodynamics to biochemical systems.

Let us consider a chemical reaction $A \rightleftharpoons B$, catalyzed by an enzyme. If the enzyme is added to a mixture of A and B, reaction will proceed until equilibrium is reached, that is until as many molecules of A are converted into B as molecules of B are converted into A, or until $d[A]/dt = -d[B]/dt$ and the net conversion of A into B, $\nu_{A\rightarrow B}$, is equal to zero. If now the system is perturbed so that it is no longer in equilibrium, more molecules of A will be converted into B than molecules of B are converted into A, or *vice versa*, so that there will be a net conversion of A into B (or *vice versa*) until a new equilibrium is reached.

For simplicity, let us consider the case where the perturbation is such that ΔG for the reaction $A \rightarrow B$ is negative, so that there will be a net conversion of A into B. If we plot the velocity of this net conversion $(\nu_{A\rightarrow B})$ against $-\Delta G$, as in Fig. 1, we now have two points: one corresponding to $\nu_{A\rightarrow B} = 0$, when $-\Delta G = 0$

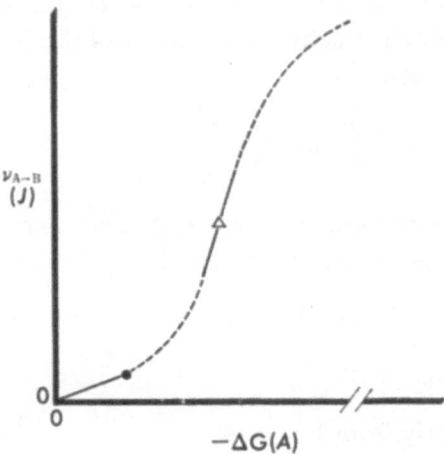

FIG. 1.   Relationship between flow (J) and force (A) in oxidative phosphorylation.

(no net reaction at equilibrium) and one, given by ●, that represents a net conversion of A into B ($\nu_{A \to B}$ positive) when $-\Delta G$ is positive.

Since for every positive value of $-\Delta G$, $\nu_{A \to B}$ is positive, we may also insert an infinite series of points between 0 and ●, *i.e.* we may draw a line between these points and, since over a sufficiently short distance any line can be approximated to a straight line (and we have given no scale in Fig. 1), we may derive the relation, for a region sufficiently close to the origin,

$$\nu_{A \to B} = \text{constant} \times (-\Delta G)$$

or, in the terminology of irreversible thermodynamics

$$\underset{\text{flux}}{J} = \underset{\text{constant}}{L} \times \underset{\text{force}}{A}.$$

What we have really said in the above paragraph is that the further the system is perturbed from equilibrium, the faster it will move towards equilibrium. Since, however, we have not derived a quantitative relationship between J and A, the conclusion that the rate of net conversion of A into B (flux) is proportional to A (the force driving the reaction towards equilibrium) is valid only when $A \; (= -\Delta G)$ is vanishingly small.

Indeed, it is known that oxidative phosphorylation does not obey this simple equation. If mitochondria are added to a mixture of oxidizable substrate (*e.g.* succinate), O_2, ADP, and P_i, O_2 is consumed rapidly until the phosphorylation potential (see above) approaches 67 kJ/mol, when it decreases sharply. The value of $-\Delta G$ for the overall reaction

$$\text{Succinate} + \tfrac{1}{2}O_2 + n\text{ADP} + n\text{P}_i \rightleftharpoons \text{fumarate} + H_2O + n\text{ATP} + nH_2O$$

depends upon the value of n. If the usual value of $n=2$ is accepted for this reaction,

$-\varDelta G = 27.6$ kJ/mol. If now ADP is added, so that its concentration is increased 10-fold, $-\varDelta G$ will be increased by only 5.7 kJ/mol, *i.e.* of the order of 20%, but the rate of respiration is increased many fold.

We must conclude that, for oxidative phosphorylation in any case, when $-\varDelta G$ becomes sufficiently large, J responds relatively more to a change in $-\varDelta G$ than expected from proportionality. We can, then, insert point $\varDelta$ in Fig. 1, and points sufficiently close to $\varDelta$ will fall on a line, the tangent of which at $\varDelta$ is given by the equation

$$J = L'A - \text{constant}$$

where L' is not the same as, but greater than, the L describing the relationship when $-\varDelta G \approx 0$.

It is obvious that the rate of oxidative phosphorylation cannot increase indefinitely as we increase $-\varDelta G$. Indeed when the enzyme systems concerned become saturated with their substrates (succinate, O_2, ADP en P_i), if $-\varDelta G$ is further increased by increasing the concentration of one of the substrates, there will be no further increase in the rate of the reaction. Thus, at sufficiently high values of $-\varDelta G$, J is independent of A.

In summary, Fig. 1 describes three regions:

(i) $-\varDelta G \approx 0$ $J = LA$
(ii) $-\varDelta G$ finite $J = L'A - \text{constant}$
(iii) $-\varDelta G = \infty$ $J = \text{constant}$.

If the three regions are joined, it is seen that the relationship between J and A is sigmoidal. It is a feature of sigmoidal curve that the middle part of it may be described, within experimental error, by a straight line.

Note that, although the relationships describing the first and third of these regions have been derived from general (essentially trivial) principles, the description of the second region is derived from experience. In the absence of a theoretical basis for this region, there is little useful that can be derived from the experimentally established relationship.

Van der Meer *et al.* (*10*) have shown that the fundamental concept of Michaelis and Menten (*11*), namely that enzyme and substrate combine, provides this theorectial basis.

If we include the enzyme in the reaction $A \rightleftharpoons B$, the latter may be written

$$A + E \rightleftharpoons AE \rightleftharpoons BE \rightleftharpoons B + E.$$

The steady-state net conversion of A to B is given by

$$J = \frac{(V_A \cdot [A]/K_A) - (V_B \cdot [B]/K_B)}{1 + [A]/K_A + [B]/K_B}$$

where V_A and V_B are the velocities of the forward and reverse reactions, at infinite concentration of A and B, respectively, and K_A and K_B are the corresponding Michaelis constants (12). The relationship between the values for V and K_m for the two reactions, and the equilibrium constant of the reactions (K_{eq}) is given by the Haldane equation (12)

$$K_{eq}=\frac{V_A \cdot K_B}{V_B \cdot K_A} .$$

The affinity A is given by the relationship

$$A=RT \ln K_{eq} \cdot [A]/[B].$$

If the reaction is conservative, i.e. [A]+[B]=constant (C), it may be derived that

$$J=\frac{V_A(e^{A/RT}-1)}{(1+K_A/C)e^{A/RT}+(V_A/V_B)(1+K_B/C)} .$$

This yields a sigmoidal relationship between J and A. Over a large part of the total velocity range, the flux is almost linearly (but not proportionally) dependent on the force, with a slope dependent on the concentration of A+B and the kinetic parameters of the system. The slope of this line is equal to $(\nu_A+\nu_B)/4\ RT$, where ν_A and ν_B are the velocities of the forward and back reactions, respectively, in the absence of B and A, respectively. The intercept of this line on the A axis is equal to

$$\frac{(\nu_A+\nu_B) \ln \nu_A/\nu_B}{4} - \frac{\nu_A-\nu_B}{2} .$$

In fact, Rottenberg (13) had found experimentally that the relationship between J an A is described by a straight line passing to the right of the origin. This was confirmed by van der Meer et al. (10) who further showed that the slope of the line is greater at higher values of C as is to be expected since $\nu_A+\nu_B$ is higher at higher values of A and B (unless they are both greatly in excess of the corresponding K_m). The intercept is dependent upon ν_A/ν_B, vanishing when $\nu_A=\nu_B$.

Van der Meer et al. (10) plotted J against A_p, the affinity of phosphorylation. Under the conditions of their experiments, the oxygen affinity was kept constant (-164 kJ/mol), and in Fig. 2 J is plotted against the total affinity, making the assumption that the P:O ratio for succinate oxidation is 2.

A striking feature of the experimentally determined relationship between J and A, shown in Fig. 2, is the steepness of the straight-line portion. This is another reflection of the sharpness of the so-called State 3-State 4 transition, to which reference has been made above. Indeed, the relationship between flows and forces in oxidative phosphorylation is very far from the proportionality between the two parameters which is one of the axioms of irreversible thermodynamics. In the

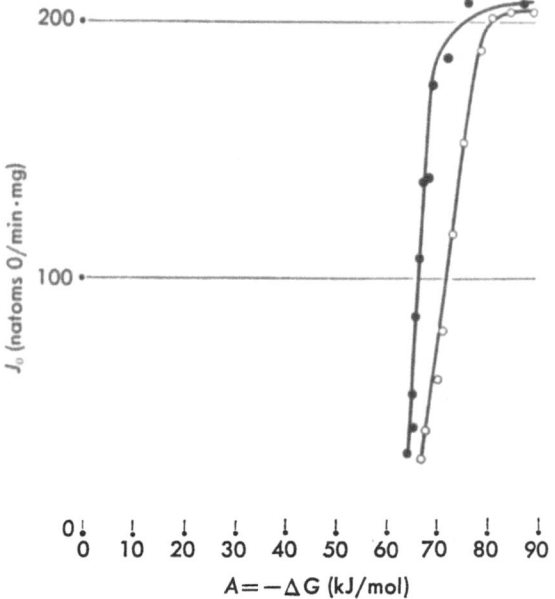

FIG. 2.   Experimentally measured rate of respiration of a mitochondrial suspension oxidizing succinate, plotted against $-\Delta G$, calculated on assumption that P:O ratio=2. Calculated from data in Ref. *10*. ○ 0.2 mM adenine nucleotide; ● 1 mM adenin nucleotide.

critical region, the relationship approaches an "all or none" phenomenon.

Although, then, in principle one might expect to obtain useful mechanistic information from the experimentally determined straight line, the very high slope of the line means that calculations of meaningful parameters from the slope will be very sensitive to experimental errors. This may well prove to limit the usefulness of non-equilibrium thermodynamics in this field.

Van Dam *et al.* (*15*) have made use of the nearly linear relationship found between flows and forces to derive equations relating the rates of respiration and phosphorylation with the phosphorylation potential and $\Delta\tilde{\mu}_{H^+}$. An important part of their treatment is to include an additional reaction describing the "leak" of protons across the membranes. The velocity of the "leak" reaction may be experimentally controlled by addition of a protonophore such as 2,4-dinitrophenol. As is to be expected from these equations, which in view of their complexity we shall not reproduce here, a linear relationship was found between the rate of net phosphorylation (J_p) and the rate of respiration (J_o) as the phosphorylation potential was lowered by addition of hexokinase and glucose. Also in agreement with these equations, the slope of the line (dJ_p/dt) and the intercept on the J_o axis increase with increasing concentration of ionophore. The lines cross at a point corresponding to a P:O ratio (n) with succinate as substrate, of 1.46.

When there was no net phosphorylation, van Dam *et al.* (*15*) measured a phosphorylation potential (ΔG_p) of 505 mV (=48.7 kJ/mol), ΔG_o=1,603 mV =155 kJ/mol, and $\Delta\tilde{\mu}_{H^+}$=218 mV=21.0 kJ/mol. The equations corresponding to

Eqs. 5 and 6 above are

$$(m+n) \, \Delta\tilde{\mu}_{H^+} \leq 154 \text{ kJ} \qquad (7)$$

$$(m+n) \, \Delta\tilde{\mu}_{H^+} \geq n \times 48.7 \qquad (8)$$

and putting $n=1.46$ and $\Delta\tilde{\mu}_{H^+}=21.0$ in Eq. 8, $(m+n) \geq 3.4$. This is consistent with Eq. 7 which yields $(m+n) \leq 7.3$. The two extremes yield values of m/n (the H^+/ATP stoichiometry of the ATP synthase reaction) of 1.3 and 4.0, respectively. Since the ATP synthase reaction is closer to equilibrium than respiration, the true value probably lies closer to 1.3 than 4.0. In fact, van Dam et al. assumed that the ATPase is in equilibrium when there is no net phosphorylation.

The new values of the stoichiometry of oxidative phosphorylation and of the proton pumps involved, derived by van Dam et al. by application of the near-linear relationship between flows and forces, have removed the inconsistencies in the accepted values that are revealed by the application of classical thermodynamics described above. However, it remains uncertain whether this procedure will be sufficient alone to settle the values of $(m+n)$ and n. The sensitivity of flows to small changes in forces, illustrated in Fig. 2, is also reflected by the small effects on $\Delta\tilde{\mu}_{H^+}$ and ΔG_p brought about by initiation of oxidative phosphorylation. For example, addition of a small amount of hexokinase doubled the respiration without any observable effect on ΔG_p and a decline in $\Delta\tilde{\mu}_{H^+}$ of less than 2%. Even the largest amount of hexokinase used by van Dam et al., which was sufficient to increase the respiration to 5.3-fold, caused a decline in ΔG_p and $\Delta\tilde{\mu}_{H^+}$ of only 13% and 12%, respectively. Thus the ratio $\Delta G_p : \Delta\tilde{\mu}_{H^+}$, which is an underlimit of $n/(m+n)$ (see Eq. 8), is not changed during a more than 5-fold change in respiratory rate. A similar insensitivity of $\Delta\tilde{\mu}_{H^+}$ to respiratory rate had been reported by Sorgato et al. (16) in malonate-inhibited sub-mitochondrial particles.

An explanation of the extreme sensitivity of the electron and phosphorylation flows to very small changes in the forces is lacking.

REFERENCES

1. De Groot, S.R.: 1961. Leven en Dood, pp. 115–121, De Erven F. Bohn, Haarlem.
2. Schrödinger, E.: 1945. What is Life, Univ. Press, Cambridge.
3. Mitchell, P.: 1966. Chemiosmotic Coupling in Oxidative and Photosynthetic Phosphorylation, Glynn Research Ltd., Bodmin.
4. La Noue, K., Mizani, S.M., and Klingenberg, M.: 1978. J. Biol. Chem. 253, 191–198.
5. Mitchell, P. and Moyle, J.: 1967. Biochem. J. 105, 1147–1162.
6. Mitchell, P. and Moyle, J.: 1966. Eur. J. Biochem. 7, 471–484.
7. Katchalsky, A. and Curran, P.F.: 1967. Non-equilibrium Thermodynamics in Biophysics, Harvard University Press, Cambridge, Mass.

8. Westerhoff, H.V. and van Dam, K.: 1979. *Current Topics in Bioenergetics*, Vol. 9 (Sanadi, D.R., ed.) pp. 1–62, Academic Press, New York.

9. Van Dam, K. and Westerhoff, H.V.: 1980. *Recueil des Travaux Chimiques des Pays-bas* 99, 329–333.

10. Van der Meer, R., Westerhoff, H.V., and van Dam, K.: 1980. *Biochim. Biophys. Acta* 591, 488–493.

11. Michaelis, L. and Menten, M.L.: 1913. *Biochem. Z.* 49, 333–369.

12. Haldane, J.B.S.: 1930. *Enzymes*, Longmans, London.

13. Rottenberg, H.: 1973. *Biophys. J.* 13, 503–511.

14. Hinkle, P.C. and Yu, M.L.: 1979. *J. Biol. Chem.* 254, 2450–2455.

15. Van Dam, K., Westerhoff, H.V., Krab, K., van der Meer, R., and Arents, J.C.: 1980. *Biochim. Biophys. Acta* 591, 240–250.

16. Sorgato, M.C., Branca, D., and Ferguson, S.J.: 1980. *Biochem. J.* 188, 945–948.

The Early Days of Polymer Science*

Cellulose and Polypeptide

H. MARK

H.F. MARK was born in Vienna, Austria on May 3, 1895. He studied at the University of Vienna, obtained his Ph.D. degree in 1921 and moved as instructor to the University of Berlin. One year later he joined the Kaiser Wilhelm Institut für Faserstoff-Chemie in Berlin-Dahlem where he worked until 1926. In 1927, he joined the I.G. Farben-Industrie in Ludwigshafen on Rhine as Research Chemist, became group leader in 1928 and assistant research director in 1930. At the same time, he was associate professor of Physical Chemistry at the Technical University in Karlsruhe. In 1932, he was appointed Professor of Chemistry and Director of the first Chemical Institute at the University of Vienna, Austria, where he stayed until 1938. After the invasion of the Nazis, he was dismissed and left Europe in 1938 to become Research Manager of the Canadian International Paper Company in Hawkesbury, Ontario, Canada. In 1940 he became adjunct professor of Organic Chemistry at the Polytechnic Institute of Brooklyn, was promoted to Full Professor in 1942, appointed Director of the Polymer Research Institute in 1946 and made Dean of the Faculty in 1961. Presently, he is Dean Emeritus at the Polytechnic Institute and an Emeritus Member of the Corporation. His principal research is on X-rays and electrons for the study of the structure of matter, on the synthesis, characterization, reactions, and properties of natural and synthetic polymers, such as cellulose, rubber, proteins, starch, and all types of synthetic products.

The expression "structure of a molecule" did not become significant and meaningful before the second half of the last century. Up to that date, organic chemists were satisfied to establish for a new molecule which they had synthesized, the chemical *composition* in terms of a stoichiometric *formula* and to describe its properties—color, specific gravity, refractive index, melting point, boiling point, *etc.*—as completely as possible.

---

* This is a shortened version of a paper published in *J. Chem. Educ.* in 1973.

When, about one hundred years ago, the establishment of a *structure* became an important part of a chemical publication, it was particularly Kekule who became the protagonist of the new approach when, in the 1850's, he had the vision of carbon atoms being bonded together and forming chains to which other atoms such as hydrogen, oxygen, or nitrogen could be attached. Later, Kekule added to the concept of an open chain that of a closed ring and explained in a global way the essential differences between aliphatic and aromatic chemistry. All formulas of these days referred to the structure and the behavior of ordinary, small molecules, but when Kekule in 1877 became Rektor of the University of Bonn, and as usual, delivered an inaugural address of general character and wider scope he advanced the hypothesis that the natural organic substances which are most directly connected with life—proteins, starch, and cellulose—may consist of very long chains and derive their special properties from this peculiar structure.

The change in emphasis from *composition* to *structure* led to the demand that any chemist would have to present in his publication the *structural formula* of the material which he was investigating. Since publications are written and printed on paper it was unavoidable that the two dimensional character of this commodity created the inclination for simplifications and distortions of what, in the well known sense of van't Hoff and Le Bel, should have been three dimensional systems. One of the greatest promoters of structural organic chemistry around the turn of the century was Emil Fischer who ingeniously used two dimensional formulas to express without any inconsistencies the most complicated three dimensional structures in the chemistry of sugars and amino acids. As early as 1893 he had already, in general terms, the structure of cellulose as a polysaccharide in mind and expressed the opinion that it might be represented as a chain of glucose units, and his later systematic work on polypeptides clearly indicated a long chain structure for natural proteins (*1*). For two decades, from the turn of the century to his death (July 1919), Emil Fischer's Institute in Berlin was the undisputed center of work on natural high polymers and their precursors.

Many of the distinguished scientists who later figured prominently in the developments of the 1920's were connected with his school for a shorter or longer period: Abderhalden, Bergmann, Delbrueck, Freudenberg, Gabriel, Harries, Leuchs, and Zemplen (*1*).

In order to get a correct impression of the general ideas prevailing in the field of such important natural substances as cellulose, starch, and proteins, early in this century, let us treat each of them separately.

Polysaccharides

French chemists—Braconnot, Payen, Fremy, and Pelouze—studies the composition of various vegetable cell walls and arrived at the conclusion that a certain substance, called "cellulose," was their most important component. Analysis showed that cellulose was a carbohydrate, isomeric with starch, and could be degraded hydrolytically into simple sugars. From these early observations it took a long time until H. Ost demonstrated in 1910 that many celluloses can be converted almost quantitatively into D-glucose by acid hydrolysis. In his "Handbook der Kohlenhydrate" (Leipzig, J.A. Barth, 1941), B. Tollens presented the concept that cellulose may be a long chain consisting of glucose units; no experimental evidence was offered for this idea.

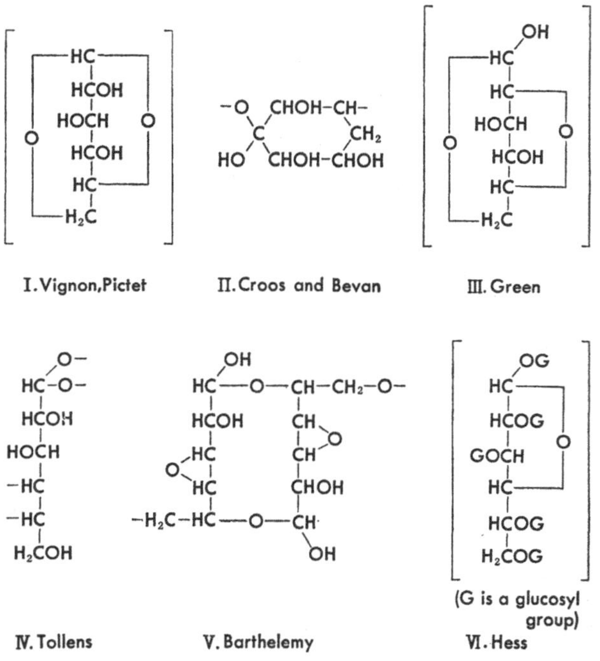

I. Vignon, Pictet   II. Croos and Bevan   III. Green

IV. Tollens   V. Barthelemy   VI. Hess

(G is a glucosyl group)

Early structural proposals for cellulose.

From 1899 to 1920 several structural proposals were offered for the "formula" of cellulose as shown in the figure; they all were based only on the fact that cellulose consists essentially of glucose but did not have the benefit of any additional or more precise evidence. Some of the proposals favored the chain concept, others that of small cyclic building units. Under these conditions it was clear that the separation of cellobiose and other oligosaccharides as intermediate degradation products would be of great importance for the experimental support of any structural concept of cellulose, and considerable efforts were spent on the accumulation of new and reliable experimental data on this point.

In 1920 and 1921 three important papers appeared which postulated long chain structure for several synthetic and natural compounds on the basis of general considerations, and offered specifically for cellulose the long chain character as a preferred alternative in comparison with other structures. The first of these papers was published by Staudinger (1920) and proposed for polystyrene, polyoxymethylene, and rubber, formulas which represented linear long chains such as

Actually, they are still accepted today.

The second paper was published by Freudenberg (1921), offered new experimental data on the yield of cellobiose during cellulose degradation, and stated that the best available data were in conformity with a long chain structure and certainly did not contradict this concept. If Freudenberg on that occasion had taken a more aggressive attitude and had said that his data *proved* the chain structure of cellulose, much confusion would have been spared during the following years; but Freudenberg was much too conservative to go with his statements beyond what he could actually prove. As a consequence this early paper did not play a very important role in the scramble for a unified and convincing formulation of macromolecular systems.

The third article refers to a lecture which M. Polanyi gave on March 7, 1921 commenting on a paper by Herzog and Jancke which presented X-ray data on various cellulosic samples. Polanyi came to the conclusion that the measured X-ray diffraction spots were in agreement *either* with *long glucosidic chains* or with rings *consisting of two glucose anhydride units*. It was made quite clear on this occasion that, on the basis of X-ray data *alone*, one could not distinguish between these two possibilities. The cautious and guarded language of this article gave rise later on the false statement that the small basic unit of the lattice of crystalline cellulose was a proof for a low molecular weight of this material. Although additional attempts were made to correct this wrong position one finds even now the erroneous opinion expressed that the molecular size of a compound cannot be larger than its crystallographic unit cell.

A few years later Haworth, Hirst, and Irvine (1923) demonstrated by a brilliant analytical technique that the hydroxyl groups 2, 3, and 6 in cellulose are still free and consequently, that the bonding between the individual glucose units must be through the carbon atoms 1 and 4. Based on this new important experimental evidence Sponsler and Dore, in 1926, made another significant step in the interpretation of the existing X-ray data by correlating the Haworth glucose ring with the Bragg atomic radii and arrived at a more detailed structure for the cellulose chain than Polanyi. Unfortunately they did not take into account the irrevocable chemical evidence of a 1–4 glucosidic bond in cellobiose and arrived, therefore, at a wrong bonding principle along the length of the chain molecules.

If Polanyi, in 1921, had known Staudinger's article of 1920 in which chain structures are postulated for several natural and synthetic materials and Freudenberg's article of 1921, he would have referred to them and probably placed more emphasis on the chain structure of cellulose, which he offered only as one of two possibilities compatible with the X-ray data alone with no additional information from chemical sources.

If Sponsler and Dore, in 1926, had seen Polanyi's article of 1921 in which he ruled out a non-polar sequence of glucose units in the chains, they would not have proposed the incorrect alternating 1-1 and 4-4 glucosidic bonds along the cellulose molecules but would have preferred the correct continuous 1-4 enchainment. Even Staudinger in his book (2) makes no mention of Polanyi's article eleven years after its appearance although it was the first correct qualitative anticipation of the ultimately accepted macromolecular chain structure of cellulose. These and many similar examples show that at that time only insufficient and slow communications between chemists existed even if they worked on the same substances.

K. Hess, for example, in his well known book on the "Chemistry of Cellulose" (3) presents a detailed account of practically all formulations up to 1928, without arriving at a completely conclusive decision between those proposals which prefer the long chain concept and those which believe that unusually strong association forces between relatively small units are responsible for the "colloidal" character of cellulose and its derivatives.

On the basis of all previously available data, with considerable additional evidence from X-ray diagrams of various cellulose derivatives and from the optical activity of cellulose and its degradation products, Meyer and Mark in 1928 accumulated convincing material for the presently accepted chain structure of cellulose and for the crystalline amorphous character of cellulosic fibers. After the acceptance of this first, approximate draft, many important details had to be settled such as the exact position of each individual atom in the unit cell, the location and direction of the intermolecular hydrogen bonds, and the relationship between the "crystalline" and "amorphous" domains. Meyer and Mark in 1928 spoke of an amorphous "bark substance." Gerngross, Hermann and Abitz in 1930 postulated that gelatin has a "fringe micellar" structure and more recently Manley (1963) has applied the chain folding principle of Keller (1955) to arrive at new concepts concerning the true supermolecular structure of cellulose.

All more recent modifications, refinement, and improvements concerning the structure of native cellulose revolved in essence around the basic formula of Freudenberg, Polanyi, Meyer, and Mark but are a classic demonstration of the fact that any new experimental evidence necessitates amendments and correction of earlier formulations.

Polypeptides

During many years of intense and systematic studies on amino acids, polypeptides, and proteins Emil Fischer never suggested or postulated anything else for the structure of his synthetic products but the character of a linear chain consisting of many amino acid units which are connected with each other by the normal –CO–NH– linkage as it occurs in all amides and peptides.

In a famous lecture which he gave in 1906 at a meeting of the *Deutsche Natur-forscher und Aerzte (1)* he declared that, in his opinion, there exists an uninterrupted line between the simplest dimeric and trimeric amino acids and the native proteins and illustrated this conviction by the demonstration of a linear synthetic polypeptide which he had synthesized step by step with a complete record of all intermediates. It had a molecular weight higher than 1,000.

Hermann Leuchs, one of Fischer's most distinguished associates, who concentrated most of his interests and efforts on the study of heterocyclics and alkaloids, made an even bolder step in the direction of true synthetic polypeptides by the preparation and investigation of the *alpha*-amino acid-N-carboxylic anhydrides; they decompose at elevated temperature, in the presence of traces of moisture, yielding CO_2 and solid bodies which he considered ploymers of a cyclic monomer.

$$\left[\text{HN–CHR–CO} \right]$$

Leuchs described the synthesis of several representative anhydrides of this class, and it is well known that compounds of this type have played and are still playing an important role in the synthesis of numerous linear polypeptides with many important ramifications concerning the structure and properties of native proteins. It is interesting that for many years, these substances did not receive much interest because of the general lack of enthusiasm for the polymer field until K.H. Meyer (1934) and R.B. Woodward (1947) put their existence and significance in the proper perspective.

K.H. Mayer also initiated the first complete X-ray diffraction study of natural silk, which was successfully carried out by R. Brill and, later, with this author, postulated a linear chain structure for natural silk and for all fibrous proteins which would explain most chemical and physical properties of these important materials. In that paper (4) one can also find the first experimentally supported proposal for a helicoidal conformation of polypeptide chains.

Adopting Staudinger's general arguments for the existence of long primary valence chains and striving for a more quantitative and precise description of their structural details, Meyer and Mark added, in 1928, chitin and starch to the list of "truly polymeric' compounds and, in 1930, published the first critical and comprehensive treatise of all natural compounds for which, in their opinion at that time, the main valence structure was an established scientific truth (5).

Lively Objection and Eventual Clarification

While, in this manner, the high polymeric or macromolecular character was first postulated and later, more and more reliably established for the most important natural substances, cellulose, rubber, proteins, and starch, there were many scientists who were unconvinced and preferred the concept that these substances consist of small building units which, however, are held together by exceptionally strong forces of aggregation or association, which were supposed to be a new and still unknown character.

The fact that all materials then under investigation are the products of *living beings*, plants, or animals, was an attractive and probably perfectly legitimate argument in favor of *something new, something which we still have to learn and to clarify* in order to understand the structure and properties of these materials which are so indispensable.

However, as often happens in science and history this somewhat romantic approach had to fade away gradually under the influence of more and better experimental evidence for the macromolecular theory. This did not happen without contradiction and opposition; on the contrary, this author remembers many meetings, symposia, and seminars in the course of which the opposing views were presented and defended with more or less emphasis and success. Strangely enough, even the champions of the long chain or macromolecular aspect, Freudenberg, K.H. Meyer, and Staudinger, did not agree with each other, as they easily could have done, but instead of concentrating on the essential principle, they disagreed on specific details and at certain occasions, they argued with each other more vigorously than with the defenders of the association theory. Of course none of them was, at that time, completely correct in all details of his approach but they all thought and worked in the right direction and, at the end emerged as the natural leaders for future developments.

There were many factors which eventually tipped the scales in favor of the concept of very long, chain-like molecules. One of them was the rapid improvement and refinement of the X-ray diffraction method which again and again not only gave answers in favor of long chains but permitted, and still permits, a progressively detailed description of every kink and twist in a macromolecule. As far as proteins are concerned, it is becoming increasingly clear that each turn and wiggle represents a significant design and has its far reaching consequences for the actions of the substance in biochemistry, biology, and medicine.

If I compare the X-ray equipment and the methods of evaluation which Brill, Katz, Polanyi, and I used in 1923—an air-filled X-ray tube, a ruler and a log table—with the present refined techniques—automatic registration of hundreds of diffraction spots with the aid of a Weissenberg goniometer and direct feed of the output (position and intensity) into a computer—it would seem that the progress is as great as that from an airplane of the early 1920's to a supersonic jet of the mid 1960's,

Another important factor was the introduction of Svedberg's ultracentrifuge (6) which played a decisive role because it was the first method which permitted a direct and absolute measurement of the molecular weight in the range between 40,000 and several millions; at the same time improved osmometers added significance and reliability to these data.

But, probably more than any other single factor, the work of W.H. Carothers and his associates contributed to the ultimate breakthrough in favor of the long chain concept (7). Their efforts extended from synthesis and characterization to ultimate properties and encompassed, with equal emphasis, condesation, and addition polymers. A careful analysis of all prior art led Carothers early to the conclusion that the macromolecular hypothesis was correct, and all his own experiments strengthened his conviction. For him the controversy, association hypothesis *versus* long chain theory, was a matter of the past, and he advanced with full scientific and industrial success on the basis of the latter.

Once the basic concepts of the new branch of chemistry were firmly established the polymer chemists settled to useful and practical work: synthesis of new monomers; quantitative study of the mechanism of polymerization processes in bulk, solution, suspension and emulsion; characterization of macromolecules in solution on the basis of statistical thermodynamics; and fundamentals of the behavior in the solid state with a resulting basic undertsanding of the properties of rubbers, plastics, and fibers.

REFERENCES

1. Hoesch, K.: 1921. Emil Fisher (A lecture on proteins given on January 6, 1906) Verlag Chemic, Berlin.
2. Staudinger, H.: 1932. *Die hochmolekularen organischen Verbindungen*, Julius Springer, Berlin.
3. Hess, K.: 1928. *Chemie der Zellulose*, Akademie-Verlag, Leipzig.
4. Meyer, K.H. and Mark, H.: 1928. Berichte 61, 1932.
5. Meyer, K.H. and Mark, H.: 1930. *Der Aufbau der hochmolekularen organischen Substanzen*, Akad. Verlag, Leipzig.
6. Svedberg, T.: 1940. *The Ultracentrifuge*, Oxford University Press, London.
7. Collected Papers of W.H. Carothers, Wiley-Interscience, New York, 1940.

Viscissitudes of "Tsuzumi" Complexes

MINORU TSUTSUI

M. Tsutsui was born March 31, 1918 in Japan, and graduated from Gifu University, 1938, received Ph.D. degree from Yale University, 1954, and D.Sci. degree from Nagoya University, 1961. After finishing the postdoctoral research at Sloan-Kettering Institute, he entered Monsanto Chemical Co. He moved to New York University where he held professorship during 1965–1968. Since 1968, he was Professor of Chemistry, Texas A&M University. He was a member of New York Academy of Sciences, Honorary Research Professor of the Chinese Academy of Sciences. Fields of his research interests were chemistry of organotransition metals, novel metalloporphyrins, semi and super conductors, catalysis, and diagnostic and chemotherapeutic tumor localizers.

The discovery of "ferrocene" (1) in 1951 brought a revolutional renaissance in organotransition metal chemistry and has expanded its influence over a broad area in science (1, 2). Although the chemical term of this complex is bis η^5-cyclopentadienyl iron, the trivial name, ferrocene, coined by Woodward, a Nobel Laureate, is now commonly used by chemists (3). The significance of ferrocene is two fold, the unusual "sandwich" type structure and the new chemical bond, the π-complex bond. In 1953, the author, who was a graduate student at Yale University, successfully reproduced "Hein's mysterious arts" which Emelius cited as only Hein being able to reproduce (4). Hein's original work was reported in 1918. The secret of Hein's initial experiment was to carry it out in an air free environment. In 1954, with the ingenious suggestion of Onsager, a Nobel Laureate, Zeiss and the author reformulated Hein's polyphenylchromium, triphenyl- and tetraphenylchromium, as benzene-diphenylchromium (2) and bis-diphenylchromium (3), (5–9). Sigma (σ)-bonded Hein's polyphenylchromium compounds were thus reformulated

Dr. M. Tsutsui passed away March 10, 1981.

(1)
Ferrocene

as π-complexes ("sandwich" type complexes). The significance of the Hein's π-complexes in chemistry was probably more serious than that of ferrocene, because this was the first π-complex of a neutral aromatic ligand, benzene, instead of an ionic aromatic ligand, cyclopentadienium ion. Later, dibenzenechromium (4) was also isolated from the reaction mixture of Hein's reaction (10). When we submitted a manuscript to the Journal of the American Chemical Society for publication in 1954, one of the reasons for the reviewer's rejection was his disbelief in the formation of the π-complex with an aromatic ligand and transition metal. I was startled when Hein showed me his lab notes indicating that this complex was first successfully isolated on March 30th, 1918. Hein told me that he must have rested on March 31st in celebration of my birth, since there was no description recorded in his lab book for that day. Although we both laughed, and shook hands, this coincidence, which could be considered a miracle, deeply impressed me as to how amusing and elaborate chemistry is. Because of the close resemblance of the structures to sandwich type π-complexes, such as ferrocene, Hein's π-complexes, and "Tsuzumi" (5), a traditional Japanese musical instrument (a hand drum), I referred to them as "Tsuzumi" complexes (11), a term which was not to last too long, because of the popularity of the term sandwich. The chemistry of "Tsuzumi" type complexes started at this time, particularly in my chemistry career. E.O. Fischer, a Nobel Laureate, independently reported the synthesis of dibenzenechromium (12). There were some talks, one in which it was stated that Zeiss and Fischer were the first

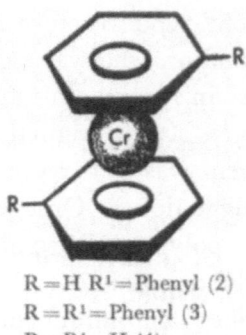

$R = H$ $R^1 =$ Phenyl (2)
$R = R^1 =$ Phenyl (3)
$R = R^1 = H$ (4)
Hein's π-Complexes

(5)

$$\underset{(6)}{\text{He}\overset{\displaystyle CH_3}{\underset{\displaystyle CH_3}{\rule{0pt}{20pt}}}\overset{O}{\underset{O}{\rule{0pt}{12pt}}}\overset{Cl}{\underset{}{Pt}}\ Ha \cdots C \overset{Hb}{\underset{\displaystyle Hc \diagdown C \diagup OH}{\rule{0pt}{12pt}}}} \underset{pK_a=3.5}{\overset{\displaystyle \longrightarrow}{\longleftarrow}} \left[\underset{(7)}{\text{He}\overset{\displaystyle CH_3}{\underset{\displaystyle CH_3}{\rule{0pt}{20pt}}} \overset{O}{\underset{O}{}}\overset{Cl}{\underset{}{Pt}}\overset{O}{\underset{CH_2-C-Hc}{\rule{0pt}{12pt}}} \atop (a,b)} \right]^{-} +H^+$$

proponents of the "Tsuzumi" structure for dibenzenechromium. It was quite certain, that we were in no way familiar with Fischer's work, nor with what he was doing at that time in this field. At the end of 1954, Zeiss told me that Fischer was working on the preparation of $Cr(CO)_6$ from the reaction of $CrCl_3$, Al powder and CO in benzene, and that he had isolated a yellow Cr complex, the structure of which he could not figure out. When Zeiss gave a lecture on our work concerning Hein's "Tsuzumi" complexes at the University of Munich, during the fall of 1954, Fischer was a member of the audience. While listening, the thought came to Fischer that his yellow complex was dibenzenechromium, which has a "Tsuzumi" type structure. Although I have not heard the story from Fischer himself, I trust that Zeiss' version is correct.

The isolation of triphenylchromium tristetrahydrofuranate ($\phi_3Cr \cdot 3THF$), and its rearrangement to Hein's π-complexes (the first example of σ-π rearrangement of an organotransition metal) (13) marked a new era in the field. It was the first to show evidence that the mechanisms of stoichiometric and catalytic aryl coupling reactions would involve σ-π (π-σ) rearrangement and π-complex intermediate (14). We have identified three different types of σ-π rearrangements, namely irreversible and reversible rearrangements, along with dynamic σ-π equilibrium (15). It was quite by accident that we found that η^2-vinyl alcohol (6) on Pt is in rapid equilibrium with η^1-CH_2CHO (7) on Pt in a protic solvent such as acetone. In the activation (conversion) of olefins, including isomerization, hydrogenation, Ziegler-Natta olefin polymerization, olefin oxidation (Wacker Process) process, hydroformylation reaction (oxo process) and others, σ-π (π-σ) rearrangements, particularly σ-π dynamic equilibrium, play essential roles. It has become clearer that σ-π rearrangements play unavoidable roles in both homogeneous and heterogeneous catalyses in the conversion of olefins. In homogeneous catalysis, the mechanism of Ziegler-Natta olefin polymerization with titanium halide and alkylaluminum involves repetition of σ-π rearrangement of an olefin ligand. It is also evident that the Wacker (olefin oxidation) and oxo (hydroformylation) processes involve π-σ rearrangement. The essential steps in hydrogenation and oxidation of olefins also go through π-σ rearrangement. In heterogeneous catalysis, it is difficult to obtain conclusive evidence in studying the reaction mechanism. However, π-σ rearrangement is an essential step in the conversion of olefins in heterogeneous catalysis such as polymerization, isomerization, oxidation, hydrogenation, and other catalytic reactions (16). While σ-π rearrangements of organotransition metals are now recognized in chemistry, we have launched a new program on unusual metalloporphyrins, in which organometallic compounds, particularly metal carbonyls, are used for insertion of metal ions into porphyrins (17). This new metal ion insertion method has produced numerous novel metalloporphyrins, which possess unusual configurations and have therefore created a new chemistry. One of our

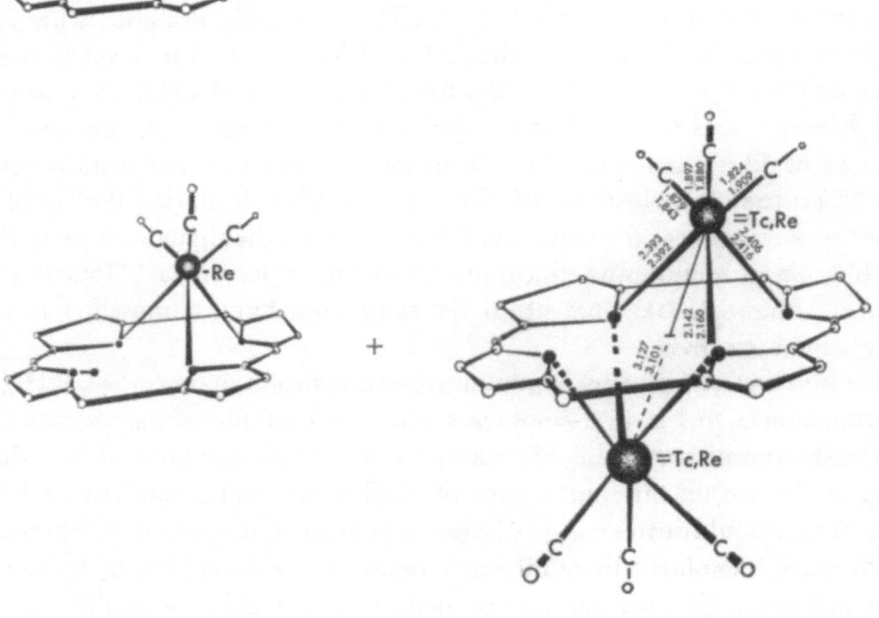

Porphyrin-tridentate ligand

(8)

Porphyrin-hexadentate ligand

(9)

most notable pieces of work is the synthesis of out-of-plane bis $Re(CO)_3$ porphyrin (8) and $Re(CO)_3$ porphyrin (9) from the reaction of $Re_2(CO)_{10}$ and porphyrins.

The structural elucidation of (8) (*17*), by X-ray diffraction analysis concludes the arguments regarding the possible existence of out-of-plane metalloporphyrins. It is interesting to note that the structure of (8) and (9) may be called half (open) and double (open face) sandwiches. We have also succeeded in synthesizing the first "skewed" complex (10) by the oxidation of (8) with $SbCl_5$ (*18*). The distance between two Re ions was 2.95 Å, which clearly demonstrates a half metal-metal bond. The structure of this "skewed" (Kushizashi) (11) complex closely resembles that of "Cupid's Arrow" (11a). However, the synthesis of (10) was truly accidental. We have not yet been able to reproduce this process. The oxidation of (8) with $SbCl_5$ has consistently produced $Re(CO)_2Cl_2PP\ 2SbCl_6$ or $Re(CO)_2\ PP\ SbCl_6$ (PP=porphyrin), which has been formulated by repeated elemental analysis.One single crystal which has been picked up from many other crystals is (10). The attempts to determine the structure of other crystals by X-ray diffraction analysis have failed.

We are also studying the chemistry and the properties of bisphthalocyarato-lanthanides $(Pc_2Ln)^-H^+$, which exhibit the unique electric properties of electro-chromism. The structure of the complex $((Pc_2Nd)^-H^+)$ (12), which had been elucidated by X-ray diffraction analysis, is shown as an exact "Tsuzumi" structure. Two phthalocyanine ligands are staggered in exact 45° angles (*19*). It is now understood that all $(Pc_2Ln)^-H^+$ and Pc_2Ac_1 normally possess a "Tsuzumi" structure. Although the structure of $Hg_3(PP)_2 2CH_3COO^-$ (13) has not been elucidated by X-ray diffraction analysis, a proposed structure has shown a modification of the "Tsuzumi" system (*17*). As the complex (12) exhibit good semiconductivity, "stacked" polymeric complexes, such as (13), are expected to exhibit good (semi)conductivity along with other electric and photo properties. In recent years, the syn-

(10) $SbCl_6^-$ (11-a)

(11)

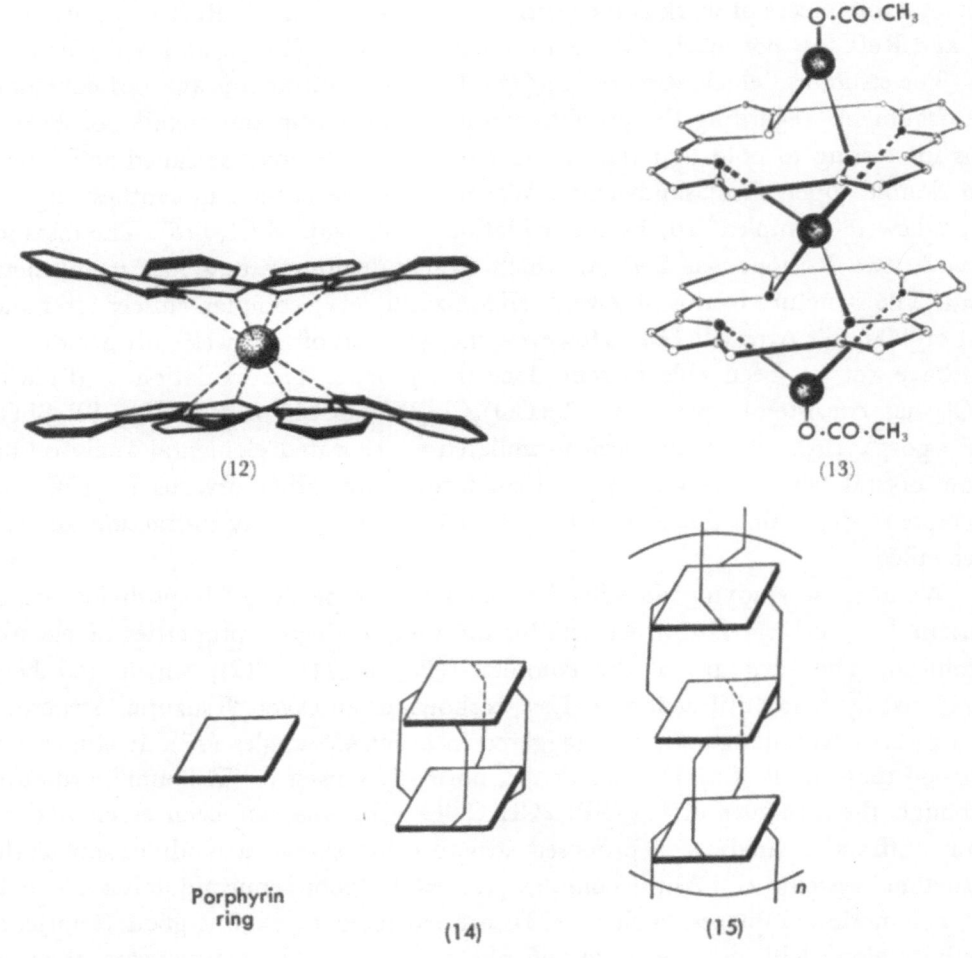

O·CO·CH₃

(12)   (13)

O·CO·CH₃

Porphyrin
ring

(14)   (15)

thesis of dimeric porphyrin (14) has attracted attention because of its potential photo and electro chemical properties.

We are currently conducting the synthesis of a stacked polymer (15) of porphyrins. Although the characterization of the compound is difficult, in the preliminary experiments, the Co complex of (15) seems to be a good dioxygen activator as a fuel cell electrode catalyst.

In summary, I have overviewed the contributions of our group in the flourishing development of organotransition metal chemistry during the past three decades. As scientists, we must be humble as human beings, knowing that much of the progress made in scientific research is due to accidental discoveries and unplanned creations, which are recognized only by dedicated researchers. I have been fortunate to have had opportunities to work with genuine scientists who brought unexpected joy to me through their experiments.

Acknowledgment

Our research programs are currently supported by the Office of Naval Research, the Gas Research Institute and the Robert A. Welch Foundation (A-420).

REFERENCES

1. Miller, A. and Tebboth, J.A.: 1952. *J. Chem. Soc.* 632.
2. Kealy, T.J. and Pauson, P.L.: 1951. *Nature* 168, 1039.
3. Wilkinson, G., Rosenblum, M., Whiting, M.C., and Woodward, R.B.: 1952. *J. Am. Chem. Soc.* 174, 2125.
4. Emelius, H. and Anderson, J.: 1938. *Modern Aspects of Inorganic Chemistry*, D. Van Nostrand Co. Inc., New York.
5. Tsutsui, M.: 1954. Ph.D. Dissertation, Yale University.
6. Zeiss, H.H., Tsutsui, M., and Onsager, L.: 1954. Abstr. 126th National Meeting of Am. Chem. Soc., pp. 29–30, New York.
7. Zeiss, H.H.: 1955. *Yale Sci. Mag.* 29, 14.
8. Zeiss, H.H.: 1955. *Angew. Chem.* 67, 282.
9. Zeiss, H.H.: 1955. *Handbook*, XIVth I.U.P.A.C., Heidelberg. pp. 262,
10. Zeiss, H.H. and Tsutsui, M.: 1957. *J. Am. Chem. Soc.* 79, 3062–3066.
11. Tsutsui, M.: 1962. *Chem. Ind. Japon.* 15, 104,
12. Fischer, E.O. and Hafner, W.: 1955. *Z. Naturforsch.* 106, 665.
13. Herwig, W. and Zeiss, H.H.: 1958. *J. Am. Chem. Soc.* 81, 4891.
14. Tsutsui, M.: 1961. *Ann. N.Y. Acad. Sci.* 93, 133–146.
15. Cotton, F.A., Francis, J.N., Frenz, B.A., and Tsutsui, M.: 1973. *J. Am. Chem. Soc.* 95, 2483–2486.
16. Courtney, A. and Tsutsui, M.: 1977. *Advances in Organometallic Chemistry*, Vol. 16 (Stone, F.G.A. and West, R., eds.) pp. 241–282, Academic Press, New York.
17. Tsutsui, M. and Taylor, G.A.: 1975. *Porphyrins and Metalloporphyrins* (Smith, K., ed.) pp. 279–313, Elsevier Sci. Publ., Amsterdam.
18. Tsutsui, M., Kato, S., Cullen, D.D., and Meyer, E.F.: 1977. *J. Am. Chem. Soc.* 99, 620.
19. Tsutsui, M., Kasuga, K., Pettersen, R.C., Tatsumi, K., Van Opdenbosh, N., Pepe, G., and Meyer, E.F.: 1980. *J. Am. Chem. Soc.* 102, 4835.

A Personal View of a Japanese Natural Products Chemist in an American Institution

KOJI NAKANISHI

K. NAKANISHI was born in Hong Kong in 1925. He studied in Department of Chemistry, Nagoya University, received Ph.D. degree in 1954 with F. Egami and Y. Hirata, and he was a post-graduate student at Harvard University (L.F. Fieser) during 1950–1952. Professor, Department of Chemistry, Tokyo Kyoiku University (1958–1963), Tohoku University (1963–1969), and Columbia University (1969–1980). He is currently Centennial Professor, Department of Chemistry, Columbia University (1980–), and Director, Suntory Institute for Bioorganic Research (1979–). He previously was Director of Research, International Center of Insect Physiology and Ecology, Nairobi, Kenya (1969–1977). His research is centered in the area of natural products, in particular: isolation and structural studies of bioactive compounds at the sub-mg level, insect antifeedants, applications of combined spectroscopy to structural studies, development of the exciton chirality method for determination of absolute configurations, and bioorganic studies in visual pigments and bacteriorhodopsin. He has received the following awards: Award in Pure Chemistry, Chemical Society of Japan (1954); Asahi Cultural Award (1968); American Chemical Society Ernest Guenther Award (1978); Chemical Society of Japan Award (1979); E.E. Smissman Medal (1979); H.C. Urey Award (Columbia University 1980); Remsen Award (1981).

It has been 12 years since I moved to Columbia University in 1969, a move which has had a strongly positive effect on myself as an organic chemist. This is despite the fact that funding in U.S.A. was starting to decline whereas that in Japan was on the uprise. Actually, this dwindling support for academic research which has ironically been one of the major beneficial factors. Namely, the stringent and severe conditions imposed on academic research by NIH and other funding agencies in the early 1970's forced me to reconsider the purpose and values of research. I was quite comfortably set up in Japan and there was no pressing need for me to leave a professor's position at a major university. When I started receiving offers from outside of Japan, I consulted several senior professors in related fields, including my mentors Professors F. Egami (then at the University of Tokyo) and Y. Hirata (then at Nagoya University). Somewhat to my surprise, the unanimous advice I received encouraged me to accept the offer of Columbia University. I was already in my early forties and a transfer at that late age would mean considerable disturbance in my life and research; more serious was my lack of confidence in moving to a major center in organic chemistry such as Columbia University and a reluctance in having to overcome the psychological barrier accompanying the move from an oriental to occidental society. I can gratefully say now that the outcome of the advice from the aforementioned professors has given me the opportunity to change the direction of my research, given me more confidence, and helped me to realize the strength in the American research system.

In those days in Japan, the bulk of federal support a researcher received came automatically with the chair. In contrast, practically all research funding in U.S.A. was based on the proposal system, and moreover, the outcome depended on the peer review system, which on the whole is fair and constructive. This forced me to move into more bio-related areas in which I am convinced lies the future of natural products chemistry. In Japan I was engaged in spectroscopy and the isolation and structure determination of natural products, mostly of plant or microbial origin. The only bio-oriented research we were carrying out was centered around studies on the insect molting hormones ecdysones (with Dr. M. Koreeda) and the determination of the meiosis inducing factor of sea urchin eggs (with Dr. Kurokawa; in collaboration with Professor H. Kanatani, University of Tokyo). Despite the fact that we had just determinated the ginkgolide structures (with Drs. M. Maruyama, A. Terahara, and others), discovered the nuclear Overhauser effect (by Dr. M. C. Woods) during the course of ginkgolide studies, and were developing the exciton chirality method based on the coupled oscillatory theory (by Dr. N. Harada), the future goals of natural products chemistry were unclear to me.

The 3rd IUPAC Symposium on Natural Products Chemistry was held in Kyoto in April 1964, the year of the Tokyo Olympics. I personally regard this year as a turning point in natural products chemistry or organic chemistry in Japan. Natural products chemistry, which unlike other countries, is carried out in the Faculties of Science, Pharmacy and Agriculture, and traditionally has been the strongest branch in organic chemistry in Japan. So called western science in Japan has a history of only one hundred years or so—besides the budding science was devastated during World War II. Moreover, Japan is geographically, language-wise and culture-wise isolated from the mainstream of science; these are inherent handicaps which western scientists do not have to bear with. The 1964 Symposium brought to Japan most of the world's leading or active natural products chemists, either as plenary lecturers or discussion leaders—the guests saw what was happening in Japan, but more importantly, the direct contact with a large number of foreign scientists gave young Japanese chemists a sense of confidence. Hence the 1964 Symposium had contributed greatly in reducing the psychological barrier to overcome the handicaps.

In spite of the explosive increase in quantity and improvement in quality of Japanese output which started in the mid 1960's, it is generally agreed that Japanese science lacks real ingenuity. One major reason for this is the fact that since science did not originate in Japan, the Japanese so far have never been exposed to the problem of creation in science. Namely, the tradition of pioneering research was lacking. The student's primary objective revolved around catching up with his Western counterpart. To learn and understand—this I recall was where the main emphasis was in the late 1950's. Japanese chemistry has by now well caught up. Therefore, the Japanese are being forced to create new science, rather than cleverly exploit, apply, improve or modify existing methods or concepts. This is a new experience for Japanese scientists. If we are to maintain the past rate of progress, there is now no alternative than to undergo this psychological molting. Practically all Japanese chemists are fully aware of the situation; in fact, a transition is occurring, but at too slow a pace. There are too many organizational as well as mental blocks which impede a smooth and healthy transition. Of course, in every country scientists are dissatisfied with the status in their respective countries. However, I shall confine myself to mentioning the shortcomings specific for Japanese society.

To put it in one word, the Japanese scientific structure is, as in practically all other cross sections of society, too homogeneous and authoritative. The hierarchy in Japanese academia was patterned after the 19th century German system is understandably much more distinct than in U.S.A. In Japan the senior professors are revered, whereas in U.S.A. the hierarchy order is rooted much more on one's activity. The hierarchy is present also in the individual research groups, which are called "kōza" (a target of frequent criticism, but if exploited properly can constitute an efficient research team). A typical kōza consists of a full professor, an assistant professor, and two assistants as faculty members, which direct research carried out by graduate students and undergraduates. A large kōza may consist of close to 50 members.

Thus, the 4th year undergraduate would join a group for his six month senior undergraduate research project, proceed on to the two year M.S. course, then the three year Ph.D. course, get promoted to the rank of Assistant, go abroad for one to two years as a post-doctoral fellow (almost invariably on grants provided by the host professor or institute and not by the Japanese government), and finally be promoted to Assistant Professor; by then he (very few she's) is at least in the midthirties. If lucky, his mentor who is the full professor may retire, and although public recruitments are announced, if the understudy is doing reasonably well, he usually is promoted.

He may of course be assigned to a full professorship at a university of lesser standing, in which case, he will transfer, often with one to two assistants. The course described here certainly is only one of a multitude, and the selection course is changing gradually. However, it is probably safe to say that it typifies the background of most current professors at major national universities.

Unlike U.S.A. graduate schools, the better undergraduates of major universities in Japan usually proceed on to the same university; there is no domestic post-doctoral system despite the universal and incessant interest for the past thirty years; there are no non-Japanese professors; there is no formalized system of the weekly seminars in which faculty members are invited to other institutes for one day to give talks and to listen to their colleagues' unpublished material. In present days, all this is typical only for Japan; only the Japanese academia has managed to retain the pre-war structure without undergoing an overhaul.

In the early Meiji era (1870–1880) promising young chemists were sent to Europe on government accounts, and also the chemistry in early Japanese universities were often started and taught by foreigners; in this respect, although in an entirely different context, the attitude of the Japanese government 100 years ago was more in common with the rest of the current world than it is today. What does the Ministry of Education do to remedy these defects with which practically every Japanese scientist is concerned? Not much—a modern wonder. Whether this is due to their lack of understanding the acuteness of the problem, their hopelessly bureaucratic mannerisms, or the cool response they receive from the professors of major universities, I do not know. However, if Japan is going to advance further in the role or influence it plays in the chemical community, it is time for action, and not for holding numerous democratic committee meetings which invade so much precious time.

The hours Japanese professors spend in committee and faculty meetings is literally astounding compared to those of U.S. chemistry professors. In the U.S., major decisions are delegated to committees, which are run in a much more informal, frank, and efficient style. The fact that the chemistry graduate students are selected in our department by a group of 3–4 young generation faculty on the basis of school records and recommendations, and that final outcomes are announced only at the semester's first chemistry faculty meeting, is for most Japanese professors a shocking but enviable revelation. Japanese professors spend countless hours at the monthly faculty meetings (all professors belonging to that faculty must attend, e.g., in the Faculty of Science this would comprise biology, chemistry, physics, mathematics, etc.). No wonder the professor, who heads the kōza, finds himself entrapped in many meetings, which steadily drains on his precious time that could be better spent on creative thinking, discussions, reading, etc.

Lack of heterogeneity in the research system and the excessive formal meetings are the two grievous impediments to Japanese science. Japanese science, both quality-wise and quantity-wise, is now all but ready to undergo a transitional molting. But the actual actions taken are slow. It is a pity.

I have benefited vastly from the interdisciplinary interaction to which I am exposed through my research at Columbia. For example, in our studies with the visual pigments and bacteriorhodopsins, we collaborate with biophysicists, physicists, biologists, electrophysiologists, medical people, biochemists, *etc*. This collaboration is also evident in the isolation-structure studies we do, and in this case the cooperation is with insect and plant physiologists, electrophysiologists, entomologists, marine biologists, molecular biologists, biochemists, *etc*. Such multi-disciplinary collaboration is more readily feasible in U.S.A. than in Japan. The major reason here again is the heterogeneous elements in the U.S. society, which is a natural outcome of the multiracial composition, the nonconformist style of college and graduate education which places values on creativity (I am not referring to the education of the "premedical cut-throat mentality") and its fluid research system.

It is usually impossible for the organic chemist to realize where an exciting problem exists in fields outside of organic chemistry proper. By the time a topic appears in the literature it is too late, or even more often, the exciting problems seldom appear in the literature because they cannot be solved without inter-disciplinary collaboration. Furthermore, frequently a bioscientist has no idea what modern isolation, structural and synthetic organic methodology can achieve. Hence exists the rationale and necessity for multidisciplinary cooperations to clarify the functions of Nature, which after all is not separated into artificial classifications such as organic chemistry, insect physiology, physics, *etc*. And if I were permitted to speak from a biased and egocentric view of a natural products chemist it is my belief that our background is ideal for unearthing and solving problems which lay dormant and unnoticed in the interdisciplinary area. An organic chemist can work in collaboration with a bioscientist, to solve the molecular structural basis of a bioscientific problem, and enable the bioscientist to proceed further. However, this is not true in the opposite case. A bioscientist can seldom help an organic chemist with an organic chemical problem.

It is my belief that many exciting problems which have a direct bearing in the daily maintenance of life or a bettering of it lie in the interdisciplinary field, an area which has only recently been tapped, and hence my optimistic views for future bioorganic chemistry, and my gratitude to my mentors, colleagues and students who have led me to realize this.

Fifty Years of Scientific Research

The Fool's Luck

EDGAR LEDERER

E. LEDERER was born in 1908 in Vienna. He studied chemistry at Vienna University and got a Ph.D. degree in 1930. After three years with Richard Kuhn in Heidelberg he worked at the Institut de Biologie Physico-Chimique in Paris and then as Director of Research at the Vitamin Institute, Leningrad. In 1938 he was back in Paris, received a D.ès.Sc. degree from Paris University and got an appointment by CNRS. After seven years in Lyon he returned to Paris in 1947, and in 1952 became Director of Research at CNRS. Since 1956, he became Full Professor at Paris University. From 1961 to 1978, he was Director of the Department of Biochemistry of the Institut de Chimie des Substances Naturelles, CNRS Gif-sur-Yvette. From 1963 to 1978, he was also Professor of Biochemistry at the Institut de Biochimie, Faculté des Sciences de Paris-Sud Centre d'Orsay. He has retired since October 1978 and is still continuing research.

Perhaps some friends and colleagues might be interested in these pages, where I shall try to describe my way from organic chemistry to natural products and biochemistry and, finally to immunology.

I have always been very lucky; indeed, I escaped the Gestapo twice (in 1933 and in 1944), once the German occupation army (Aug. 1st, 1944), several times the French Vichy police and milicia; I also escaped the allied bomb which fell on the Institut de Chimie at Lyon on May 25th 1944, killing several people there.

I was lucky, because the choice of nearly all of my research topics were due to chance observations, or contacts. Lucky also to be able to work with many brilliant and dedicated young colleagues.

My interest in biology is due to my uncle, Hans Przibram (Professor of Biology at the University of Vienna) who initiated me, as a boy, to the mysteries of nature and to the immense joy of scientific discovery.

During my last years in high school, I spent most of my free time with a microscope, looking at the wonders of the unicellular world. I was especially fascinated by the amoebae, by their phagocytic activities and by Metchnikoff's macrophages. My first review article was a "digest," published in 1925 in "Mikrokosmos" ("Amoeboide Zellen im Tier und Pflanzenreich"). It is amusing, and perhaps not quite fortuitous that I am co-author, 55 years later, of a paper describing the activation of macrophages (1).

Why did I study chemistry? Simply because it was the only non-mathematical science promising a living to its students (at that time at least).

After receiving my Ph.D. with a thesis on the synthesis of indole alkaloids under the direction of E. Späth, in July 1930, I had the great luck of being accepted by Richard Kuhn at the Kaiser Wilhelm Institut für Medizinische Forschung, Heidelberg. I arrived there at a crucial moment, when it was essential to develop a method for separating and purifying natural carotenoids (carotene had just been shown to be a provitamin A). This led me to the first preparative applications of Tswett's chromatography and the discovery of the isomeric α- and β-carotenes and several other carotenoids (2). Tswett's method became an essential tool in my further work, for the isolation of new natural compounds.

In June 1932 I was married in Paris with Hélene Fréchet whom I had met six months before in Heidelberg. But since January 1933 the political situation became more and more dangerous and we left Heidelberg "à toute vitesse" on March 25th 1933. Four days later the Gestapo came to fetch me...

We settled in Paris, near the popular rue Mouffetard, where I bought all sorts of sea food to isolate new carotenoids. But despite my French wife and two children, born in Paris, I could get neither fellowship nor position in France. I therefore accepted a 3-year contract as director of research at the Vitamin Institute, Leningrad. But after two most interesting years there (Nov. 1935–Nov. 1937), the political situation became very dangerous, especially for foreigners (the sadly known "Yeshowchina") and we returned to Paris, where I at last got a position as attaché de recherches at the CNRS (Centre National de la Recherche Scientifique).

I first worked in one of Professor Robert Lévy's Labs at the Ecole Normale Supérieure where I isolated pigments of marine invertebrates (echinochrome, the spinochromes, bonelline, etc.), then Professor Eugène Aubel and René Wurmser offered me a laboratory at the Institut de Biologie Physico-Chimique where Dr. Yvonne Khouvine gave motherly help to the arriving refugees.

In March 1939 I signed a contract with Max Roger, Director of a perfumery factory near Paris, to work on "natural perfumes." I chose "animal perfumes" and especially ambergris and castoreum, of which nearly nothing was known. This choice had most beneficial consequences for me some ten years later.

In December 1938 I became French Citizen and when the war broke out I was drafted into the army. In March 1940 the "soldat de 2ème classe" was demobilized after the birth of his 4th child. But on May 10th 1940 the "real war" started for France, and soon the German armies were approaching Paris. I put a precious box of 1 kg ambergris and extracts of castoreum in the trunk of an old car and we moved southwards until the armistice was signed.

From September 1940 to March 1947 I was most lucky again to be able to continue work in the Laboratoire de Chimie Biologique of the Université de Lyon, due to the great kindness of Professor Claude Fromageot. Space does not permit the description of all the personal problems of the four difficult war years until May 1945: the battle for food and money, dodging the police etc... Fortunately Max Roger turned up when all funds were exhausted (I had been excluded from the CNRS in application of the Nuremberg laws) and renewed our contract. I could engage a technician, Daniel Mercier and an excellent chemist, Dr. Judith Polonsky who both stayed with me for nearly 40 years.

Our work on ambergris led to the conclusion that most odorous constituents were oxidation products of the triterpene ambrein which we formulated as a cyclized squalene (3).

When the Swiss Chemical Society invited French chemists to a meeting in Basle, in October 1945, I submitted an abstract on the chemistry of ambrein, and was immediately called to Zürich by Professor Leopold Ruzicka who had a paper on ambergris odorants unpublished since 1939, because of its importance for the perfumery industry. At the Basle meeting I was contacted by Dr. Max Stoll, Head of Research, Firmenich, Geneva, with whom Ruzicka was working and have ever since enjoyed a priviledged relationship with this firm, especially with Dr. Roger Firmenich and Dr. Max Stoll and in recent years with his successor Dr. Günther Ohloff. My contract with Firmenich, signed with the CNRS in 1950 has been essential over the last 30 years for the development of my laboratory in Paris and later in Gif, in procuring technical help, research workers and laboratory funds.

In March 1947 we moved back to Paris and I started work again at the Institut de Biologie Physico-Chimique. There I met Dr. Nine Choucroun, who had described in 1939 a biologically active paraffin oil extract of tubercle bacilli. Her "PmKO" gave the impetus for a detailed study of the chemistry of Mycobacteria which started with the excellent doctor's thesis of Jean Asselineau and led to more than 150 papers on the chemistry biosynthesis and biological activity of mycobacterial lipids, glycolipids, peptidolipids and, finally in recent years to the development of synthetic immunostimulants, which will be mentioned below.

From 1947 to 1960 my group at the Institut de Biologie Physico-Chimique expanded rapidly, working on a rather large variety of problems, such as constituents of medicinal plants (4), the pheromone of the queen bee (5), the volatile constituents of jasmin essence (6) or of roasted cocoa (7) etc...

A typical example of a "small problem" leading by chance to more important results might be described under the heading: "From ascaryl alcohol to immunospecific determinants of bacterial endotoxins."

The embryologist Fauré-Frémiet had isolated in 1913 from the eggs of the parasitic nematode *Ascaris equi* a lipid which he called "alcool ascarylique." I received a sample of the original material and we prepared larger quantities by extracting *Ascaris* from the slaughter house of Paris. Dr. Judith Polonsky with Claudine Fouquey separated the crude unsaponifiable extract into three "alcools ascaryliques" A, B, and C. All three were characterized as glycosides of long chain fatty alcohols or diols containing a new sugar, ascarylose, which was shown to be a 3,6-dideoxy-hexose. Two isomers had been previously described by Westphal and Lüderitz, abequose and tyvelose, as immunodeterminant constituents of endotoxins of *Salmonella* species. A branched chain structure had been proposed for tyvelose, but after a discussion with Otto Westphal and Otto Lüderitz in Freiburg, it was clear that ascarylose, abequose, and tyvelose (as well as the more recently discovered paratose and colitose) where all stereoisomers of the same linear 3,6-dideoxyhexose. Abequose, tyvelose, and paratose were synthesized in a joint effort; ascarylose was shown to be the optical antipode of tyvelose, and colitose the antipode of abequose (*8*). The fortuitous isolation of ascarylose had thus led to the structural determination of the immunodeterminant constituents of *Salmonella* species.

In 1958 I was most lucky and proud to be appointed Director of the Department of Biological Chemistry of the Institut de Chimie des Substances Naturelles of the CNRS which was being built south of Paris, at Gif-sur-Yvette. The University of Paris had also recognized the importance of Natural Products Chemistry and had created a chair for me in 1956. This I abandoned at the end of 1958 to occupy the chair of Biochemistry, vacant after the death of Claude Fromageot. The CNRS Institute in Gif opened in 1961 and in 1963 I moved part of the biochemistry group of the old Paris lab to a new Institute on the campus of Orsay (now Université Paris-Sud), 4 km from Gif.

Our "excursion into the mysterious world of *Mycobacteria*" (*9*) was extended to *Corynebacteria* and *Nocardia* and brought many unexpected fruits. One of these can be described under the heading: "From fortuitine to mass spectrometry of permethylated peptides."

During an analysis of the lipids of *Mycobacterium fortuitum* Dr. E. Vilkas noted a precipitate gathering at the interface of ether and an aqueous layer. It turned out to be a nearly pure peptidolipid and was called "fortuitine." Chemical degradation gave the following structure (MW 1047):

$$\text{Ac}$$

$$CH_3(CH_2)_{20}CO\text{-Val-Val-Val-}\overbrace{\text{Thr-Thr}}\text{-Ala-Pro-OCH}_3$$

not quite in agreement with some analytical data. At that time Dr. Michael Barber of AEI Manchester, was looking for high molecular weight well defined compounds for testing the capacity of the new MS9 mass spectrometer. I sent him a few mg of fortuitine and received 2 weeks later a letter giving the correct, full structure with a molecular ion at 1359:

$$\text{CH}_3(\text{CH}_2)_{20}\text{CO-Val-MeLeu-Val-Val-MeLeu-}\overset{\overset{\displaystyle\text{OAc}}{|}}{\text{Thr}} - \overset{\overset{\displaystyle\text{OAc}}{|}}{\text{Thr}}\text{-Ala-Pro-OCH}_3.$$

Here we had the proof that mass spectrometry not only could give molecular ions above 1,300, but, more importantly, that the "sequence ions" obtained could easily lead to the discovery of unsuspected amino acid residues and to the complete structure of acylated peptide methyl esters (10).

The two N-methylated amino acids and one proline in fortuitine seemed to be the reason for its relative volatility. This led to the idea (first proposed by Dr. Jean Van Heijenoort at Orsay) of developing a method of permethylation of acyl-oligopeptides to increase volatility for mass spectrometry. With Drs. S.D. Géro, B.C. Das, and later D. Thomas and E. Vilkas a permethylation method was developed, allowing us to sequence a series of natural peptidolipids and peptide antibiotics (11). The fortuitous isolation of fortuitine had thus led to a new and useful method of peptide sequencing.

Another unexpected and rather important fruit from our excursion into the mysterious world of Mycobacteria can be described under the title: "From cord factor and cell walls to synthetic immunostimulants."

A "toxic lipid" extracted in 1950 by Hubert Bloch in New York, was found to be a 6,6'-dimycolate of trehalose, by H. Noll and J. Asselineau (12). Several years later, Adam Bekierkunst in Jerusalem revived interest in cord factor by showing its strong immunostimulant activities (13). Independently, Edgar Ribi in Montana isolated a biologically active glycolipid "P_3" which was nothing else than cord factor; today natural and synthetic trehalose diesters are recognized as strong immunoadjuvants and in particular as activators of macrophages (1, 14).

The first mycobacterial lipid which Robert White in London had found adjuvant active was Jean Asselineau's wax D (15); when the chemical similarity of active wax D fractions with cell walls became evident, we got interested in cell wall chemistry. Here we were again very lucky to have several excellent junior colleagues at the Institut de Biochimie at Orsay. Dr. Jean-François Petit had learnt the delicate art of dissecting bacterial cell walls enzymatically and showed that the immunoadjuvant activities of whole mycobacterial cells which are used in Freund's complete adjuvant, are an intrinsic property of the cell wall (16); then in 1972 Dr. Arlette Adam treated purified cell walls of Mycobacterium smegmatis with lysozyme and isolated the first water soluble adjuvant (WSA) (17); careful enzymatic and chemical degradations lead then to the conclusion that a muramyl-dipeptide is the minimum active structure of the whole mycobacterial cells, in Freund's complete adjuvant. The first few mg of synthetic N-acetyl-L-alanyl-D-isoglutamine (MDP) prepared by Pierre Sinaÿ at Orléans arrived in May 1974, and, as expected, were found to be strongly active in the guinea pig for stimulating antibody production and producing delayed hypersensitivity (18).

Since this first paper on MDP several hundred analogues and derivatives have been prepared and many large pharmaceutical firms have filed patents in this field. Our own work has developed rapidly due to the chemical expertise of Pierre Lefrancier at the Institut Choay and the enthusiasm and competence of Louis Chedid and his group at the Pasteur Institute. MDP and many of its derivatives not only stimulate antibody production but also non-specific resistance to bacterial infection (19). More recently, "desmuramyl" peptidolipids have been synthesized, which are not adjuvant active, but which strongly stimulate non-specific resistance (20), a property which might be interesting in certain clinical situations.

Considering, however, the large number of new and quite different natural and synthetic immunomodulators described recently, it is certainly hazardous to predict that MDP or its derivatives will become widely used drugs; as tools for experimental studies they have certainly already yielded a large amount of new information in cellular immunology (21–23).

Let us quote one last, more recent, example of luck, leading—hopefully—to a new approach to chemotherapy of some viral and parasitic diseases: it started as a rather academic exercise: to determine if in the biosynthesis of tuberculostearic acid and the side chain of phytosterols the methyl group of S-adenosyl-methionine is transferred with 3, or only 2 of its hydrogen atoms to a double bond of a precursor (24, 25). But we then wanted some "useful applications."

It was known that in some tumors hypermethylation of nucleic acids occurred; so why not try to inhibit these? In 1971 Jean Hildesheim first started in Gif to synthesize analogues of S-adenosyl homocysteine (SAH), the universal inhibitor of biological transmethylations; in his *in vitro* experiments, however, SAH was always more active than the synthetic compounds; so we were disappointed. Luckily, in September 1974 Dr. Malka Robert-Géro came to work with us; she had great experience, in particular, with chick embryo fibroblasts infected with Rous sarcoma virus (RSV); very soon she made the discovery that the synthetic SAH analogues prepared in our lab, especially 5'-deoxy-5'-S-isopropyl-thioadenosine (SIBA) are strong inhibitors of RSV induced cell transformation, whereas SAH is only weakly active. This finding was rapidly extended to other oncogenic RNA as well as DNA viruses (26). More recently an antifungal antibiotic Sinefungin, a "carba-analogue of SAH" isolated by Lilly Research (Indianapolis) was shown to be a potent inhibitor of methyltransferases and of viral transformation (27) and has, in particular, a strong antiparasitic activity (28, 29).

Will any of these compounds be of future use? Clearly, more active and less toxic SIBA-and Sinefungin-analogues should be found. I am expecting a new "coup de chance" in this field.

In conclusion, let me state that I am now especially happy to continue some work after my official retirement. I have been able to realize the dreams of my youth, in getting closer and closer to biology and to medical applications. As Goethe's Faust already said "Der gute Mensch in seinem dunklen Drange ist sich des rechten

Weges wohl bewusst." And, finally I wish to quote Einstein: "Imagination is more important than knowledge."

Acknowledgment

The research of my laboratory has been generously supported by CNRS and by grants from DGRST, la Fondation de la Recherche Médicale, La Ligue Française contre le Cancer, National Institutes of Health, Cancer Research Institute New York, and a contract from Laboratoires Choay and Institut Pasteur, Paris.

REFERENCES

1. Tenu, J.P., Lederer, E., and Petit, J.F.: 1980. *Eur. J. Immunol.* 10, 647–653.
2. Lederer, E.: 1972. *J. Chromatogr.* 73, 361–366.
3. Lederer, E.: 1949. *J. Chem. Soc.* 2115–2125.
4. Polonsky, J. and Lederer, E.: 1954. *Bull. Soc. Chim. France* 924–932.
5. Barbier, M. and Lederer, E.: 1960. *C. R.* 250, 4467–4469.
6. Demole, E., Lederer, E., and Mercier, D.: 1962. *Helv. Chim. Acta* 45, 675–685, 685–692.
7. Dietrich, P., Lederer, E., Winter, M., and Stoll, M.: 1964. *Helv. Chim. Acta* 47, 1581–1590.
8. Fouquey, C., Polonsky, J., Lederer, E., Westphal, O., and Lüderitz, O.: 1958. *Nature* 182, 944.
9. Lederer, E.: 1964. *Proc. Plenary Sess., IUB* 33, 63–78.
10. Barber, M., Jollès, P., Vilkas, E., and Lederer, E.: 1965. *Biochem. Biophys. Res. Commun.* 18, 469–473.
11. Das, B.C. and Lederer, E.: 1973. *Peptides 1971* (Nesvadba, H., ed.) pp. 253–264, North-Holland, Amsterdam.
12. Noll, H., Bloch, H., Asselineau, J., and Lederer, E.: 1956. *Biochim. Biophys. Acta* 20, 299–309.
13. Bekierkunst, A., Levij, I.S., Yarkoni, E., Vilkas, E., and Lederer, E.: 1971. *Science* 174, 1240–1242.
14. Lederer, E.: 1979. *Springer Semin. Immunopathol.* 2, 133–148.
15. White, R.G., Bernstock, L., Johns, R.G.S., and Lederer, E.: 1958. *Immunology* 1, 54–66.
16. Lederer, E., Adam, A., Ciorbaru, R., Petit, J.F., and Wietzerbin, J.: 1975. *Mol. Cell. Biochem.* 7, 87–104.
17. Adam, A., Ciorbaru, R., Petit, J.F., and Lederer, E.: 1972. *Proc. Natl. Acad. Sci. USA* 69, 851–854.
18. Ellouz, F., Adam, A., Ciorbaru, R., and Lederer, E.: 1974. *Biochem. Biophys. Res. Commun.* 59, 1317–1325.
19. Chedid, L., Parant, M., Parant, F., Lefrancier, P., Choay, J., and Lederer, E.: 1977. *Proc. Natl. Acad. Sci. USA* 74, 2089–2093.
20. Parant, M., Audibert, F., Chedid, L., Level, M., Lefrancier, P., Choay, J., and Lederer, E.: 1980. *Infect. Immun.* 27, 826–831.

21. Lederer, E.: 1980. *J. Med. Chem.* 23, 819–825.
22. Lederer, E.: 1980. *Immunology 1980* (Fougereau, M. and Dausset, M., eds) pp. 1194–1212, Academic Press, New York, San Francisco, and London.
23. Adam, A., Petit, J.F., and Lederer, E.: 1981. *Mol. Cell. Biochem.*, in press.
24. Lenfant, M., Ellouz, R., Das, B.C., Zissmann, E., and Lederer, E.: 1969. *Eur. J. Biochem.* 7, 159–164.
25. Lederer, E.: 1969. *Q. Rev.* 23, 453–481.
26. Robert-Géro, M., Blanchard, P., Lawrence, F., Pierré, A., Vedel, M., Vuilhorgne, M., and Lederer, E.: 1979. *Transmethylation* (Usdin, R.T., Borchardt, R.T., and Creveling, C.R., eds.) pp. 204–214, Elsevier/North-Holland, Amsterdam.
27. Vedel, M., Lawrence, F., Robert-Géro, M., and Lederer, E.: 1978. *Biochem. Biophys. Res. Commun.* 85, 371–376.
28. Bachrach, U., Schnur, L.F., El-On, J., Greenblatt, C.L., Pearlman, E., Robert-Géro, M., and Lederer, E.: 1980. *FEBS Lett.* 121, 289–291.
29. Lederer, E.: 1981. *Pre-FEBS Meeting on "Biochemistry of Parasites,"* Pergamon Press, New York, in press.

The Problem of Chemical Convergence in Secondary Metabolism

KURT MOTHES

K. MOTHES was born on November 3, 1900 in Germany. He was educated at University of Leipzig: pharmacy, biochemistry, botany. Dr. Phil. Leipzig; Dozent Halle 1928. Professor of Botany and Pharmacognosy 1935–1945: Königsberg, Head of Chemical Physiology, Institute für Kulturepflanzenforschg. Gatersleben; 1949–1957; Director of Botanical Institute, University of Halle; Director of Plant Biochemistry Institute of Academy of Sciences, Berlin at Halle 1958–1967; President of Deutsche Akademie der Naturforscher Leopoldina, Halle 1954–1974; Honorary Member: Acads. of Science Vienna, Tokyo, Budapest. Foreign member: Royal Society (U.K.), Acad. Nauk. Moscow. His scientific works have been done in nitrogen metabolism of plants, in alkaloids, in senescence, and in cytokinin.

The enormous progress in bioorganic chemistry enables us to make a comparison of the chemical capabilities between various classes in the plant and animal kingdoms and those of microorganisms. It should be noted that the data obtained have been selected mainly for the practical aspects which stimulated the search for new substances. Thirty years ago microorganisms were still considered to be mostly unable to synthesize secondary substances, although the experiences with lichen substances by Asahina and Shoji Shibata on the one hand and with fungi by Raistrick on the other were published.

Fleming's and Waksman's discoveries stimulated research for substances which are characterized by a specific activity against pathogenic bacteria and fungi. The number of known antibiotics amounts to almost 6,000. How many compounds may exist which are not recognized by the biological screening tests? Twenty years ago about 2,000 alkaloids were described; today the number 6,000 is estimated. The chance discovery of therapeutic substances such as reserpine was important, because it resulted in a profound search for similar alkaloids among the *Gentianales*. At present we know of about 1,000 alkaloids biochemically related to reserpine.

Only occasionally did scientists systematically look for certain types of substances. Bohlmann (*1*) discovered 650 new acetylenes in Compositae and Umbelliferae. Recently he reported on 1,000 newly-discovered sesqui- and diterpenes in 2,000 to 3,000 tested species of Compositae.

I will restrict this paper to alkaloids of N-heterocyclic compounds which until recently were regarded as characteristic products of the highly-evolved angiosperm plants. Their alkaloids are fairly uniform; proteinogeous amino acids provide their α-N atom and C atoms (with the exception of the carboxyl group) for the heterocyclic ring system.

We know at present that this type of alkaloid is lacking in prokaryotes and rare in filamentous fungi but real alkaloids with other structures do occur in these organisms. Here nature demonstrates a greater variability. It is incorrect to state that lower plants are primitive in regard to alkaloid biosynthesis. They are only unlike higher plants.

Of course, the occurrence of alkaloids is not determined by the ability to synthesize these substances. The producers need the equipment to excrete toxic substances or to protect themselves by alkaloid accumulation in protected cell compartments (storage excretion).

Similar aspects are observed in animal cells. Alkaloids as defense substances play an occassional role, but only where mechanisms have been developed for the prevention of selfintoxication. Alkaloids in plant cells do not in general protect plants from becoming animal foodstuff. On the other hand, there are animals which eat specific alkaloid-containing plants. They use such plants avoided by other animals as an ecological niche.

Among the alkaloids occurring in vertebrates, tetrodotoxin is one of the best-known animal poisons. Its chemical constitution has been elucidated by Tsuda (*2, 3*), Goto (*4*), and Woodward (*5*). It is a guanidine derivative but the biosynthesis and hence the origin of the guanidine group are unclear. In plants guanidine derivatives are rare. The guanidine group in galegin derives from arginine; this is a rare case. Normally arginine is cleaved by arginase, yielding ornithine which in the form of a diamine is converted to amine-aldehyde, which often appears in cyclic structures.

Interestingly the biochemically complicated and unusual tetrodotoxin has been detected in different animals, in bowlfishes (*Speroides ruripes*) and some other fish, in the Californian salamander (*Taricha toroso*), in gobi fish (*Gobius criniger*), in the eggs of the frog *Atelopus chiriquensis* and in the cephalopode *Hapalochlaena maculosa*. It is difficult to imagine that such a substance has been invented several times during evolution.

The case reminds us of saxitoxin (6). As a neurotoxin it causes serious and epidemic diseases in the inhabitants on the west coast of North America, however, the diseases occur only in short periods during the year and always after having eaten mussels. This period coincides with the appearance of a red "sea flower" caused by *Gonyaulax catanella*, a dinoflagellate. Indeed this algae contains saxitoxin, too. This alga is an important nutritional source for various marine animals and its poison is a diguanidine. Here chemical convergence is clearly related to a nutritional correlation. Does a similar correlation exist also for tetrodotoxin?

In 1972 Kupchan *et al.* (7, 8) reported the occurrence of maytansine, a novel antileukemic ansamacrolide from *Maytenus ovatus* which is a cyclic oligopeptide. Almost the same substance was found in 1977 by Higashide *et al.* (9) within the actinomycete *Nocardia*. It appears unlikely that maytansine has been invented twice during evolution. Are there other possible explanations (10)?

Cyclic oligopeptides are known to occur in the families of Rhamnaceae, Celastraceae, Sterculiaceae, Pandaceae, Urticaceae, Euphorbiaceae, Hymenocardiaceae, and Loranthaceae. The first two families grow as pioneer plants in sterile soils, because they can fix atmospheric nitrogen due to symbiosis with *Actinomycetes*. According to Klemmedson (11), 145 of 290 species live in symbiosis with Actinomycetes and they are able to fix N_2. Most of the species belong to the genus *Ceanothus*, but *Ceanothus* and *Alnus* dominate in respect to the number of individuals. Of 52 *Ceanothus* species in western North America, 25 fix N_2, but it is unknown whether the remaining 27 species likewise grow in symbiosis with root nodules causes by *Actinomycetes*.

Thus it is possible that a Nocardia symbiosis exists within *Maytenus ovatus*, which is responsible for the presence of maytansine. It is likewise possible that during evolution an *Actinomycete* has transiently coexisted in the manner of a mycorrhiza with a higher plant but later abandoned this symbiosis. The gene complex for synthesis of its peptide alkaloid may have been left within the higher plant. This explanation would be rendered more probable if the peptide synthetases were programmed on plasmids.

A similar problem concerns lyngbyatoxin, which according to Cardellina (12) occurs in a shallow-water variety of *Lyngbya majuscula* and which chemically resembles Teleocidin B of a *Streptomyces* (13) so that one is inclined to assume an impurity by a *Streptomyces*, or the presence of a plasmid. The importance of plasmids for the distribution of secondary substances has been emphasized by Okami (14) and Hopwood (15). It seems to be necessary to study whether plasmids during evolution of green plants have been incorporated in addition to chloroplasts and mitochondria.

Most of the alkaloids in higher plants are programmed by Mendelian inheritance. Convergence, however, does not necessarily mean the same enzyme species or the same type of precursors occur. Unfortunately the reaction mechanisms of the biosynthesis of alkaloids are still obscure. Hence, we do not know whether convergence in the occurrence of quinolicidines, benzylisoquinolines, anabasine, β-carbolines, pyrrolicidines, or nicotine is caused by a complete identity of their synthesis. These seemingly simple types of alkaloids are by no means proper chemotaxonomic characters. Their variabilities due to secondary reactions increase their taxonomic value by complexity. For example, benzylisoquinolines are present in 7 different orders, derived aporphine in 6, protoberberines in 4, morphinans in 2, and morphine as such in 2 species.

On the other hand, nicotinic acid as a coenzyme in primary metabolism is an important precursor for alkaloids in plants and animals; it has been distributed in all organisms and can be synthesized in different ways: from degradation of tryptophan, during synthesis of aspartic acid (or N-formyl aspartic acid), *via* a glycerol derivative and from nicotine.

There are complicated examples of convergence, *e.g.* the occurrence of ergoline alkaloids in *Claviceps* species, but also in a few other *Ascomycetes*, in Mucorineae and in Convolvulaceae. Since the synthesis of this type of alkaloid can be regarded as very difficult, a more detailed study of the enzymes involved should provide information about the evolution of this specific chemical quality (*16*).

REFERENCES

1. Bohlmann, E., Burkhardt, T., and Zdero, C.: 1973. *Naturally Occurring Acetylenes*, 547 pp., Academic Press, New York.
2. Tsuda, K.: 1966. *Naturwissenschaften* 53, 171.
3. Tsuda, K. and Kawamura, M.: 1952. *J. Pharm. Soc. Japan* 72, 721.
4. Goto, T., Kishi, Y., Takahashi, S., and Hirata, Y.: 1965. *Tetrahedron* 21, 2059.
5. Woodward, R.B.: 1964. *Pure Appl. Chem.* 9(1), 49–74.
6. Shantz, E.J., Chazarossian, V.E., Schnoes, H.K., Strong, F.M., Springer, J.P., Pezzamite, J.V., and Chardy, J.: 1975. *J. Am. Chem. Soc.* 97, 1238–1239.
7. Kupchan, S.M., Komoda, Y., Court, W.A., Thomas, G.J., Smith, R.M., Karim, A., Gilmore, C.J., Halliwanger, R.C., and Bryan, R.R.: 1972. *J. Am. Chem. Soc.* 94, 1354–1356.
8. Kupchan, S.M., Kodamo, Y., Branfman, A.R., Dailey, R.G., and Zimmerly, V.A.: 1974. *J. Am. Chem. Soc.* 96, 3706–3708.
9. Higashide, E., Asai, M., Ootsu, K., Tanida, S., Kozai, Y., Hasegawa, I., Kishi, I., Sugino, Y., and Yoneda, M.: 1977. *Nature* 270, 721–722.
10. Komoda, Y. and Isogai, Y.: 1978. *Sci. Pap. Coll. Gen. Educ. Tokyo* 28, 129–134.
11. Klemmedson, J.V.: 1979. *Bot. Gaz.* 140 (Suppl.), 91–96.
12. Cardellina, J.H., Marner, Fr-J., and Moore, R.: 1979. *Science* 204, 193–195.
13. Takashima, M., Sakai, H., and Arima, K.: 1962. *Agric. Biol. Chem.* 26, 660.
14. Okami, Y.: 1979. *J. Nat. Prod.* 42, 583–595.
15. Hopwood, D.A.: 1979. *J. Nat. Prod.* 42, 596–602.
16. Gröger, D.: 1975. *Planta Med.* 28, 269–288; 1978. *Antibiotics* (FEMS Symp. No. 5) pp. 201–217, Academic Press, New York.

Seed-Transmitted Crop Diseases

Diagnostics, Control, International Co-operation

PAUL NEERGAARD

P. NEERGAARD was born on February 19, 1907, and was awarded the degrees of B.Sc. (Horticulture) 1932, Ph.D. 1935, Sc.D. 1945, Copenhagen. He was Seed Pathologist at J.E. Ohlsens Enke, Copenhagen, 1935–1951; Head of Department, Government Plant Protection Service, Copenhagen, 1952–1966; Professor of Plant Pathology and Seed Technology, American University of Beirut, Lebanon, 1959–1960. Since 1966 he has been Director of Danish Government Institute of Seed Pathology for Developing Countries, Copenhagen. In 1979, he was Sir M. Visveswaraya Visiting Professor at University of Mysore, India. During 1956–1974 he was Chairman of the Committee on Plant Diseases of International Seed Testing Association, and of 18 international workshops on Seed Pathology, 1958–1974. President of Danish Phytopathological Society, 1973–1978. Since 1963 he has been Vice-President of the International Esperanto Academy for Language and Literature; he has been elected fellow of Indian National Science Academy, New Delhi, National Academy of Sciences, India (Allahabad), L'Academie d'Agriculture de France, Paris, and Explorers Club, New York. He published several books and about 200 research papers on topics of plant pathology and mycology, several books on horticulture, and several books in Esperanto on various topics of science and of esperantology.

In following the suggestion of the Editor to present a recollection of my past research and a review of my present work, I have been given an opportunity to express my deep gratefulness for decades of close co-operation with colleagues throughout the world. My essay will be about the joint efforts in developing *seed pathology* as a new cross-breed discipline of plant pathology and seed technology, and I shall take into consideration that my readers will be scientists outside of my own field of research. First some background information.

Importance of Crop Diseases

The losses to world agriculture due to plant diseases are estimated at 50,000 million US dollars/annum or 550 million tons of production, 25% of this being cereals. To compare, in 1975 the food deficit of the developing countries was about 37 million tons and it is expected to be 120–145 million tons by 1990 (*1*). It is indicative that the total number of malnourished children in the age group 0 to 4 years in the world is around 100 million (*2*)!

Approximately 90% of all food crops are propagated by seed. More than a thousand host/pathogen combinations of seed-borne diseases are known, and this number is rapidly increasing as our knowledge grows (*3*). Some of these diseases cause devastating crop epidemics resulting sometimes in famine conditions.

The technology of the Green Revolution aims at (1) increase of yield potentials by breeding new high yielding varieties, and (2) increase of the actual yields in the farmer's field by improving agricultural practices such as use of better, disease-free seed, fertilizers, irrigation, *etc*. It is much easier to increase theoretical yield potentials than to increase actual yields—there is a long way from the experimental field to the poor farmers' fields in the Third World. The two most pressing problems are provision of water and control of pests and diseases.

Diagnostics

In order to control an epidemic disease we must be able to recognise the disease condition, to reveal the causal agent and further to define the behaviour of the pathogen. Only for less than 150 years science has understood the true nature of epidemic disease in man, animals, and plants. Still for many diseases our understanding is poor and sometimes primitive. A recent example from my work in the developing countries:

Bangladesh is the world's biggest exporter of jute, and this crop is fundamental for the economy of the country. The most devastating disease of this crop is a yellowing of the jute, still widely believed to be a genetic disorder. The average national losses are about 20% of the production, in many fields they go up to 50%. In 1979 co-operation was established between the Bangladesh Jute Institute and our Institute of Seed Pathology for Developing Countries to study the aetiology of the disease. By infection experiments, sowing of jute seed from Bangladesh in Copenhagen, and by electron microscopic studies we are now convinced that the causal agent is a seed-transmitted viroid. This opens the way for controlling the disease efficiently by testing and screening the propagative material because the disease starts with infected seed.

Some 45 years ago my studies on a "wrong" celery disease introduced me into the aetiology of plant disease. As a graduate student I was given some celery roots badly attacked by scab, a disease well known to be caused by the fungus *Phoma apiicola*, and I was told: go ahead. This was my luck. A very different fungus came out in my isolations, and following Koch's postulates I could prove that *Alternaria radicina* was the causal agent: I had a new disease (*4*). And this disease was transmitted by seed.

This work qualified me after graduation to join a horticultural seed company, J.E. Ohlsens Enke (the biggest exporter of horticultural seed in the world), where I was given a free hand to do independent research as a plant pathologist. During the 15 years of my appointment here I analysed about 40,000 samples of horticultural seed for seed-borne infections (*5*), and concentrated my research on two of the most harmful and widespread parasitic fungus genera, *Alternaria* and *Stemphylium*. The seed enabled me to define the distribution of the pathogens in the crops, and by infection experiments I could define their importance as pathogens. The work resulted in a monograph now used internationally as a standard handbook (*6*). In these years, working in close contact with the seed industry, I became increasingly involved in the study of this fundamental aspect in plant disease epidemiology: seed as a common carrier of inoculum, a paramount vehicle of trouble and losses to the farmer. I continued detecting seed-borne pathogens, when from 1952–1966 I was associated with the Danish Plant Protection Service responsible for quarantine inspection of plants, including seed, in import and export.

Seed Testing and Plant Protection

There are two kinds of institutions taking care of checking seeds for infection with the aim of preventing crop epidemics. They are (1) *seed testing stations*, testing seed for planting value (germinability, purity, moisture content, health condition), and (2) *plant protection services*, inspecting plants for quarantine to prevent international spread of crop epidemics from infested areas (regions, countries, continents) to non-infested areas.

By necessity, there is intensive international co-operation and co-ordination between these testing and inspecting agencies. The International Seed Testing Association (ISTA) is an intergovernmental organisation having the responsibility of authorising competent seed testing stations to carry out tests using methods prescribed or recommended by ISTA and to issue certificates according to the ISTA International Rules for Seed Testing (*7*). Testing and certifying seed for germination and purity, being rather simple, have been done for more than 100 years while testing for seed health is a rather new development, implying the complications of many different pathogens, including fungi, bacteria, nematodes, and viruses on a huge range of hosts, and regarding a considerable range of testing methods to be elaborated and tried out in international comparative tests before they can be standardised.

Plant quarantine regulations and procedures of plant inspection for quarantine are obviously also based on a system of international agreements and co-operation. The International Plant Protection Convention, under the aegis of FAO, Rome 1951 (the Rome Convention), provides for establishment of national organisation for plant protection, for issuance of standard phytosanitary certificates, moreover, it specifies basic quarantine requirements in relation to import, and defines agreement or international co-operation and establishment of regional plant protection organisations. Such organisations have been established, and cover most of the world, for example the European and Mediterranean Plant Protection Organisation (EPPO), the Inter-African Phytosanitary Council (IAPSC), and Plant Protection Committee for the South-East Asia and Pacific Region (SEAPPC). Still two important countries, Japan and the Peoples Republic of China, have not yet entered formalised regional organisational co-operation.

By now 77 countries have signed or acceded to the Convention, and they have thus adopted the Model Phytosanitary Certificate, given in the Convention, the so-called "Rome Certificate." This certificate, however, is now in general use and accepted not only by the contracting countries but by practically all other countries who have not yet signed the Convention.

I have participated actively in this international network of co-operation for about 25 years, namely within ISTA and with some of the Regional FAO Plant Protection Organisations (in particular EPPO and IAPSC). In 1956 I was appointed chairman of the Plant Disease Committee of ISTA. This Committee is responsible for international co-operation in development of routine seed health testing methods.

A routine test must produce repeatable results within a testing laboratory and reproducible results between different laboratories. Therefore, comparative seed health testing schemes were needed in order to develop reliable testing procedures. The first such scheme was organised in 1957, comprising cabbage, carrot, flax, barley, and oats, and the first workshop on seed pathology was convened in Cambridge in 1958. Here the results were discussed and joint laboratory studies carried out on incubated samples duplicate to those tested at home by the participating stations. Thus the ground was laid for further action.

Up to now 16 such international workshops on seed pathology have been held under the aegis of the Plant Disease Committee of ISTA, all based primarily on discussions on results of comparative tests carried out by the workshop participants at their home institutions. In these international schemes a total of about one hundred comparative tests, involving seeds of 28 crop plants and half a hundred seed-borne pathogens, have been studied and restudied, for details see Ref. 8. In most of the tests 25–30 stations from 20 or more countries have participated. In recent years the number of participating stations and countries has increased and testing activities are now organised within the Committee by a number of working groups, specialised according to categories of crops and pathogens. I have not counted the number of colleagues who have co-operated in these activities—now hundreds of plant pathologists are involved. These joint studies of experimentation and practical observations have led to increasing reliability of seed health testing procedures now available for a considerable number of crop diseases, and to increasing practical application of them. This empirical development of testing techniques is increasingly supported by research, and considerable progress has been made in recent years, perhaps particularly in testing seed for bacteria and viruses.

Testing of seed for quarantine has almost entirely derived advantage from the experience of the above development work organised by the ISTA Committee on Plant Diseases. EPPO has convened several Working Parties (Reports published in 1954, 1959, 1964, and 1966) which urged development of testing techniques and uniformity of regulations. In recent years IAPSC has taken the first steps to set up a network of seed health testing laboratories for quarantine in Africa.

The Third World

The problem of producing healthy seeds for sowing is very much connected with the potentials of the Green Revolution. Pests and diseases of crops are far more predominant under the climatic conditions of the tropical than under those of the temperate zones. Due to more advanced technology in pest control the industrial countries in the temperate zones have substantially reduced crop losses during the last decades, whereas the tropical countries still have to develop proper technology, and to do this under conditions much more unfavourable, due to the heavy epidemiological pressure to which tropical crops are subject. For this reason seed pathology is a major issue in the Third World's agricultural development.

In 1965 I was invited by the Danish Ministry of Foreign Affairs to set up an Institute of Seed Pathology for Developing Countries, and in 1967 we received the first scholars from developing countries in Copenhagen.

In order to introduce seed pathology directly to the agricultural authorities and plant pathologists in the developing countries themselves we arranged a series of two weeks introductory courses in different regions of the Third World: in August 1969 in the Philippines with delegates from ten South Asian countries, in November 1970 in Ghana with delegates from eleven African countries, in April 1972 in Turkey with delegates from nine countries from the Mediterranean area and the Middle East, and in September 1973 in Argentina with delegates from eleven Latin American countries, all expenses being paid by the Danish Government. The workshops gave us an excellent opportunity to introduce the subject not only to the 80 delegated scientists and some 20 observers but also to senior authorities in many countries. In about 35 developing countries in the four regions I visited in connection with the scheme, I established personal contacts with senior agricultural authorities in governments and institutions—with the decision makers, and this achievement turned out to be invaluable for the further development.

The Institute in Copenhagen has so far received 242 scholars and research fellows from 59 developing countries, and 20 from 8 industrialised countries, making a total of 262 from 67 countries. To qualify for admission the candidate must have a M.Sc. degree or a comparable university degree, though exceptions may be given in respect to some developing countries.

The course, given for one or two (to four) semesters, includes practical and theoretical training. Practical laboratory testing of seed for fungi, bacteria and viruses is carried out using seeds brought to Copenhagen by the scholars from their home country. The theoretical course (9) deals in detail with the epidemiology of seed-borne diseases and includes the mechanism of seed transmission of diseases, location of inoculum in the seed, taxonomy of seed-borne organisms, principles of control, and of assessing the harmful effects of these diseases, general methodology and ecology of seed health testing, including testing for fungi, bacteria, viruses, and nematodes, and principles of organising seed health certification schemes and for establishing disease tolerances and prescribing quarantine measures.

The research aspect of the activities of the Institute is important. We give priority to problems pertaining to seed-transmitted pathogens occurring in the developing countries, which means in the tropical and subtropical zones. We have particularly concentrated on two areas of research: (1) identification of seed-borne organisms and (2) development of routine seed health testing methods for fungi, bacteria and viruses, but we also take up more basic problems such as those of mechanisms of seed transmission; ecology of fungi during incubation, the influence of light period, light quality and quantity, temperature, humidity, and pH; taxonomy studies on growth characters of fungi as they occur on seeds that have been incubated; anatomical studies on infected seeds, location of fungi within the seed tissues of crop plants; preliminary surveys on the distribution of seed-borne fungi for a number of developing countries, including Argentina, Egypt, some states in India, Nepal, Pakistan, the Philippines, Turkey, and Uganda. We are working on world mapping of distribution of seed-borne fungi.

What has been the response of the Third World to our efforts? We have contact with the majority of our scholars, and a recent review of their activities indicates that about 95% of them are working in teaching, research or practical testing, directly or indirectly related to studies or control of seed-borne diseases. After their return they have initiated a considerable amount of activities. Special seed health testing laboratories have been set up in many developing countries, the exact number is unknown but I estimate between 50 and 100. The Institute in Copenhagen has enquired, how many universities are teaching seed pathology. More than 50 universities in 24 developing countries have introduced seed pathology in their research and teaching programmes, and about 40 of these offer credit courses in this subject. This important development was followed up by a Seminar on University Teaching of Seed Pathology in Developing Countries of Africa and Asia, organised by the Institute in co-operation with Indian authorities in 1976, and in this connection a book on this subject, to be used in organising university courses, has recently been published (*10*).

Other training activities are now going on in many developing countries, such as short courses in seed health testing, seminars on quarantine for seed, and on control of seed-borne diseases in seed production, the primary aim here being to introduce health testing in seed certification schemes, *i.e.* to organise strict health control to be implemented during the process of seed multiplication, from breeders seed to certified seed.

Research activities are also going on in many countries. A recent specified list of research projects in Brazil includes 95 projects, involving 137 scientists.

In the efforts of our Institute to promote the science and technology of seed pathology in the Third World we have encountered serious language difficulties. To be admitted for studies at the Institute good working knowledge of English is required. Obviously this discriminates against scientists from countries with a different language background, and this has had a definite impact on the development of seed pathology in such countries. We have had proportionally far fewer scholars from developing countries with French language background, and when we do get scholars from such countries they have considerable language difficulties during their studies in Copenhagen. Consequently, as a group, they have obtained lower marks at the final examination than the categories of students with English language background. Finally, as a further consequence, there are fewer related projects and initiatives in French than in English speaking developing countries. In these, almost 50 universities have introduced seed pathology while only 3 universities in French speaking countries have introduced courses in this subject. Fortunately, to overcome these difficulties in so far as at least French is concerned, an institute of seed pathology is likely soon to be set up in Morocco for scientists from French speaking developing countries. However, remains the fact that differences of language is a major problem interfering in the urgent process of developing the Third World.

REFERENCES

1. James, C.: 1980. Preprint No. 28. The 19th International Seed Testing Congress, Vienna, Austria, 1980. 8 pp., Vienna.
2. Bengoa, J.M.: 1975. *Man, Food and Nutrition* (Miloslav, R., ed.) 344 pp., Cleveland, Ohio.
3. Richardson, M.J.: 1979. *An Annotated List of Seed-borne Diseases*, 3rd ed., 320 pp., ISTA and Commonwealth Mycological Institute, Kew, England.
4. Neergaard, P.: 1937. *R. Vet. Agric. Coll. Yearb. 1937*, 1–42.
5. Neergaard, P.: 1935–1966. (Series of Annual Reports on seed health testing): *1–15 Årsberetning fra J.E. Ohlsens Enkes plantepatologiske laboratorium 1935–1950*. With General Index of 1–10 Annual Report and General Index of 11–15 Annual Report.

(Summaries in English and Esperanto) Copenhagen 1936–1951. 280 pp. *6–11, 13-17 Årsberetning vedrørende frøpatologisk kontrol, Statens Plantetilsyn, 1953–1959, 1960–1965.* Summaries in English. Copenhagen 1956–1960, 1962–1966. 183 pp.

6. Neergaard, P.: 1945. *Danish Species of Alternaria and Stemphylium. Taxonomy, parasitism, economical significance*, 560 pp. (Summaries in Danish and Esperanto), Diss. Copenhagen.

7. ISTA: 1976. *International Rules for Seed Testing*; Annexes 1976, 177 pp.

8. Neergaard, P.: 1980. *Seed—A Horse of Hunger or a Source of Life?* University of Mysore, Mysore, India.

9. Neergaard, P.: 1977. *Seed Pathology*, Vol. 2 (revised ed. 1979) 1191 pp., The Macmillan Press Ltd., London and Basingstoke.

10. Neergaard, P. and Mathur, S.B.: 1980. *University Teaching of Seed Pathology*, 161 pp., University of Mysore and Danish Government Institute of Seed Pathology for Developing Countries, Mysore, India.

Some Personal Recollections and Reflections

JOHN T. EDSALL

J.T. EDSALL was born in Philadelphia, Pennsylvania, USA on November 3, 1902. He received the A.B. degree from Harvard University in 1923 and the M.D. degree to 1928. From 1924 to 1926 he studied at Cambridge University. In the Department of Physical Chemistry at Harvard Medical School he worked on the physical chemistry of proteins, amino acids, and peptides, and was also a tutor in Biochemical Sciences, later becoming Professor of Biochemistry in Harvard University. He is an editor of "Advances in Protein Chemistry." He is a member of the United States National Academy of Sciences, and was President of the American Society of Biological Chemists in 1957–1958. His current interests are in the history of science, especially biochemistry, and in problems of scientific freedom and responsibility.

My general interest in science and medicine began very early, for I grew up with parents of strong intellectual interests, my father having been one of the leaders in American medicine, first in Philadelphia and later in Boston, as Professor of Medicine and later Dean at Harvard Medical School (1). As a student at Harvard College I was much influenced by Professor Lawrence J. Henderson, both in his writings on the fitness of the environment (2) and his deep studies of blood as a highly integrated and regulated system for the performance of its biological functions (3). He was, for me, an inspiring teacher and later a most kind and helpful friend. After one year of medical school at Harvard, I spent two years in Cambridge, England (1924–1926) where I took advanced biochemistry in the Sir William Dunn Institute—which had just opened a few months before my arrival—headed by the great Sir Frederick Hopkins. ("Hoppy"). He was and is an inspiration in my life, and the same is true of David Keilin, who was at that time rediscovering the cytochrome system, and starting on that long series of studies on biological oxidation systems which is one of the great achievements of our time.

Returning to Harvard in 1926, to finish my medical studies, I sought to do extra work in research in what free time I had. Thanks to a wise adviser, Dr. Alfred C. Redfield, I came to the Department of Physical Chemistry headed by Edwin J. Cohn, which was devoted to the study of proteins. That laboratory was the center of my activity for the next quarter century with only two or three sabbatical years spent far away (4). I have recently written of the work of Alexander von Muralt and myself 50 years ago on double refraction of flow in the muscle protein that we then called "myosin" (5).

Under Cohn's direction and influence, our laboratory was involved for a decade, from 1930 on, in studies of the physical chemistry of amino acids and peptides. We were a fairly small and intimate group. From the laboratory at the Medical School, in addition to Cohn himself, it included Arda A. Green (6), T.L. McMeekin (7), and Jesse P. Greenstein (8). Later came J. Lawrence Oncley, John W. Mehl, and John D. Ferry. My intimate friend Jeffries Wyman, in the biological laboratories at Harvard College, worked in extremely close collaboration with our group in his important studies on the dielectric constants and dipole moments of amino acids, peptides, and certain proteins. George Scatchard (9) and John G. Kirkwood (10) at the Massachusetts Institute of Technology, extended the Debye-Hückel theory of interionic forces to deal with dipolar ions, such as the amino acids and peptides, and also with large multipolar ions such as proteins. Greenstein synthesized peptides, studied their ionization constants, and other physical properties, with immense energy and enthusiasm. I worked on Raman spectra, to characterize the state of ionizing groups and various side chains in teams of characteristic molecular vibrations. McMeekin, in constant interplay of ideas with Cohn and the rest of us, did extensive solubility studies on amino acids and related compounds, in water and organic solvents. These studies served to define how the free energy of transfer of these compounds was influenced by charged, polar, and nonpolar groups. We were concerned, not only with forces between electrically charged and polar groups, but also with what are now known as hydrophobic interactions. In the course of the work I discovered, by correlating some of scattered data in the literature, the remarkable effect of nonpolar groups in organic compounds in increasing their partial molal heat capacities (11); an effcet now extensively studied by modern techniques (see for instance Edsall and McKenzie, Ref. 12, for a survey of modern work on the subject). Of course it was very hard to understand some of the effects we recognized; I puzzled for years over the meaning of these heat capacity data, without being able to come up with any good ideas. Indeed it was the work of Frank and Evans (13) and then that of Kauzmann (14) first began to make sense out of all this; though indeed hydrophobic interactions are still far from being adequately understood.

In all this work we were sustained by a great intellectual excitement; we began to perceive coherent relations between structure and properties of proteins, amino acids, and peptides, and we finally set down a systematic account of it all in a book (15). Although Cohn was the initiator of this project, he was too busy with the direction of the department to do more than a limited number of chapters, and most of the writing fell to my lot. Other colleagues were also involved; important chapters by Scatchard, Kirkwood, Mehl, Oncley, and Hans Mueller formed an essential part of the book. We finished it just before the United States was engulfed in the war, which changed our lives and activities profoundly, as I explain below. Much of the book is now out of date; yet other portions, especially the solubility data on amino acids and related compounds, and the calculation of the specific volumes of proteins, have still not been altogether superseded. Working scientists still consult the book and cite it, sometimes to my surprise.

The work of the laboratory in the 1930s had seemed to most of our medical colleagues remote from any practical applications. Yet from 1940 on, throughout the war years, we became deeply involved in collaboration with clinicians in the development of new uses of blood plasma proteins in medicine and surgery. Before 1940 our interests were already shifting from the amino acids and peptides to the proteins; the demands of the war enormously accelerated the process. It was Edwin Cohn who had the vision to see the possibility of a large scale fractionation process for obtaining clinically useful fractions of blood plasma, each separated from the others so that each could be applied to its own specific use. Other proteins, which at the moment could not find practical use, would become available for study and possible later use. To make that vision a reality required the talents of dozens of chemists, biochemists and clinicians, financial support on a previously unprecedented scale from government sources, and the cooperation of scientists with the industrialists and laboratory scientists of seven major manufacturers in the pharmaceutical industry. It led to the large scale production of serum albumin for transfusion, of γ-globulin for temporary immunization against measles and hepatitis, of fibrin foam and thrombin to stop bleeding in neurosurgical operations, and of fibrin film to replace the dural membrane, also in neurosurgery. It led also to new insights into other proteins, notably the plasma lipoproteins and transferrin. The story is far too long to tell here; Cohn (16) has given a history of the development of the project as a whole, and I (17) have described the results in their relation to the basic physical chemistry of proteins. It brought us all into a new and different world, involving the combination of science with industrial development. Personally we could be thankful that our war work was devoted to the healing and protection of the sick and wounded, and that the results could continue to be useful in peace as in war.

The Post-war Years: Return to Basic Science and to New Concerns

The ending of the war brought a gradual return to basic scientific problems, combined for me, as for many others, with a much deepened concern for the role of science and scientists in the larger problems of the world. I was appalled by the implications of atomic weapons for the future of mankind, for I had little faith in the ability of the governments and peoples of the world to deal wisely with such revolutionary developments. The results appear in the terrible and essentially insane arms race that is going on today. We can indeed be thankful that no nuclear weapons have been used in war since 1945, and I hope (but find it difficult to believe) that the nations will exert sufficient wise restraint to avert catastrophe, in a world torn by so many wars and revolutions, with rapid population growth accompanied by shortages of energy supplies and (in many countries) of food and basic necessities. I have been an active supporter of two organizations that have tried to do something about these problems: the Federation of American Scientists (in which I have just finished serving a term as Secretary) and the Council for a Livable World, founded by the late Leo Szilard. My own contribution to the work for a more stable world order has been a very small one, but I applaud all those who have devoted a major part of their time and effort to these supreme concerns.

However science, and the advancement of biochemistry, and protein chemistry in particular, remained a central and absorbing interest. I have written elsewhere (18) of my teaching and research activities during those years, and of the many colleagues, younger associates and students whose work was closely intertwined with mine. A turning point in my life was the move in 1954 from Harvard Medical School to Harvard College in Cambridge, and the formation there of a new Committee on Higher Degrees in Biochemistry—a Committee that later grew into a Department. In 1958 came a major change in my life, when I became Editor-in-Chief of the Journal of Biological Chemistry (JBC) and held the job for ten years, with one year's leave of absence. Recently I have written a history of the first 75 years of the JBC for its anniversary in 1980 (19). Strenuous as the editorship was, I still had time and energy to direct research on carbonic anhydrases of red cells —a class of enzymes that had always fascinated me—and to continue teaching on a more limited scale. The splendid group of young researchers in the laboratory had plenty of initiative, and were able to pursue their research with only a modest amount of guidance from me.

Twice I was a Fulbright Lecturer—in Cambridge, England in 1952, and at the University of Tokyo in 1964. The students in my class in Tokyo, about 35 in number, were an excellent group, fully comparable to the students at Harvard, and also, to my admiration, were able to understand and assimilate the words of a lecturer in a foreign language. Professors H. Noda and K. Maruyama provided me with constant help, advice and encouragement, both in my teaching and in becoming acquainted with life in Japan, and the beauties of the cities and the countryside. Having attended all my lectures they volunteered, with my full approval, to use them to compose a book in Japanese, on protein chemistry and related matters, listing me as coauthor with them. This is the only one of my books that I cannot read myself. Apparently it has been a success in Japan, and has recently appeared in a revised edition, including a section on nucleic acids by a newly recruited author.

In 1958 Jeffries Wyman and I completed our book on "Biophysical Chemistry" after several years of close collaboration. We had planned a second volume which, alas, was never completed, largely because of my heavy involvement with the Journal of Biological Chemistry. We remain constantly in touch, through letters and occasional visits, and our friendship has remained unbroken over 60 years.

Work in the History of Biochemistry

I have always had a deep interest in history, including the history of science; for we cannot fully understand where we are today unless we have some appreciation of our origins. For me, in any case, the study of history is inherently of great interest, regardless of its possible practical use in widening our perspectives. In recent years this interest has become active; I have explored aspects of the history of blood and hemoglobin, and biochemical regulatory processes (*20, 21*). Moreover, the Committee on the History of Biochemistry and Molecular Biology of the American Academy of Arts and Sciences, of which I was Chairman, recognized the great importance for historians of preserving personal correspondence and other unpublished papers of scientists who have played a significant role. Too many of these papers are carelessly thrown away. These considerations have led us to a Survey of Sources for the History of Biochemistry and Molecular Biology (*22*) which in four years of work has described nearly 600 significant archival collections of papers of scientists working in this area (*23*). The major work on this project was done by my young colleagues David Bearman, Margaret Miller, and Matthew Konopka; to them and especially to David Bearman, goes the chief honor for completion of the project. Actually the completed book represents only a beginning, though I think an important one; we know that great quantities of important archival material are in existence, but not yet known to us, and often in danger of being lost. We earnestly hope that others will continue the work we have begun.

Scientific Freedom and Responsibility

One of my major recent activities has been the work of the Committee on Scientific Freedom and Responsibility of the American Association for the Advancement of Science. The AAAS set up this Committee because of increasing concern over the responsibility of scientists, including engineers, physicians, and public health workers, to monitor the consequences and implications of developments in applied science and technology, to warm of risks of which the public may not be aware, and to point out opportunities for improvement. Sometimes the scientist's concern for these matters brings him into conflict with his employer, who may prefer to avoid facing embarassing evidence of risks that it would be expensive to correct. The scientist must endeavor first to persuade his employer to take the necessary corrective action; but if such efforts fail he may feel obliged to raise the issue more publicly, often with great risk of losing his job. The problems are complex and difficult, but they are vitally important for the protection of the public. The Committee's reports set forth the problems in much more detail (24, 25). Our concern extends also to the defense of fellow scientists in foreign countries whose rights have been violated by oppressive regimes. This has indeed been a major part of our concerns, and it must remain so, in a world so turbulent and violent as that of today.

I have mixed views concerning the place of science in the world today. I marvel at, and delight in, the amazing achievements of modern biology in unraveling the nature of heredity, and the biological controls that direct the harmonious activity of living organisms, revealing knowledge that, as a young researcher, I would never have expected to see in my lifetime. On the other hand the applications of science, often so beneficent, can be, and often are, productive of new dangers for mankind. The creative scientist derives immense delight and satisfaction, mixed with periods of confusion and anguish, from the process of discovery; but increasingly we must be concerned with the wise use of science which is now a major force—probably the major force—in the transformation of the complex modern world.

Acknowledgment

I am indebted to the National Science Foundation (grant SOC7912543) for the support of my current research.

REFERENCES

1. Abu, J.C. and Hapgood, R.: 1970. *Pioneer in Modern Medicine: David Linn Edsall of Harvard*, Harvard University Press, Cambridge, Mass.
2. Henderson, L.J.: 1913. *The Fitness of the Environment*, Macmillan; reprinted by Beacon Press, Boston, 1958.
3. Henderson, L.J.: 1928. *Blood: a Study in General Physiology*, Yale University Press, New Haven.
4. Edsall, J.T.: 1961. *Biog. Mem. Natl. Acad. Sci. USA* 35, 47–84.
5. Edsall, J.T. and von Muralt, A.: 1980. *Trends Biochem. Sci.* 5, No. 8, 228–230.
6. Colowick, S.P.: 1958. *Science* 128, 519–521.
7. Edsall, J.T.: 1980. *Nature* 285, 58.

8. Edsall, J.T. and Meister, A.: 1960. *Amino Acids, Proteins, and Cancer Biochemistry* (Edsall, J.T., ed.) Academice Press, New York. pp. 1–8, New York.

9. Edsall, J.T. and Stockmayer, W.H.: 1980. *Biog. Mem. Natl. Acad. Sci. USA* 52, 335–377.

10. Scatchard, G.: 1960. *J. Chem. Phys.* 33, 1279–1281.

11. Edsall, J.T.: 1935. *J. Am. Chem. Soc.* 57, 1506–1507.

12. Edsall, J.T. and McKenzie, H.A.: 1978. *Adv. Biophys.* 10, 137–208.

13. Frank, H.S. and Evans, M.W.: 1945. *J. Chem. Phys.* 13, 507–532.

14. Kauzmann, W.: 1959. *Adv. Protein Chem.* 14, 1–63.

15. Cohn, E.J. and Edsall, J.T.: 1943. *Proteins, Amino Acids and Peptides*, 700 pp., Reinhold, New York.

16. Cohn, E.J.: 1948. *Advances in Military Medicine*, Vol. I (Andrus, E.C. *et al.* eds.) pp. 365–443, Little Brown, Boston.

17. Edsall, J.T.: 1947. *Adv. Protein Chem.* 3, 383–479.

18. Edsall, J.T.: 1971. *Annu. Rev. Biochem.* 40, 1–28.

19. Edsall, J.T.: 1980. *J. Biol. Chem.* 255, 8939–8951.

20. Edsall, J.T.: 1972. *J. Hist. Biol.* 5, 205–257.

21. Edsall, J.T.: 1980. *Fed. Proc.* 39, 226–235.

22. Edsall, J.T. and Bearman, D.: 1979. *Proc. Am. Philos. Soc.* 123, 279–292.

23. Bearman, D. and Edsall, J.T. (eds.): 1980. *Archival Sources for the History of Biochemistry and Molecular Biology*, xii+338 pp., American Philosophical Society, Philadelphia.

24. American Association for the Advancement of Science: 1975. *Scientific Freedom and Responsibility*. A report for the Committee by J.T. Edsall, AAAS, xiii+50 pp., Washington, D.C.

25. American Association for the Advancement of Science: 1979. Annual Report of Committee on Scientific Freedom and Responsibility, 77 pp., Washington, D.C.

From Enzymes to Insulin and Back

C. L. TSOU

C.L. Tsou was born on May 17, 1923 in Tsingdao, China. He had his college training during the war and graduated from the Chemistry Department of the then National South Western Associated University in Kunming in 1945. Later he won a scholarship after a keenly competed examination to go to England and had the good fortune to work with the late Professor Keilin in Cambridge to whom he owed his training as a research biochemists. After he obtained his Ph.D. degree with a thesis on the cytochromes and succinic dehydrogenase, he returned to China. During his 20 years in the Shanghai Institute of Biochemistry, apart from spending most of his time on enzymes, he also took part in the collective effort culminating in the total synthesis of crystalline bovine insulin, the first ever synthesis of a protein. He moved to Peking in 1970 and now holds the position of the Vice Director of the Institute of Biophysics of the Academia Sinica.

I began my research career on the study of cytochromes under the late Professor D. Keilin. My friends outside China perhaps still know me best for my work on the respiratory chain. However, in China, with the possible exception of a few older biochemists, most people do not even know that I had worked in this field.

Work Leading to the Total Synthesis of Insulin

The year 1958 will always be remembered in China as the year of the Great Leap Forward. In those days, everyone in China was talking about how to accomplish great tasks and scientists were no exception. I was then working in the Shanghai Institute of Biochemistry of the Academia Sinica. Being the head of the Enzyme Division, I was busily engaged in the study of oxidizing enzymes. Invited to speak on succinic dehydrogenase, I nearly paid my first visit to Japan to attend a Symposium in 1957, however political upheaval and the subsequent Great Leap Forward made this quite impossible.

Together with a group of young people, we proposed that we could tackle the problem of the first ever synthesis of a protein. Sanger had solved the sequence of insulin only two years previously and insulin was still the only protein whose sequence was known. It would be difficult for people outside China to imagine what a difficult task we were facing. In those years, biochemistry in China was still in its infancy and we were very short of all kinds of reagents. Just consider that we had to make every one of the amino acids ourselves in the Institute (fortunately bovine insulin has only 17) and we were to synthesize a protein!

It is known that insulin is composed of two chains linked by two disulfide bridges and in addition, the A chain has an intrachain disulfide. The ideal way to synthesize insulin would be to block the 6 SH groups by 3 different protecting groups and after the synthesis of the chains, these can be removed one at a time to insure the correct pairing of the disulfides. This looks easy now, but it was quite impossible 20 years ago. The easiest way to synthesize insulin was to synthesize the chains separately with the same SH protecting group which could be removed after the completion of the chains, and then the correct structure of insulin would be formed simply by oxidizing the SH form of the chains. This job was given to me and I had the good fortune to have a group of very able young people to help me, namely: Du, Zhang, Lu, and Xu, all of whom are now professors and are making valuable contributions in different fields, and Jiang, who joined us a few years later.

When we looked up the literature, we were very much saddened by the finding that all previous attempts to reduce insulin and to get the active hormone back by reoxidation had failed. Furthermore, in a model study, Rydon (1) reached the conclusion that by oxidizing two peptide chains, each with two cysteine residues, the most stable product would be that with the chains antiparallel to each other. Undaunted by these findings, we started to separate the chains by sulfitolysis, put them together again, reduce and then reoxidize them by simply leaving the solution in air. Soon we had the good results: an activity recovery of nearly 10%. At first we did not publish our results, not so much because we had heard that some other laboratories were also actively engaged in the synthesis of insulin, but due to the simple fact that *Scientia Sinica* was not being published at that time. However, after a report by Dixon and Wardlaw appeared in 1960 (2) with a lower activity recovery, we published our results in 1961 in the first issue of *Scientia Sinica* which had just resumed publication (3). We not only had a higher activity recovery but we also succeeded in the purification and crystallization of the reoxidized product, and proved by fingerprint mapping that it indeed had the correct structure. After 1962, I was able to resume my enzyme work; however, I kept a small group working on this problem with Du and Jiang and we were soon able to increase the activity recovery to 50% (*4*).

I have often been asked by foreign friends why in the final synthesis of insulin from the synthetic chains, the yield was only 1–2%. It has also often been written in the literature about the "low yield" of insulin from recombination of the chains, citing one of the synthetic references.

I wish to point out that everyone who is familiar with peptide synthesis knows the difficulty in purifying heavily protected chains of this length owing to their extremely low solubility in most solvents. The best way would be to use the crude chains directly and, after deblocking and reoxidation, purify the regenerated insulin instead. This is the reason for the low yield in combining the synthetic chains. Due to similar reasons, the yield for the half synthesis of insulin with a natural chain and a synthetic chain was usually about 10%. When both natural chains are used, a yield of 50% is not difficult to obtain and this has later been confirmed (*5, 6*). We suggested in our 1961 paper that "among all the possible structures that can be formed on reoxidation of reduced A and B chains, that with the insulin structure is one of the most stable."

My attention has recently been drawn to the production of insulin by the recombinant DNA technique (*7*). I was of course greatly interested to read that human insulin was produced from synthetic genes for the two chains and after expression of these genes in *Escherichia coli,* insulin could be obtained from the chains with a yield of 80% when the A chain was present in excess. I wish to point out that we reported the fact that a nearly quantitative yield could be obtained with the A chain in excess as early as 1964 (*8*). Katsoyanis's results came several years later (*9*). It is interesting to note that even without the connecting C chain, correct pairing of the A and B chains can thus be easily achieved.

Number and Type of Essential Groups in Enzymes

By 1962, I was able to come back to my old favourite, the study of enzymes, after a short diversion in protein chemistry. However, the experience of the last three years had had some effect on me and I became interested in the effect of chemical modification of protein side chain groups on their activities. I was struck by the loose interpretation of experimental results in the literature. This was further aggravated by the fact that the specificity of the reagents then available left much to be desired.

In this connection, I found the kinetic method of Ray and Koshland (*10*) most interesting, in that they made an important contribution in relating quantitatively the number of essential groups involved and the biological activity lost on the protein modified. However, the method of Ray and Koshland, although a big step forward, is still seriously limited in its applications. Some reagents may react so fast as to make accurate determination of the rate constants a very difficult matter. In some other cases, the reagent employed may not be in great excess of the protein and the first order rate law does not hold. If a set of more generally applicable quantitative relations can be worked out, it will not only give greater impetus to future work on the determination of the number and type of essential groups but will also render a large number of data already in the literature amenable to such an analysis so that more definite conclusions can be obtained.

This led to the publication of a paper in *Scientia Sinica* in 1962 proposing such a method (*11*). I do not wish to go into the details of the method, especially as it has now find its way into the third edition of the classical textbook on enzymes by Dixon and Webb and a recently published book by Cornish-Bowden (*12, 13*). I wish only to point out that after examination of a fairly large number of papers then available with sufficient data amenable to such an analysis, it appeared most remarkable that although the number of the side chain groups modified by a certain reagent may sometimes be very large, the number really essential for the protein to express its biological activity is usually rather small. I do not wish to quarrel with the authors of the above mentioned famous textbook who have done me an honour by including my paper in that respected volume. Nevertheless, I might suggest that a cumbersome method is better than no method at all and perhaps computers can now take over the routine calculations. I can only hope that more authors will become familiar with both the method of Ray and Koshland as well as that of myself and that loose interpretations of the effect of chemical modification of essential groups of proteins and enzymes will become less and less frequent in the literature.

Irreversible Enzyme Inhibition

If only a limited number of people know about the works mentioned above, I am inclined to think that probably no one outside China knows about the two papers (*14, 15*) I published in Chinese in 1965. I would have translated these two papers and published them in the English version of *Scientia Sinica* but for the Cultural Revolution which swept over China shortly thereafter.

Enzyme inhibition has always appeared to me to be an important field of study because it is not only relevant to the understanding of enzyme mechanisms but is also indispensable to a rational approach to pharmacology and toxicology. In both these respects, I think irreversible inhibition may be even more important than reversible. However, the kinetic aspect of irreversible inhibition was rather neglected, at least compared to reversible inhibition, at that time. A vast literature already existed for the latter. I think a unified scheme can be proposed based on the following:

$$\begin{array}{ccccc}
\overset{\text{Y}}{\underset{}{+}} & & \overset{\text{Y}}{\underset{}{+}} & & \\
\text{S+E} & \overset{k_{+1}}{\underset{k_{-1}}{\rightleftharpoons}} & \text{ES} & \overset{k_{+2}}{\longrightarrow} & \text{E+P} \\
k_{+0}\Big\|k_{-0} & k'_{+1}\Big\|k'_{+0} & k'_{-0} & k_{-1}\Big\|k'_{+2} & \\
\text{S+EY} & \overset{}{\underset{k'_{-1}}{\rightleftharpoons}} & \text{ESY} & \longrightarrow & \text{EY+P} .
\end{array} \tag{1}$$

The rate of decrease of the native forms of the enzyme E_T including both E and ES can be easily derived:

$$-\frac{d[E_T]}{dt} = \left\{ \frac{[Y](k_{+0}+k'_{+0}\bar{K}[S])}{1+\bar{K}[S]} + \frac{k_{-0}+k'_{-0}\bar{K}'[S]}{1+\bar{K}'[S]} \right\}[E_T]$$

$$- \frac{[E_0](k_{-0}+k'_{-0}\bar{K}'[S])}{1+\bar{K}'[S]} . \tag{2}$$

Where: $\bar{K}$ and $\bar{K}'$ are the respective inverted Michaelis constants and $[E_0]$ the initial enzyme concentration. For irreversible inhibition, it is a special case of the general scheme with both k_{-1} and k'_{-1} equal to zero. If we now write Eq. (2) as:

$$-\frac{d[E_T]}{dt} = (A[Y]+B)E_T - B[E_0] \tag{3}$$

we can now present unified conditions describing competitive, noncompetitive and uncompetitive inhibitions for both reversible and irreversible inhibitors as shown in Table 1.

The terms in the last column of Table 1, a_∞, easily derived from Eq. (3) by integration to $t=\infty$, represent the fractional activity left at equilibrium. It can be seen that they are exactly the same as the terms derived in the usual way as given in various textbooks of enzyme kinetics for reversible inhibitors. For irreversible inhibitors, the terms k_{-0} and k'_{-0} vanish and hence the coefficient B. Only coefficient A need be considered.

TABLE 1. General Conditions for Different Types of Inhibition

Types of inhibition	Conditions	A $\dfrac{k_{+0}+k'_{+0}\bar{K}[S]}{1+\bar{K}[S]}$	B $\dfrac{k_{-0}+k'_{-0}\bar{K}'[S]}{1+\bar{K}'[S]}$	a_∞ $\dfrac{B}{A[Y]+B}$
Competitive	EYS does not form	$\dfrac{k_{+0}}{1+\bar{K}[S]}$	k_{-0}	$\dfrac{1+\bar{K}[S]}{1+\bar{K}S+\bar{K}_0[Y]}$
Non-competitive $\bar{K}_0=\bar{K}'_0$	$\alpha k_{+0}=k'_{+0}$ $\alpha k_{-0}=k'_{-0}$	$\dfrac{k_{+0}(1+\alpha\bar{K}[S])}{1+\bar{K}[S]}$	$\dfrac{k_{-0}(1+\alpha\bar{K}[S]}{1+\bar{K}[S]}$	$\dfrac{1}{1+\bar{K}_0[Y]}$
	$k_{+0}=k'_{+0}$	k_{+0}	k_{-0}	
Un-competitive	EY does not form	$\dfrac{k'_{+0}\bar{K}[S]}{1+\bar{K}[S]}$	k'_{-0}	$\dfrac{1+\bar{K}[S]}{1+\bar{K}[S]+\bar{K}'_0\bar{K}[S][Y]}$

The usefulness of Eq. (3) can be further demonstrated by converting it to another form. The product P formed at any time t is given as:

$$[P]=\frac{v}{A[Y]+B} \left\{ Bt+\frac{A[Y]}{A[Y]+B} \ (1-e^{-(A[Y]+B)t}) \right\} \tag{4}$$

and when the inhibition is irreversible, $B=0$:

$$[P]=\frac{v}{A[Y]}(1-e^{-A[Y]t}) . \tag{5}$$

Since the exponential term vanishes when t is sufficiently large, a plot of $[P]$ against t can then give the steady state inhibited rate in the straight-line portion of the curve. From the slopes and the intercepts at the abscissa at different Y concentrations, coefficients A and B can also be obtained.

I should perhaps also point out that the above equations can be easily adapted to the study of enzyme activation.

Owing to the unexpected interruption of my research program, the above approaches have yet to be verified experimentally. We are now, after 15 years, coming back to this problem.

REFERENCES

1. Rydon, H.N.: 1958. *Ciba Foundation Symposium on Amino Acids and Peptides with Anti-metabolic Activity* (Wolstenholme, G.E.W. and O'Connor, C.M., eds.) pp. 192–204, Churchill, London.
2. Dixon, G.H. and Wardlaw, A.C.: 1960. *Nature* 188, 721–724.
3. Du, Y.C., Zhang, Y.S., Lu, Z.X., and Tsuo, C.L.: 1961. *Sci. Sin.* 10, 84–104.
4. Jiang, R.Q., Du, Y.C., and Tsou, C.L.: 1963. *Sci. Sin.* 12, 452–453.
5. Zahn, H., Gutte, B., Pfeiffer, E.F., and Ammon, J.: 1966. *Liebig's Ann.* 691, 225–231.
6. Katsoyannis, P.G. and Tometsko, A.: 1966. *Proc. Natl. Acad. Sci. USA* 55, 1554–1561.
7. Geoddel, D.V., Kleid, D.G., Bolivar, P., Heyneker, H.L., Yansure, G.G., Crea, R., Hirose, T., Kraszewski, A., Hakura, K., and Riggs, A.D.: 1979. *Proc. Natl. Acad. Sci. USA* 76, 106–110.
8. Du, Y.C., Jiang, R.Q., and Tsou, C.L.: 1964. *Acta Biochim. Biophys. Sin.* 4, 665–672; 1965. *Sci. Sin.* 14, 229–236.
9. Katsoyannis, P.G., Trakatellis, A.C., Johnson, S., Zalut, C., and Schwartz, G.: 1967. *Biochemistry* 6, 2642–2655.
10. Ray, W.J. and Koshland, D.E., Jr.: 1961. *J. Biol. Chem.* 236, 1973–1979.
11. Tsou, C.L.: 1962. *Sci. Sin.* 11, 1535–1558.
12. Dixon, M. and Webb, E.C.: 1979. *Enzymes*, 3rd ed. pp. 376–379, Longman Group Ltd., London.
13. Cornish-Bowden, A.: 1979. *Fundamentals of Enzyme Kinetics*, pp. 94–96, Butterworths, London.

14. Tsou, C.L.: 1965. *Acta Biochim. Biophys. Sin.* 5, 400–408.
15. Tsou, C.L.: 1965. *Acta Biochim. Biophys. Sin.* 5, 409–417.

The Chemist's Magic Bullets: Selectivity as a Guiding Principle in Biomedical Research

BERNHARD WITKOP

B. WITKOP was born in 1917 in Freiburg, Germany. After graduation from the University of Munich in 1938, he started his research career under the guidance of H. Wieland. He received a doctor's degree in 1940. He moved to the United States in 1947. Since 1950, he has been associated with the National Institutes of Health. His research covers a wide range of bioorganic chemistry, comprising alkaloids, arrow poisons, and various kinds of naturally occurring toxins, pharmacologically active amines, *etc*. Not only in low molecular weight substances but also in biopolymers was he interested. Nucleic acid components and polypeptides were also investigated. The discovery and development of the cyanogen bromide method for the cleavage of polypeptides is particularly appreciated in protein chemistry. He was awarded the Paul Karrer Medal in 1971. He is interested not only in science but also in arts and literature, old and new, west and east.

The seventieth birthday of Fujio Egami, the seventy seventh (Ki-Ju) of Munio Kotake (*1*) as well as the hundredth of my teacher, Heinrich Wieland (*2*) not only serve as reminders that we are just links in the chain of scientific tradition, but prompt an inquiry into the aims and accomplishments of organic chemistry, especially within a mildly mission-oriented biomedical research organization, such as the Mitsubishi-Kasei Institute of Life Sciences or the National Institutes of Health, where *research priorities* are a constant topic for discussion (*3*).

In looking for a valid criterium for the usefulness of a method or an agent,

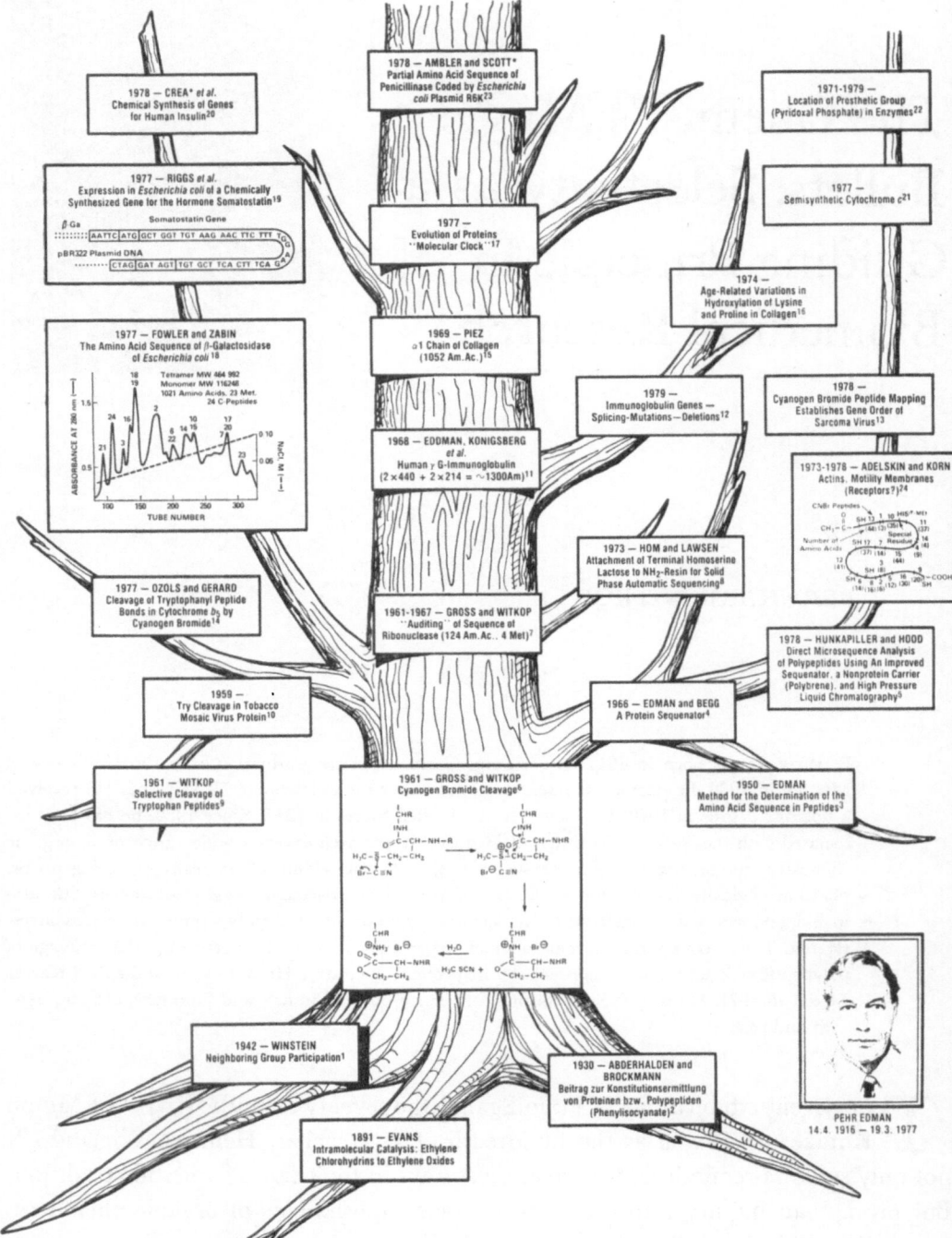

FIG. 1. Neighboring group participation as the guiding principle for the discovery and effectiveness of two major methods that enabled sequencing of large proteins: sequential degradation by the Edman method and selective cleavage of methionine (and tryptophan) peptide bonds by the cyanogen bromide reaction.

I personally felt guided by the principle of selectivity. In that respect the organic chemist, like Weber's Freischütz, is looking for magic bullets ("Freikugeln") that he may direct to known targets or that nature directs to unpredictable or still unknown targets. *Cyanogen bromide*, discovered in 1961 (*4*), is one of the smallest and most effective of these bullets, and its success rests on the principle of neighboring group participation (*5, 6*), in particular on the phenomenon that the late Saul Winstein liked to call *anchimeric assistance* (*7*), while Myron Bender refers to *intramolecular catalysis* (*8*). The beginnings of this concept, before it was generalized and properly recognized, go back to 1891 when W.P. Evans in Giessen described a kinetic investigation of the base-promoted intramolecular cyclization of ethylene chlorohydrins to ethylene oxides (*9*).

The usefulness of cyanogen bromide consists in its exclusive *search and cleave* action on methionyl peptide bonds. The longest protein sequence established so far by this method is the single chain of 1,021 amino acids, MW 116,248, of β-galactosidase of *Escherichia coli* (*10*). The 24 C-peptides formed by selective cleavage of the 23 methionyl bonds, separate easily on a column of O-carboxymethylcellulose, and then are accessible to successive Edman degradations, nowadays mostly done automatically on a "sequenator" with an accuracy that permits, under optimal conditions, including high pressure liquid chromatography, the sequencing of 47–77 residues on a nanomole to picomole scale, depending on the nature of the protein (*11*).

The late *Pehr Edman* (1916–1977) (*12*), shortly before his death, took pride and satisfaction in the accomplishment that more than 80,000 amino acids have been properly positioned in sequence (*13*). The 200,000 mark should be reached within a few years, but this number is dwarfed by the coding capacity of the mammalian genom of 2 billion residues.

But the cyanogen bromide method, beyond its value in sequencing (*14*), resynthesis of fragments (*15*), improvement by attachment to carriers for combination with solid phase Edman degradation (*16*), localization of prosthetic groups in enzymes (*17–19*) or of points of selective oxidation in (pro)-collagen (*20*), has proven its usefulness most recently and most dramatically in the *genetic*

←1. *cf.* refs. *5, 6*, and *7* in the text. 2. Abderhalden, E. and Brockmann, H.: 1930. *Biochem. Z.* 225, 386–404. 3. Edman, P.: 1950. *Acta Chem. Scand.* 4, 283–293. 4. Edman, P.: 1966, *Eur. J. Biochem.* 1, 80–91. 5. *cf.* ref. *11* in the text. 6. Gross, E. and Witkop, B.: 1961. *J. Am. Chem. Soc.* 83, 1510. 7. Gross, E. and Witkop, B.: 1962. *J. Biol. Chem.* 237, 1856; 1967. *Biochemistry* 6, 745. 8. *cf.* ref. *16* in the text. 9. *cf.* ref. *4* in the text. 10. Ramachandran, L.K. and Witkop, B.: 1959. *J. Am. Chem. Soc.* 81, 4028. 11. Waxdal, M.J., Konigsberg, W.H., Morley, W.L., and Edelman, G.M.: 1968. *Biochemistry* 7, 1959. 12. Franklin, E.C., Prelli, F., and Frangione, B.: 1979. *Proc. Natl. Acad. Sci. USA* 76, 452–456. 13. Oskarsson, M.K., Elder, J.H., Gantsch, J.W., Lerner, R.A., and Vande Woude, G.F.: 1978. *Proc. Natl. Acad. Sci. USA* 75, 4694–4698. 14. *cf.* refs. *32* and *33* in the text. 15. *cf.* ref. *20* in the text. 16. Barnes, M.J., Constable, B.J., Morton, L.F., and Royce, P.M.: 1976. *Biochem. J.* 139, 461. 17. *cf.* ref. *36* in the text. 18. *cf.* ref. *10* in the text. 19. *cf.* ref. *21* in the text. 20. *cf.* ref. *22* in the text. 21. *cf.* ref. *15* in the text. 22. *cf.* refs. *17, 18*, and *19* in the text. 23. *cf.* ref. *27* in the text.

synthesis of important hormones. A purely chemical reaction was the final step in recovering somatostatin from its precursor made by inserting a synthesized gene for somatostatin into *E. coli*. The gene sequence for the human hormone somatostatin, preceded by the codon ATG for methionine, was fused to the gene for β-galactosidase (*10*), a very large protein (1,021 residues). The *E. coli* thus produced the short peptide methionyl-somatostatin linked to β-galactosidase. Cyanogen bromide cleavage, which breaks the polypeptides at methionine liberated the tetradecapeptide which itself does *not* contain methionine. This overall approach thus not only provided a safety factor in that the *E. coli* could not make the hormone directly, but it protected the somatostatin produced from being degraded by the cells' own enzymes, since *E. coli* does not normally produce peptides as small as somatostatin. This marks the first successful use of recombinant DNA techniques to get a bacterium to produce a substance from a gene of a high organism (*21*). More recently this approach has been extended to the chemical synthesis of genes for the A (77 base pairs) and B (104 base pairs) chains of human insulin (*22*), neither containing methionine. Therefore the liberation of the two chains, or, conceivably, of the larger (pre)-proinsulin (*23*) requires the addition of the methionine triplet (ATG) to the insulin gene, so that cyanogen bromide will yield the peptide hormones or their precursors.

While proteins of more than thousand amino acids still are laborious to sequence, the astonishing speed and accuracy of DNA sequencing techniques (*24*) are rapidly establishing DNA sequences, such as the nucleotide sequence of bacteriophage φx 174 with 5,386 nucleotides (*25*). The "magic bullet" in this case is simply *dimethyl sulfate* which, under properly controlled conditions, selectively cleaves at guanine, adenine, cytosine, or thymines, preferentially, indiscriminately or selectively. Such methods will make it possible eventually to sequence a complete DNA, *e.g.*, adenovirus DNA with about 35,000 nucleotides (Richard Roberts, Cold Spring Harbor). An equally small selective molecule is *sodium borohydride* which, under proper conditions, photo-reduces exclusively bound uridine to dihydrouridine residues (*26*).

The accuracy of the DNA technique is so good that, in favorable cases, protein sequences have been audited and corrected (*27*) in the same way as the introduction of the cyanogen bromide cleavage—historically—served to audit the primary sequence of many proteins, such as ribonuclease (*28*).

The most recent and most glamorous example of this new approach is the deduction of the complete sequence of 166 amino acids of human fibroblast interferon by isolation and sequencing of the corresponding gene (*29*).

We had noticed earlier that the indole chromophore of tryptophan is slightly affected by prolonged contact with cyanogen bromide. Bovine β-lactoglobulin-AB with 612 amino acids and 5 methionine residues (*30*) not only yields the expected C-peptides, but also 3 minor peptides resulting from noticeable cleavage at the two tryptophan residues in positions 19 and 61 (*31*). The yields of this tryptophan cleavage can be made quantitative by working in more acidic medium, hexafluoro-butyric, rather than 70–100% formic acid, and be made selective by suppressing methionyl peptide cleavage through prior photooxidation to the sulfoxide (*32*, *33*). In this form cyanogen bromide may compete with, or be superior to, cleavage of the tryptophanyl peptide bond by dimethyl sulfoxide and hydrobromic acid (*34*), but still not as elegant as the (non-concerted) selective cleavage by *o-iodosobenzoic acid* (*35*).

A word about the mechanism of these cleavages: in the pseudohalogen cyano-gen bromide, polarization and induction of dipole may operate in two ways. Normally in formic acid the strong nucleophilic methylmercapto group of methio-nine approaches the dipole at the cyano group leading, eventually, to methyl thiocyanate (CH_3–S–C$\equiv$N) and Br$^-$. In a more acid medium the less nucleophilic β-position of the indoleninium form of tryptophan may approach the bromine end of the dipole to form the requisite β-bromoindole intermediate. In this case the energetically favorable cyanide anion may be expected to be formed.

In summary, two fundamental methods, both based on neighboring group participation, are available for rapid sequencing of protein chains, the (automated) Edman degradation, preceded by the cleavage of methionyl (and/or tryptophanyl) peptide bonds by cyanogen bromide (or *o*-iodosobenzoic acid). Both methods have made possible major advances, such as an evolutionary pattern of protein molecules, an evaluation of the "neutral mutation theory" and the establishment of sub-stitution rates, *i.e.* substitution/year/nucleotide as an indication for the existence of a "*molecular clock*" (*36*). Beyond that, the cyanogen bromide cleavage, because of its great selectivity, has found numerous new applications, especially in the production of peptide hormones by genetic engineering. *Figure 1 presents these advances as ramifications of a tree that is still growing.*

Magic bullets with unknown targets are supplied by nature. They are normally discovered by their high toxicity, whose mechanism of action eventually reveals the target(s).

My first contact with such a bullet was with calabash curare, the arrow poison used by many South American Indians. That it was related to a degradation pro-duct from strychnine, the Wieland-Gumlich aldehyde (*2*) was established by the late Hans Schmid in the Laboratory of Paul Karrer (*37*). Likewise, Wieland was the pioneer of the venoms of the toads (*2*). The combination of the two subjects led me to the study of arrow poisons prepared from frogs by the Indians of Western Colombia (*38*).

One of the strongest toxins known to man is the steroidal ester *batrachotoxin*. The target of this magic bullet are the "gates" of sodium channels (*39*) in nerve and muscle (*40*). Other toxins, such as saxitoxin, tetrodotoxin or scorpion venom help to establish a connection between multiple recognition and receptor sites, between (non)cooperative and heterogeneous binding sites (*41*). Voltage clamp techniques, electral stimulation and the use of agonists are required to find out that batrachotoxin, in order to have an effect, requires a certain population of open channels, as the trans-membrane potential is lowered and reclosure of the channel is prevented more or less irreversibly (42). As a blocker of axonal transport of proteins and vital enzymes batrachotoxin is about one thousand times more potent than vinblastine or colchicin.

Another magic bullet is *histrionicotoxin* (43) and its congeners (44) which selectively blocks cholinergic postsynaptic effects, a phenomenon dependent on open channels, *i.e.*, a critical minimum frequency of stimulation (44, 45). Such stimulation may be electrical or iontophoretic, but at concentrations of histrionicotoxin (10^{-12}–10^{-7} M) which block the effects of iontophoretic acetylcholine, the responses to spontaneously released acetylcholine from presynaptic vesicles (micro-endplate currents measured under voltage clamp conditions) are unaffected (46). Such a discovery has a profound effect on the validity of observations and conclusions arrived at by direct and indirect stimulation, as is customary in neuro- and electrophysiology.

As a potential ligand for the ion conductance modulator of the acetylcholine receptor (47) perhydrohistrionicotoxin has proved its usefulness (48) in separating the acetylcholine receptor from the ion translocation site as two distinct protein fractions. Perhydrohistrionicotoxin also helped to clarify the voltage-dependent action of tetraethylammonium ions on endplate conduction and inhibition of ligand binding to postsynaptic proteins (49). Successful recombination of such solubilized fractions with the aid of artificial membranes has been reported from three laboratories (50–52).

Studies on the inhibiting effect of dihydroisohistrionicotoxin (H_2HTX) on ACh-receptor-dependent $^{22}Na^+$ uptake by cultured chick embryo muscle cells have led to the suggestion that H_2HTX inhibits the nicotinic receptor by causing an increase in the affinity of the desensitized form of the receptor for agonists and thereby stabilizing the desensitized state (53). In addition, H_2HTX has been found to inhibit the muscarinic receptor system; it, as well as tetracaine, inhibited noncompetitively [^3H]-scopolamine binding to neuroblastoma glioma hybrid cells (54).

The subject of such magic bullets is as fascinating as inexhaustible. In order to recognize the target, the time has often to be right and the stage to be set by the requisite advances in molecular biology.

A man-made molecule made possible selective cleavage of proteins as elaborated in the first part of this overview. Selective cleavage of the less diversified DNA sequences is possible, of course, by special enzymes. But a natural metal-binding glycopeptide antibiotic, *bleomycin*, under appropriate conditions, cleaves double- and single-stranded DNA at G-T and G-C sequences only. The mechanism of such a cleavage is intriguing and provides new directives for more "magic bullets" (55) and new challenges for greater selectivity, adumbrated as the guiding principle at the outset.

Returning to the central nervous system we notice that *6-hydroxydopamine*, isolated as an autoxidation product (*56*) and urinary metabolite of dopamine 20 years ago (*57*), has turned out to be capable of selective destruction of peripheral and central adrenergic nerve terminals (*58*). This magic bullet has found many applications, for instance, in the production of Parkinson's disease in animal models (*59*) or in studying (presynaptic) dopamine receptors with tritiated apomorphine in nigrostriatal pathways before and after 6-hydroxydopamine lesions (*60*). Attempts to utilize 6-hydroxydopamine, attached to analgesic agents, for the construction of affinity-directed irreversible inhibitors of the narcotic receptor (*61*) have not been as successful as the synthesis of irreversible inhibitors ("irazepine") of the diazepam (inosine) (*62*) receptor (*63*). Analogously hydroxyserotonins have opened up a wide field of selective neurotoxins affecting serotinergic receptors (*64*).

In a related fashion, but by a different mechanism, *kainic* acid (*65*), a naturally occurring rigid analog of the neurotransmitter glutamic acid, is the most powerful substance known that produces selective neuronal lesions in the central nervous system while sparing axonal passages. It is a potent neuroexcitatory agent which destroys cells containing glutamic acid receptors (*66*). The reader interested in these advances may find the latest review in "Neurotoxins: Tools in Neurobiology" (*67*).

Our most recent "magic bullet" is an unusual tropane alkaloid, *anatoxin-a*, isolated from toxic blue algae (*68*). It is an ideal agonist or nicotinic acetylcholine receptors, about 20 times more active than carbamylcholine and much closer to acetylcholine, the natural neurotransmitter, with regard to the time of opening (~1.5 msec) of the ion conductor channel (*69*).

In conclusion we may modify Montaigne's aperçu and say that we find the desired magic bullets more by chance than by systematic search.

REFERENCES

1. Witkop, B.: 1977. *Kagaku* 8, 605–619 (in Japanese).
2. Witkop, B.: 1977. *Angew. Chem. Int. Ed.* 16, 559–572.
3. Brooks, H.: 1978. *Daedalus* 20, 171–190.
4. Witkop, B.: 1961. *Adv. Protein Chem.* 16, 221–321.
5. Winstein, S. and Buckles, E.: 1942. *J. Am. Chem. Soc.* 64, 2780.
6. Capon, B. and McManus, S.P.: 1976. *Neighboring Group Participation*, Plenum Press, New York and London.
7. Winstein, S., Lindegren, C.R., Marshall, H., and Ingraham, L.L.: 1953. *J. Am. Chem. Soc.* 75, 148.
8. Bender, M.L.: 1957. *J. Am. Chem. Soc.* 79, 1258.
9. Evans, W.P.: 1891. *Z. Phys. Chem.* 7, 337.
10. Fowler, A.V. and Zabin, I.: 1977. *Proc. Natl. Acad. Sci. USA* 74, 1507–1510.
11. Hunkapiller, M.W. and Hood, L.E.: 1978. *Biochemistry* 17, 2124–2133.
12. Blombäck, B.: 1977. *Thromb. Res.* 11, 695–698, Pergamon Press, London.

13. Edman, P.: 1976. *Carlsberg Res. Commun.* 42, 1–9.
14. Dayhoff, M.O.: 1976. *Atlas of Protein Sequence and Structure*, Vol. 5 (Suppl. 2) (Dayhoff, M.O., ed.) pp. 1–345, National Biochemical Research Foundation, Washington, D.C.
15. Barstow, L.E., Young, R.S., Yakali, E., Sharp, J.J., O'Brien, J.C., Berman, P.W., and Harbury, H.A.: 1977. *Proc. Natl. Acad. Sci. USA* 74, 4248–4250.
16. Laursen, R.A. and Horn, M.J.: 1977. *Advanced Methods in Protein Sequence Determination* (Needleman, S.B., ed.), Springer-Verlag, Berlin, Heidelberg, and New York.
17. Smith, E.L., Landon, M., Piszkiewicz, D., Brattin, W.J., Langley, T.J., and Melamed, M.D.: 1970. *Proc. Natl. Acad. Sci. USA* 67, 724–730.
18. Sygusch, J., Madsen, N.B., Kasvinsky, P.J., and Fletterick, R.J.: 1977. *Proc. Natl. Acad. Sci. USA* 74, 4757–4761.
19. Titani, K., Koide, A., Hermann, J., Ericsson, L.H., Kumar, S., Wade, R.D., Walsh, K.A., Neurath, H., and Fischer, E.H.: 1977. *Proc. Natl. Acad. Sci. USA* 74, 4762–4766.
20. Piez, K.A.: 1976. *Biochemistry of Collagen* (Ramachandran, G.N. and Reddi, A.H., eds.) pp. 4–9, Plenum Publishing Corp., New York.
21. Itakura, K., Hirose, T., Crea, R., Riggs, A.D., Heyneker, H.L., Bolivar, F., and Boyer, H.W.: 1977. *Science* 198, 1056–1063.
22. Crea, R., Kraszewsky, A., Hirose, T., and Itakura, K.: 1978. *Proc. Natl. Acad. Sci. USA* 75, 5765–5769.
23. Villa-Komaroff, L., Efstratiadis, A., Broome, S., Lomedico, P., Tizard, R., Naber, S.P., Chick, W.L., and Gilbert, W.: 1978. *Proc. Natl. Acad. Sci. USA* 75, 3727–3731.
24. Maxam, A.M. and Gilbert, W.: 1977. *Proc. Natl. Acad. Sci. USA* 74, 560–564.
25. Sanger, F., Coulson, A.R., Friedmann, T., Air, G.M., Barrell, B.G., Brown, N.L., Fiddes, J.C., Hutchison III, C.A., Slocombe, P.M., and Smith, M.: 1978. *J. Mol. Biol.* 125, 225–245.
26. Witkop, B.: 1968. *Photochem. Photobiol.* 7, 813–828.
27. Ambler, R.P. and Scott, G.K.: 1978. *Proc. Natl. Acad. Sci. USA* 75, 3737–3741.
28. Witkop, B.: 1968. *Science* 162, 318–326.
29. Derynck, R., Content, J., DeClercq, E., Volckaert, G., Tavernier, J., Devos, R., and Fiers, W.: 1980. *Nature* 285, 542–547.
30. Braunitzer, G., Chen, R., Schrank, B., and Stangl, A.: 1972. *Hoppe-Seyler's Z. Physiol. Chem.* 353, 832–834.
31. Wootton, J.C., Chambers, G.K., Holder, A.A., Baron, A.J., Taylor, J.G., Fincham, J.R.S., Blumenthal, K.M., Moon, K., and Smith, E.L.: 1974. *Proc. Natl. Acad. Sci. USA* 71, 4361–4365; Braunitzer, G. and Aschauer, H.J.: 1975. *Hoppe-Seyler's Z. Physiol. Chem.* 356, 473–747.
32. Ozols, J. and Gerard, C.: 1977. *Proc. Natl. Acad. Sci. USA* 74, 3724–3729.
33. Ozols J. and Gerard, C.: 1977. *J. Biol. Chem.* 252, 5986–5989.
34. Fontana, A., Savige, W.E., and Zambonin, M.: 1980. *Peptide and Protein Sequence* (Proc. 3rd Int. Conf.) (Birr, C., ed.) pp. 309, Elsevier/North-Holland, Amsterdam.
35. Mahoney, W.C. and Hermodson, M.A.: 1980. *Peptide and Protein Sequence* (Proc. 3rd Int. Conf.) (Birr, C., ed.) p. 323, Elsevier/North-Holland, Amsterdam.

36. Matsubara, H. and Yamanaka, T. (eds.): 1978. *Evolution of Protein Molecules*, Japan Sci. Soc. Press, Tokyo.

37. Isler, O.: 1978. *Biogr. Mem. Fellows Roy. Soc.* 24, 245–321.

38. Witkop, B.: 1971. *Experientia* 27, 1121–1138.

39. Khodorov, B.I., Peganov, E.M., Revenko, S.V., and Shrishkova, L.D.: 1975. *Brain Res.* 84, 541–546.

40. Albuquerque, E.X. and Daly, J.W.: 1977. *A Selective Probe for Channels Modulating Sodium Conductances in Electrogenic Membranes*, Vol. 1 (Ser. B) (Cuatrecases, P., ed.) pp. 297–338, Chapman and Hall, London.

41. Catterall, W.A. and Morrow, C.S.: 1978. *Proc. Natl. Acad. Sci. USA* 75, 218–222; Sastre, A. and Podleski, T.R.: 1976. *Proc. Natl. Acad. Sci. USA* 73, 1355–1359.

42. Garrison, D.L., Albuquerque, E.X., Warnick, J.E., Daly, J.W., and Witkop, B.: *Mol. Pharmacol.* 14, 111–121.

43. Witkop, B.: 1979. *La Recherche* 6, 528–539.

44. Daly, J.W., Brown, G.B., Mensah-Dwumah, M., and Myers, C.W.: 1978. *Toxicon* 16, 163–188.

45. Albuquerque, E.X., Barnard, E.A., Chiu, T.H., Lapa, A.J., Dolley, J.O., Jansson, S.E., Daly, J.W., and Witkop, B.: 1978. *Proc. Natl. Acad. Sci. USA* 70, 949–953; Albuquerque, E.X., Kuba, K., and Daly, J.W.: 1974. *J. Pharmacol. Exp. Ther.* 189, 513–524.

46. Albuquerque, E.X. and Cage, P.W.: 1978. *Proc. Natl. Acad. Sci. USA* 75, 1596–1599.

47. Eldefrawi, A.T., Eldefrawi, M.E., Albuquerque, E.X., Oliveira, A.C., Mansour, N., Adler, M., Daly, J.W., Brown, G.B., Burgermeister, W., and Witkop, B.: 1977. *Proc. Natl. Acad. Sci. USA* 74, 2172–2176.

48. Eldefrawi, M.E., Eldefrawi, A.T., Mansour, N.A., Daly, J.W., Witkop, B., and Albuquerque, E.X.: 1978. *Biochemistry* 17, 5474–5483.

49. Adler, M., Oliveira, A.C., Eldefrawi, M.E., and Eldefrawi, A.T.: 1979. *Proc. Natl. Acad. Sci. USA* 76, 531–535.

50. Epstein, M. and Racker, E.: 1978. *J. Biol. Chem.* 253, 6660–6662.

51. Changeux, J.-P., Heidmann, T., Popot, J.-L., and Sobel, A.: 1979. *FEBS Lett.* 105, 181–187.

52. Gonzalex-Ros, J.M., Paraschos, A., and Martinez-Carrion, M.: 1980. *Proc. Natl. Acad. Sci. USA* 77, 1796–1800.

53. Burgermeister, W., Catterall, W.A., and Witkop, B.: 1977. *Proc. Natl. Acad. Sci. USA* 74, 5754–5758.

54. Burgermeister, W., Klein, W.L., Nirenberg, M., and Witkop, B.: 1978. *Mol. Pharmacol.* 14, 751–767.

55. Takeshita, M., Grollman, A.P., Ohtsubo, E., and Ohtsubo, H.: 1978. *Proc. Natl. Acad. Sci. USA* 75, 5983–5987.

56. Senoh, S. and Witkop, B.: 1959. *J. Am. Chem. Soc.* 81, 6222–6231.

57. Senoh, S., Witkop, B., Creveling, C.R., and Udenfriend, S.: 1959. *J. Am. Chem. Soc.* 81, 1768.

58. Creveling, C.R., Rotman, A., and Daly, J.W.: 1975. *Chemical Tools in Catecholamine*

Research. I. 6-Hydroxydopamine as a Denervation Tool in Catecholamine Research (Jonson, G., Malmfors, T., and Sachs, C., eds.) pp. 23–32, North-Holland, Amsterdam.

59. Creese, I., Burt, D.R., and Snyder, S.H.: 1977. *Science* 197, 596.

60. Nagy, T., Lee, T., Seeman, P., and Fibiger, C.: 1978. *Nature* 274, 278–280.

61. Rice, K.C., Shiotani, S., Creveling, C.R., Jacobson, A.E., and Klee, W.A.: 1977. *J. Med. Chem.* 20, 673–675.

62. Tallmann, J.F., Paul, S.M., Skolnick, P., and Gallager, D.W.: 1980. *Science* 207, 274–281.

63. Rice, K.C., Brossi, A., Tallman, J., Paul, S.M., and Skolnick, P.: *Nature* 278, 854–855.

64. Jacoby, J.H. and Lytle, L.D. (eds.): 1978. *Serotonin Neurotoxins*, Vol. 305, The New York Academy of Sciences, New York.

65. Mauger, A.B. and Witkop, B.: 1966. *Chem. Rev.* 66, 47–86.

66. McGeer, E.G., Olney, J.W., and McBeer, P.L. (eds.): 1978. *Kainic Acid as a Tool in Neurobiology*, Raven Press, New York.

67. Ceccarelli, B. and Clementi, F. (eds.): 1979. *Neurotoxins: Tools in Neurobiology (Advances in Cytopharmacology*, Vol. 3), Raven Press, New York.

68. Carmichael, W.W. and Gorham, P.R.: 1978. *Int. Ver. Limnol.* 21, 285–295.

69. Spivak, C.E., Witkop, B., and Albuquerque, E.X.: 1980. *Mol. Pharmacol.* 18, 384–394.

On the Chief Factors of Historical Evolution in Biochemistry

JEAN ROCHE

J. ROCHE was born on January 14, 1901, in Sorgues (Vaucluse, France) and educated at Universities of Montpellier (1917–1923) and Strasbourg (1923–1925). M.A. (1921), D.M. (1925), D.Sc. (1934). Reader of Biochemistry, Strasbourg (1925–1930); Professor of Biochemistry, Lyon (1930–1931), Marseilles (1931–1947). Professor of General and Comparative Biochemistry, Collège de France, Paris (1947–1972); Directeur du Laboratoire de Biologie marine du Collège de France in Concarneau (Finistère) (1948–1972); Rector of the University of Paris (Sorbonne) (1961–1969); general delegate for international university relationship of the Ministry of Education (1969–1972). Member of the Academy of Sciences (Institute of France) and of the Nationale Academy of Medicine and of various foreign Academies. President of the Society of Friends of Universities of Paris (since 1975). Research work on proteins (hemoglobins and respiratory pigments of invertebrates), on enzymes (phosphatase, arginase), on thyroid hormones (identification of 3,5,3'-triiodothyronine, hormone metabolism) and comparative biochemistry of iodine.

The only privilege of being a biochemist having started research work nearly sixty years ago is to have been a witness of the development of modern biochemistry through many unexpected steps (*1*). Therefore, I should like to discuss shortly on the importance and efficiency respectively of creative spirit and of experimental results, on the rapid evolution and on the extension to new fields of one of biological sciences highly promising for the future.

The most striking steps of the evolution of biochemistry have been unforseen; they were partly due to discoveries stimulating the creativity more than to experiments planned in order to reach a strictly defined answer to a limited interrogation. The viewpoint expressed in 1867 by Claude Bernard on this problem in physiology can be applied to all biological sciences of our time. For him, the chief reason of the rapid progress of physiology from the middle of nineteenth century is that experimental researches lost their "empirical character" (2). In France, under the influence of mathematicians, like Laplace, the best physiologists of this time, as Magendie, consider the results of their experiments as just limited to facts. Their interpretation open to speculative spirit became later more and more usual, opening the way to new and fruitful conceptions. I should like to illustrate by a few examples how biochemistry reached its actual prominent position by successive impulses originating from this tendency, even in the initial period of its differentiation between other biological sciences, all formerly united in physiology. Only a few significant steps of the evolution of biochemistry will be evoked to illustrate it: but it will be easy to present many others, even if the lack of chemical and biological knowledges in many fields has longly been a heavy limiting factor of possibilities open to creativity.

The origin of researches on metabolism is undoubtedly linked to discovery by Lavoisier, at the end of the eighteenth century, of the biochemical significance of respiratory exchanges. For him, the fixation of oxygen from air and the excretion of carbon dioxyde by lungs was the global expression of combustion of organic molecules in living animal organisms ("as a burning candle." The first extrapolation of these findings to a definite metabolism concept has been possible only fifty years later, when Liebig showed the presence of lactic acid—called sarcolactic acid—in muscle and its formation from glucose. The possibility of a non aerobic degradation of this carbohydrate, established by Pasteur in the chemical study of alcoholic fermentation, has led to another step in research. Progresses remained very slow during a long time, because they were linked to the poor knowledge on intermediary products of carbohydrate metabolism. One of the most important discoveries in this field has probably been the discovery of fructosediphosphate by Harden and Young at the beginning of the present century. However, it is significant that, even shortly after 1910, the work of Fletcher and Hopkins on the biogenesis of lactic acid in muscle brought a sum of analytical results more than a precise metabolic interpretation of these.

Completely new concepts had to be elaborated in order to further a step in the understanding of metabolic pathways. This occurred by the study of association of degradative and synthetic reactions in biological cycles and of the energy transfers involved in these reactions, initially studied by A.V. Hill and Meyerhof. One of the consequences of the introduction of energetic studies in metabolism has been the discovery in cells of chemical compounds containing an energy-rich link of phosphoric acid. For example, the adenosine triphosphate or ATP, phosphocreatinine, phosphoarginine, and a series of phosphoamide of various guanidine derivatives were described by us with Nguyen van Thoai and coworkers in invertebrates and the most important compound of this type present in nearly all cells. The synthesis of such a link, as this ATP from ADP and phosphate, leads to a storage of energy and its splitting, such as during the degradation of ATP in ADP and phosphate, to the liberation of energy at the disposal of various metabolic reactions. From these findings, due in great part of Lipmann, and from the work of Lynen on acetyltransfer, as from the large extension of knowledge on general metabolic cycles due to Krebs, the way has been largely open to progresses. These were rapidly reached since 1940–1942 by the use of molecules labeled by radioactive or heavy isotopes. This exceptional technical advance allowed to study the metabolism under physiological conditions of concentration of products participating to cell life and to identify many new compounds of great biological interest, especially hormones. During the same period the introduction of electron microscope allowed biologist to identify elements of subcellular structure and to study the biochemical functions of these, according to the new concepts of molecular biology. The results obtained between 1940 and 1965 in the study of cell metabolism would probably have taken more than a century without the help of these new techniques. Therefore, the remarks reminding the evolution of metabolic studies show clearly the part due to experimental findings and this due to the spirit of creativity in the progresses.

The need of a precise fitness between experimental findings and creative spirit to stimulate the evolution of sciences is evident from may examples; it acts positively when it is reaches and negatively when not. This will be illustrated specificaly by the development of enzymology as by that of endocrinology and hormones study. At the start of each of these fields, physiological researches were the unique possibility of an approach able to lead to biochemistry in the future. Each of these had a very different evolution from the other, because the biological interpretation of experimental results was objective in the case of enzymology, and limited to a dogmatic erroneous viewpoint on "internal secretions" in endocrinology.

At the end of the eighteenth century, Spallanzani demonstrated that gastric digestion of meat could be observed *in vitro* by stomach juice. This was the first experimental proof that an enzymic action was not linked to its behaviour in a living organism, under the direct and necessary influence of a so-called "vital principle." The first chemical synthesis of a product elaborated by mammals and excreted in urine, the urea, by Wohler shortly after 1820, about thirty years after its discovery by Vauquelin and Fourcroy, was considered as another definite proof of a possibility to escape the "vital principle," to interprete all events linked to life. The preparations of numerous crude enzymes during the nineteenth century strengthened this viewpoint. However, many biologists remained convinced that cells, or at least cell particles, were necessary to complex enzymic processes, like the production of alcohol by yeast. Thus, the observation at the beginning of the twentieth century by the brothers Buchner, that alcoholic fermentation of glucose could proceed in presence of cell-free yeats juice was a very important step to eliminate the myth of "vital principle" from enzymology. However, ten years later, such a distinguished biochemist as Bayliss—considered officially as a physiologist —wrote a classical monograph on "The nature of enzyme action," in order to discuss if this action could be considered as a catalytic one and concluded positively. In 1912, Michaelis and Menten act as true precursors in their kinetic studies of enzyme action. They showed the formation of enzyme-substrate combination and they characterized an enzyme by its affinity constant for a substrate. In 1920, the catalytic action of enzymes was practically admitted without restriction, and applied to degradation as to synthetic biological reactions. Only the complete darkness on the mechanism of biosynthesis of proteins seemed to justify some doubts about the general character of this viewpoint. At the same time, the nature of the enzymes was entirely unknown.

Prominent biochemists as Willstätter and Hans von Euler remained convinced in 1925 that enzymic catalysis was due to colloids containing an "active group," specific of its chemical substrate and included in a relatively small molecule, associated to an unspecific colloid bearer. This was the so-called "Trägertheorie," admitting that the colloidal support could be indifferently an unspecific protein or a carbohydrate. The isolation by Sumner of the first pure crystallysed enzyme, the urease from soja beans, and a few years after from 1925–1927 this of pure proteinases from digestive glands and that of inactive precursors or proenzymes by Northrop and Kunitz cleared the situation, allowing Warburg to declare, "all enzymes are proteins." From this time, the study of enzyme structure became a chapter of protein chemistry, taking account of the specificity of constitution linked to a catalytic activity: nature of the active part of the molecule, transformation of proenzymes in enzymes, fixation of the substrate on the protein in a position allowing its activation. One can conclude from this short survey, that the evolution of enzymology has been delayed by a lack of creativity due to the appeal of a "vital principle" to explain their properties. The opening of biochemical decades of their study has been due to the concept they were all proteins.

The last examples I will remind is related to the beginning of hormone biochemistry from endocrinology, to show how a misunderstanding in the interpretation of results of a remarkable experimental work has delayed the birth of the concept of hormone. There is no doubt that Claude Bernard demonstrated in 1852 the regulation of blood sugar level through the formation and hydrolysis of liver glycogen, according to the excess or the too low level of glucose circulating in blood. The secretion of this hexose in blood from liver was called by him "an internal secretion" and he considered, as all physiologists until 1891, that such a process could be only nervous, as all regulations of biological functions. Glycosuria in human diabetics was known since centuries, but attributed at this time only to excretory troubles ruled by nervous factors. In 1891, Brown-Sequard understood, after poorly demonstrative results, that substances secreted in blood by specialized cells or organs, exert a physiological action on target cells they reach as "chemical messenger" of other cells. An entirely new concept of what a product of internal secretion is, was created by this discovery of hormonal function of testis, later definitively established and located in the interstitial cells by Ancel and Bouin on the beginning of the twentieth century. The first direct experimental demonstration of the action "at distance" of a chemical messenger was given by Bayliss and Starling in 1902 by the secretion if pancreatic juice after injection of an extract of duodenal mucosa containing an unknown active product, called secretion, protein the structure of which has been established nearly sixty years later. The word hormone for products of internal secretion has been introduced by Starling in 1905 and the expression endocrine gland in 1912 by the pathologist Pende. Biochemistry of hormones started much later, even if pure adrenalin has been isolated in 1901 by Aldrich and by Takamine, but its vascular action only was known during more than 20 years. During the three first decades of the present century, important progresses in physiology of endocrine organs allowed the identification of numerous new internal secretions and the coordination of some of the secretary glands. This physiological step of researches had to be overpassed to start the biochemical study of hormones, partly initiated by this of steroidic hormones of corticoadrenal gland and of sex organs, due on its beginning to Kendall, to Allen and Doisy and to Butenandt. Every family of hormones had its own biochemical evolution, dominated by advances reached in the field of chemistry of substances related to them: sterols, peptides, proteins, or products of specific structure deriving from amino acids or from fatty acids.

The scope of all these remarks is to show how the creativity of Brown-Sequard in 1891 has been necessary to modify an initial conception of Claude Bernard on internal secretions, from which the endocrinology could not spring. The work of Claude Bernard on the regulation of blood sugar level led to an important discovery, due to a precise coordination of remarkable planned experiments. That of Brown-Sequart is typically creative, because it opened the way to a completely new general and fruitful conception from very limited and poorly defined experiments results.

The potentialities of a continuously growing science as Biochemistry are limited to discoveries unexpected or inspired by the creativity of scientists. In some cases, a discovery opens a new field by itself. For example, it was sure nearly sixty years ago that an endocrine product secreted by pancreas plays an important role on glucose metabolism, but the extraction of crude insulin by Banting Best and MacLeod was necessary to open the way to biochemistry of this protein. The role of a new concept is quite different, as shown by two examples: first, the understanding of antibiotic action, and secondly the need of the maintenance of a relatively stable "milieu intérieur" to secure the continuity of cell life. When Fleming understood that penicillin acts on bacterial growth by its inhibition and not by destroying the bacteria, he opened an extremely fruitful way to researches and medical applications. The concept of Claude Bernard of a "milieu intérieur," elaborated at about 1860, is still surviving; it lead van Slyke and L.J. Henderson to consider as necessary a strict regulation of blood pH in 1920. These examples are significant of the permanent value of some creative concepts, but the opening of completely new field is another great sources of potentialities for the future of biological sciences.

During the last decades, the molecular biology has probably been the best example of this viewpoint on the evolution of biochemistry not only in the field of structure, biogenesis and functions of proteins and nucleic acids, but in the development of genetics and of immunochemistry. Therefore, if the present extension of biochemistry is chiefly linked to the factors reminded above, we can be confident in its future from the same principles. The opening of new fields from previously known but not clearly interpreted results and the discovery of entirely original and unexpected facts will remain the source and the chief consequence of creativity in experimental sciences.

REFERENCES

1. Roche, J.: 1974. *J. Comp. Biochem. Physiol.* 478, 521–529.
2. Bernard, Cl.: Rapport historique sur les progrès et la marche de la Physiologie en France, Vol. 1, p. 226, publie sous les auspices du Ministère de l'Instruction publique, Imprimerie imperiale, Paris, 1867.

Control of Differentiation Pathways—Ontogenesis of an Embryologist

TUNEO YAMADA

T. YAMADA was born in Japan on January 20, 1909. He graduated from Tokyo Imperial University (Department of Zoology) in 1931, where he received Ph.D. degree in 1941. After working as a guest research worker under supervision of Walther Vogt at Anatomical Institute, University of Zürich and then University of Munich (1934–1937), he returned to Tokyo Imperial University (1937–1946). In 1946, he organized his group at Department of Biology, Nagoya University, where he became Professor (1951) and Chairman (1958). He accepted the offer of the research position as a member of Biology Division at Oak Ridge National Laboratory. After he retired by age (1974), he became a member of Swiss Institute for Experimental Cancer Research. Now he retired (1980).

When I started the zoology course at the University of Tokyo in the late twenties, experimental embryology had been introduced in Japan by Professor Naohide Yatsu. The causal analysis of embryogenesis as opposed to descriptive embryology began to attract the interest of young biologists. In this atmosphere, I enjoyed my student life under the warm guidance of this unassuming teacher who accomplished beautiful studies of the cytoplasmic localization in the invertebrate egg cells. Among many approaches of experimental embryology flourishing around that time, the determination of developmental pathways in the early stages of amphibian embryos fascinated me, because it was directly related to developmental diversification and emergence of new levels of organization, the most basic problems of embryology. It was the time when this area was high-lighted by the School of Hans Spemann with his doctrine of organizer. For preparing my research in this area, new in Japan, I learned the technique of surgical operation on amphibian embryos from literature, and did some experiments on determination of the sucker gland (*1, 2*). In the postgraduate period I intended to go to Freiburg to work under Professor Spemann. My dream was not realized, because the plan evoked an opposition from a Japanese scientist working with Professor Spemann. Since I was overwhelmed with the brilliant publications coming out of Spemann's laboratory, the failure of the plan was a heavy disappointment. In retrospect, however, I feel that it was fortunate that I took another route. This allowed me to develop my own idea and look at the problem from an angle different from that of Spemann and his coworkers. I decided to go to Zürich to work under Professor Walther Vogt whose analysis of morphogenesis of amphibian embryos with the localized vital staining method was already well known. In 1934 I was received by Professor Vogt at the Anatomical Institute of University of Zürich. The dynamic personality of this great embryologist quickly captured my mind. He suggested me to follow the segregation of the trunk mesoderm after gastrulation with the transplantation technique. The experiment started in Zürich was continued in Munich as Professor Vogt moved back to Munich in 1935 (*3*). The experiment was further continued in Tokyo, after I left Munich for home in 1937 under a menace of the world war. In Tokyo the experiment was expanded by introducing the method of culture *in vitro* (*4–6*). The unexpected result was the formation of pronephros by somite material either grafted in the ventral area or isolated *in vitro*. The existing theories could not explain the finding. This and other results were interpreted by the idea that the control of developmental pathways is not done by signals specific for each pathway but by a shift of levels along the dorso-ventral axis and each level is connected with specific pathways. I am pleased to note that the interest in this type of approach has been recently revived. The old results are confirmed and expanded, and the concept of dorsalization is adopted (*7*). Around the end of the thirties, it became obvious to me that the concept of dorsalization is applicable to the organizer action, although such concept is incompatible with the theory of specific inducing substances which was dominating at that time in the area of organizer research (*8*). In 1941, as some experiments on the inductive effect of muscle protein on isolated ectoderm were being carried out (*9*), I was called out for military service. Subsequently I had to stay in service on the Pacific War front for 5 years and then in captivity. It was a miracle that I could come back alive after the ordeal. During this long period, whenever it was possible, I made an effort to formulate a theoretical model for the organizer action along the line of my earlier concept of the control of developmental pathways, taking consideration of facts and ideas published in the area of regional induction. Basically the model (*10*) proposed that two different activities occurring in ectodermal cells determine, in combination, the differentiation pathway of those cells. One of such activities controls the development along the dorso-ventral axis and the organizer shifts the level toward dorsal (dorsalization), and the other activity controls the development along the cephalo-caudal axis, and the organizer shifts the level towards caudal (caudalization). On the basis of the observation that caudalization is closely associated with a specific type of morphogenetic movement, stretching-convergence, it was speculated that this type of morphogenetic movement is involved in caudalization of dorsalized ectoderm. The model was proposed as an alternative to the theory of specific inducing substances for the organizer action (8). In this respect the model was successful, because two theories of the organizer action, subsequently published (11, 12), adopted the idea of the combinative effect of two factors as the basis of regional induction by the organizer. As to the proposal that morphogenetic movement is involved in the organizer action, the general evaluation remained negative. However, the new situation in the area of organizer research to be mentioned in the next paragraph appears to raise the question of involvement of morphogenetic movement anew in the organizer action (13).

Meanwhile a research group was built up in Nagoya, which was interested in approaching the control of differentiation in amphibian embryogenesis with new biological techniques and ideas. We were aware that the aim cannot be achieved easily and quickly, but believed that it was the time to start a long-range program. Yujiro Hayashi, Shozo Osawa, Reizi Okazaki, Shuichi Karasaki, Kenzo Takata, and Chinami Takata were among the original members of the group which changed its composition during the course of time. Beside other projects we were interested in chemical factors responsible for regional induction in newt gastrula ectoderm as an approach to the organizer problem. We took the standpoint that such factors do not operate in normogenesis, but they should provide powerful tools for the analysis of the organizer action once isolated in pure form (14). From guinea pig liver, a ribonucleoprotein fraction was isolated which induced almost exclusively archencephalic structures at a high frequency in isolated ectoderm (15, 16). This effect was reduced by treatments of the sample with trypsin and pepsin (17), but not by treatment with ribonuclease (18). Attempts were made to isolate the mesoderm inducing factor of guinea pig bone marrow. The factor was found not in the ribonucleoprotein fraction, but in a fraction non-dial0izable and precipitable with acid or ammonium sulphate (19–21). In the last series of experiment made in this direction, the acid precipitable fraction was submitted to DEAE-cellulose column chromatography. A fraction called E15 had the strongest mesoderm inducing effect when tested with the nylon technique which allows a study of morphogenetic effects of macromolecular samples in solution on gastrula ectoderm (22). Five μg/ml of E15 gave the minimum detectable effect and 50 μg/ml the optimum effect. In another approach, Chinami Takata made a long term culture of ectodermal sandwich containing bone marrow tissue and showed that various endodermal tissues are formed from ectoderm (23). It was quite probable that also some samples separated from bone marrow had the ability to induce endodermal tissues in ectoderm. Considering this aspect, the morphogenetic effect of bone marrow on ectoderm appeared to be vegetalization. However, at that time we interpreted the results of our bone marrow experiments as indicating involvement of specific polypeptides in caudalization by the organizer in normogenesis, because this tissue and its fractions produced abundant trunk and tail ectodermal structures. Now the subsequent development in the organizer research has shown that in normogenesis the mesodermal pathways are laid down during cleavage and blastula stages under the vegetalizing influence of the endoderm in the marginal zone (24, 25). Hence it appears reasonable to assume that the bone marrow factor and other factors of heterogenous inductors, which can be characterized as vegetalizing agents, mimic this influence of the endoderm in normogenesis, and do not have much to do with the organizer action (13). Trunk and tail ectodermal structures induced by these factors could be due to secondary inductions by the mesoderm which is primarily produced by the vegetalizing factor. For understanding caudalization one should start a new approach avoiding heterogenous inductors as a tool. However, the archencephalic effects of heterogenous inductors may have a value for understanding the dorsalization by the organizer in normogenesis. In contrast to claims made in the fifties that RNA transmitts the inductive effect of the organizer, our studies showed that different samples of RNA either do not have any detectable morphogenetic effect or possess very weak archencephalic and non-regional neural effects on ectoderm (21, 26). Also the measurement of RNA distributed in different areas of the gastrula in the process of induction of neural plate did not reveal the alleged massive migration of RNA from the organizer to reacting ectoderm (27).

One result of ultrastructural studies of developing amphibian embryonic cells carried out by Karasaki (28, 29) in Nagoya, was the discovery of crystalline structure of yolk platelets. This finding later led to the establishment of a new system of cell differentiation under participation of a number of laboratories: synthesis of yolk protein precursor (vitellogenin) in liver cells as a response to estrogen, its discharge into blood stream, its uptake by oocytes, and elaboration of yolk crystalline structure. Karasaki later worked in this area with Robin Wallace in Oak Ridge.

In 1961 I accepted a research position at the Oak Ridge National Laboratory and started a group called "Cell Growth and Differentiation." During 12 years of its existence, the group received participation of the following scientists: Rudolf Achazi, Samir K. Brahma, Günther Cleffmann, Tom G. Connelly, Jim Dumont, John Eppig, Aida Goldstein, Horst Grunz, Fritz Jauker, Arthur Jurand, Shuichi Karasaki, Jussi Kohonen, David S. McDevitt, Isaura Meza, Christian Michel, Oscar Miller, Sohan Modak, Rudolf Nöthinger, José Ortiz, John Pappaconstantinou, Edigio Puccia, David H. Reese, Chinami Takata, Victor Idogaya-Vargas, Robin Wallace, and Sara Zalik (Eisenberg). In the framework of this group, I first continued the investigation on embryonic induction together with Karasaki (30–32). But soon our work faced an obstacle. The supply of Japanese newts suddenly stopped, and the similar species available in Oak Ridge were found unsuitable for our purpose. The fact that the urodele species most abundant around Oak Ridge was best suited for lens regeneration prompted us to shift our research project to conversion of iris epithelial cells into lens cells, which was believed to occur during lens regeneration in adult newts. If this conversion is real, it represents the extreme case of reprogramming of cell differentiation, and its mechanism must be valuable for basic understanding of the control of cell differentiation. The shift in research subject was favoured by the situation that Heinz Tiedemann had started a long range project for chemical characterization and isolation of the chick embryo factor which is equivalent to the bone marrow factor in morphogenetic effects. Concerning the lens regenerating system, its experimental morphological studies had been done in a large number of laboratories, but almost nothing had been done from the cell-biological or biochemical angle. Our approach to the lens regenerating system was started around 1962 in Oak Ridge and carried over to Switzerland, when I retired from the Oak Ridge position, and moved to Lausanne. Since the results of our research on this subject have been repeatedly reviewed until recently (33–37), they will be only briefly summarized.

Iris epithelial cells (IECs) of adult newts are fully differentiated and withdrawn from cell cycles. However, they can be shifted into cell cycles *in situ* by lens removal (38–40) and by putting in culture (41). *In situ*, proliferating IECs located in the dorsal pupillary margin give rise to lens cells (the conversion pathway), while other proliferating IECs go back to the original state of differentiation when they retreat from cell cycles (the retrieval pathway). Estimation of the cell cycle time demonstrated that the conversion pathway has a significantly shorter cell cycle time than the retrieval pathway (42). Knowing the average cell cycle time, the time of first appearance of DNA synthesis (38, 39), and the time of appearance of lens specific antigens (43, 44) one can conclude that more than 6 cell cycles are needed for the progenies of IECs to express the lens phenotype in the conversion pathway. A number of evidence suggest that in the retrieval pathway *in situ*, cells fail to stay more than 6 cell cycles in proliferation. Supporting the earlier report of Eguchi *et al.* (45) it was found that in primary culture of IECs, whether it is derived from dorsal or ventral IECs, a certain proportion of cells express lens phenotype (36). The time needed for appearance of lens cells is longer in primary culture than *in situ* after lentectomy. The cell cycle time measured in primary culture of IECs is longer than that of the conversion pathway (46), and no difference was found in the cell cycle time between cultures derived from dorsal IECs and ventral IECs. The number of cell cycles needed for appearance of lens antigens in primary culture was calculated to be 6–7 (47). In an attempt to clone IECs, it was found that high motility of IECs in culture makes the application of the conventional cloning method unreliable. A new method was deviced to overcome the difficulty. Cell count and immunofluorescence were used to determine the cell cycle number needed for the appearance of lens antigens in IECs cloned in this way. The results indicate that the number is 5–7. These data as a whole support the working hypothesis that conversion of IECs into lens cells is controlled by the extent of proliferation (36), and are in conformity with the observation that X-ray inhibits lens regeneration (48). Earlier it was generally agreed that the presence of neural retina is essential for the cell-type conversion, and speculated that an inductive influence of neural retina is the basis for the conversion. Available data however show that neural retina is needed for conversion *in situ* and in organ culture (*49, 50*), but not in cell culture. (The role of neural retina in organ culture can be replaced by pituitary *etc. 51.*) Since the presence of a growth factor in retina has been demonstrated (*52*), one can speculate that neural retina shortens the cell cycle time of IECs, so that cells can pass through more than 6 cell cycles during the limited proliferation period after lentectomy *in situ* or in organ culture. In cell culture this shortening of the cell cycle time is not needed for obtaining lens cells, because the time for proliferation is much longer than *in situ* or in organ culture. If our interpretation is right, we have here an interesting situation that, although *in situ* the appearance of lens specificity is dependent on retina, the direct control of developmental pathway is done by a cellular activity not by a retinal factor. The situation is basically in accord with my concept of embryonic induction discussed above.

While IECs are in proliferation, they go through changes in ultrastructure (Jurand cited in *35, 53–55*) and in biochemical parameters (*56–62*). The most prominent change is the loss of melanosomes which fill up the cytoplasm of IECs. Our studies show that when IECs enter into cell cycles, lysosomes are activated and attack melanosomes (*63*), and that those "activated" melanosomes are discharged by elevation of intracellular concentration of Ca^{2+} (*64*). Evidence shows that this elevation of Ca^{2+} is mediated by cAMP, whose role in dedifferentiation has been suggested by its effect on configuration of IECs (*65, 66*). During the period of extensive exocytosis of melanosomes *in situ*, the level of cAMP of iris was found to be maximum by the group of José Ortiz (*67*). Those events occur in parallel with the changes in electrophoretic mobilities of IECs as demonstrated by Eisenberg-Zalik and her coworkers (*68–70*). Available data, as a whole suggest that dedifferentiative events are associated with cell cycles and as cells traverse a series of cell cycles, alterations in cytoplasm and on cell surface occur, which initiate a new pattern of transcriptional and post-transcriptional events. My personal speculation is that this last event is mediated by alterations in chromatin proteins.

REFERENCES

1. Yamada, T.: 1933. *J. Facult. Sci., Tokyo Imp. Univ. Sec. IV*, 3, 239–254.
2. Yamada, T.: 1938. *J. Facult. Sci., Tokyo Imp. Univ. Sec. IV*, 5, 133–163.
3. Yamada, T.: 1937. *Roux. Arch. Entwicklungsm. Org.* 137, 151–270.
4. Yamada, T.: 1939. *Folia Anat. Japon.* 18, 565–568.
5. Yamada, T.: 1939. *Folia Anat. Japon.* 18, 569–572.
6. Yamada, T.: 1940. *Folia Anat. Japon.* 19, 131–197.
7. Slack, J.M.W. and Forman, D.: 1980. *J. Embryol. Exp. Morphol.* 56, 283–299.
8. Toivonen, S.: 1940. *Ann. Acad. Sci. Fenn. Ser. A*, IV, 55 (No. 6), 1–150.

9. Yamada, T.: 1942. *Proc. Imp. Acad. Tokyo* 18, 86–88.
10. Yamada, T.: 1950. *Embryologia* 1, 1–20.
11. Nieuwkoop, P.D. *et al.*: 1952. *J. Exp. Zool.* 120, 1–108.
12. Saxén, L. and Toivonen, S.: 1961. *J. Embryol. Exp. Morphol.* 9, 514–533.
13. Yamada, T.: 1981. *Netherlands J. Zool.* 31, 78–98.
14. Yamada, T.: 1958. *Chemical Basis of Development* (McElroy, W. and Glass, B., eds.) pp. 217–238, Johns Hopkins Press, Baltimore.
15. Hayashi, Y.: 1956. *Embryologia* 3, 57–67.
16. Hayashi, Y. and Takata, K.: 1958. *Embryologia* 4, 149–160.
17. Hayashi, Y.: 1958. *Embryologia* 4, 33–53.
18. Hayashi, Y.: 1959. *Dev. Biol.* 1, 247–263.
19. Yamada, T.: 1958. *Experientia* 14, 81–87.
20. Yamada, T.: 1958. *Ann. Biol.* 34, 165–178.
21. Yamada, T.: 1960. *Advances in Morphogenesis*, Vol. 1 (Abercrombie, M. and Brachet, J., eds.) pp. 1–53, Academic Press, New York and London.
22. Yamada, T. and Takata, K.: 1961. *Dev. Biol.* 3, 411–423.
23. Takata, C. and Yamada, T.: 1960. *Embryologia* 5, 8–20.
24. Nieuwkoop, P.D.: 1973. *Advances in Morphogenesis*, Vol. 10 (Abercrombie, M., Brachet, J., and King, T.J., eds.) pp. 1–39, Academic Press, New York and London.
25. Nakamura, O.: 1978. *Organizer—A Milestone of Half-century from Spemann* (Nakamura O. and Toivonen, S. eds.) pp. 179–220, Elsevier/North-Holland, Amsterdam.
26. Yamada, T., Takata, K., and Osawa, S.: 1954. *Embryologia* 2, 123–131.
27. Takata, K.: 1953. *Biol. Bull.* 105, 348–353.
28. Karasaki, S.: 1959. *Embryologia* 4, 247–272.
29. Karasaki, S.: 1959. *Embryologia* 4, 273–282.
30. Yamada, T.: 1962. *J. Cell. Comp. Physiol.* 60 (Suppl. 1), 49–64.
31. Karasaki, S.: 1963. *J. Ultrastr. Res.* 9, 225–247.
32. Karasaki, S.: 1963. *J. Cell Biol.* 18, 135–151.
33. Reyer, R.W.: 1976. *Handbook of Sensory Physiology*, Vol. VII/5 (Crescitelli, F., ed.) pp. 309–390, Springer-Verlag, Berlin.
34. Yamada, T.: 1966. *Am. Zool.* 6, 21–31.
35. Yamada, T.: 1967. *Current Topics in Developmental Biology*, Vol. 2 (Moscona, A.A. and Monroy, A., eds.) pp. 247–283, Academic Press, New York and London.
36. Yamada, T.: 1977. *Monographs in Developmental Biology*, Vol. 13 (Wolsky, A. ed.) pp. 1–126, S. Karger, Basel.
37. Yamada, T.: 1981. *Bollettino Zool.*, in press.
38. Eisenberg, S. and Yamada, T.: 1966. *J. Exp. Zool.* 162, 353–367.
39. Yamada, T. and Roesel, M.E.: 1969. *J. Exp. Zool.* 171, 425–431.
40. Yamada, T. and Roesel, M.E.: 1971. *J. Exp. Zool.* 177, 119–128.
41. Horstman, L.P. and Zalik, S.E.: 1974. *Exp. Cell. Res.* 84, 1–14.
42. Yamada, T., Roesel, M.E., and Beauchamp, J.J.: 1975. *J. Embryol. Exp. Morphol.* 34, 497–510.
43. Takata, C., Albright, J.F., and Yamada, T.: 1964. *Dev. Biol.* 9, 385–397.
44. Takata, C., Albright, J.F., and Yamada, T.: 1966. *Dev. Biol.* 14, 382–400.

45. Eguchi, G., Abe, S., and Watanabe, K.: 1974. *Natl. Proc. Acad. Sci. USA* 71, 5052–5056.

46. Yamada, T. and Beauchamp, J.J.: 1978. *Dev. Biol.* 66, 275–278.

47. Yamada, T. and McDevitt, D.S.: 1979. *Experientia* 35, 985.

48. Michel, C. and Yamada, T.: 1974. *Differentiation* 2, 193–201.

49. Yamada, T., Reese, D.H., and McDevitt, D.S.: 1973. *Differentiation* 1, 65–82.

50. Yamada, T. and McDevitt, D.S.: 1974. *Dev. Biol.* 38, 104–118.

51. Connelly, G.T., Ortiz, J.R., and Yamada, T.: 1973. *Dev. Biol.* 31, 301–315.

52. Arruti-Mizrali, C.: 1979. Thèse, Université René Descartes, Paris V.

53. Karasaki, S.: 1964. *J. Ultrastr. Res.* 11, 246–273.

54. Dumont, J.N., Yamada, T., and Cones, M.V.: 1970. *J. Exp. Zool.* 174, 184–204.

55. Dumont, J.N. and Yamada, T.: 1972. *Dev. Biol.* 29, 385–401.

56. Yamada, T. and Karasaki, S.: 1963. *Dev. Biol.* 7, 595–604.

57. Yamada, T. and Takata, C.: 1963. *Dev. Biol.* 8, 358–369.

58. Reese, D.H., Puccia, E., and Yamada, T.: 1969. *J. Exp. Zool.* 170, 259–268.

59. Reese, D.H.: 1973. *Exp. Eye Res.* 17, 435–444.

60. Achazi, R. and Yamada, T.: 1972. *Dev. Biol.* 27, 295–306.

61. Idoyaga-Vargas, V. and Yamada, T.: 1974. *Differentiation* 2, 91–98.

62. Idoyaga-Vargas, V., Yamada, T., and Michel, C.: 1976. *Dev. Biol.* 49, 563–568.

63. Yamada, T., Dumont, J.N., Moret, R., and Brun, J.P.: 1978. *Differentiation* 11, 133–147.

64. Patmore, L. and Yamada, T.: 1980. *Experientia* 36, 755.

65. Ortiz, J.R., Yamada, T., and Hsie, A.W.: 1973. *Proc. Natl. Acad. Sci. USA* 70, 2286–2290.

66. Ortiz, J.R. and Yamada, T.: 1975. *Differentiation* 4, 135–142.

67. Velázquez, F.M. and Ortiz, J.R.: 1980. *Differentiation* 17, 117–120.

68. Zalik, S.E. and Scott, V.: 1972. *J. Cell Biol.* 55, 134–146.

69. Zalik, S.E. and Scott, V.: 1973. *Nature New Biol.* 244, 212–214.

70. Zalik, S.E., Scott, V., and Dimitrov, E.: 1976. *Progress in Differentiation Research* (Müller-Bérat, N., ed.) pp. 361–367, North-Holland, Amsterdam.

On Becoming an Exact Science

JAMES D. EBERT

In nearly three decades as an experimental embryologist, J.D. EBERT has opened new approaches to the organization and specificity of cells and their arrangement in living organisms. Following his graduate training at Johns Hopkins, he served on the faculties of Massachusetts Institute of Technology and Indiana University, and in 1956 he assumed leadership of the Department of Embryology of Carnegie Institution of Washington. Under his guidance, the Department made great strides in the study of genes, their structure, evolution, and control, and in the study of membrane development and function. During those years, his joint appointments at Johns Hopkins as Professor of Biology and Professor of Embryology in the School of Medicine were indicative of the interactions between the Department and the University. He served the Carnegie Institution and the Marine Biological Laboratory at Woods Hole as President and Director over nearly a crucial decade. In 1978 he assumed his present responsibility as President of Carnegie Institution of Washington. He holds honorary degrees from Yale and Indiana Universities, and Moravian College, as well as from his Alma Mater, Washington and Jefferson College, and is a member of the American Philosophical Society, the National Academy of Sciences, the American Academy of Arts and Sciences, and the Institute of Medicine, and served as the first Chairman of the Assembly of Life Sciences of the National Research Council. He has just been elected Vice President of the National Academy of Sciences.

"That the fundamental aspects of heredity should have turned out to be so extraordinarily simple, supports us in the hope that nature may, after all, be entirely approachable. Her much-advertised inscrutability has once more been found to be an illusion due to our ignorance. This is encouraging, for if the world in which we live were as complicated as some of our friends would have us believe we might well despair that biology could ever become an exact science."   *Thomas Hunt Morgan (1)*

Throughout the world there is widespread apprehension about both the costs and possible social consequences of contemporary scientific inquiry and discovery, exemplified by the dreary refrain in the *New York Times* (2), "Genetic engineering is controversial. Some critics fear it may be used unwisely to tinker with human heredity or to produce changes in animals or plants that may have unforeseen bad results." As I wrote in a recent editorial (3), "To a cynical and dispirited public, no news is indeed good news. Accentuating the negative has become a national disease." But I would accentuate the positive, emphasizing the resilience of science and the richness of new opportunity in fields ranging from astronomy and planetary science to geophysics, from environmental science to cellular and molecular biology (3, 4).

In the latter fields alone, the recombinant DNA methodology and the hybrid-oma techniques for producing monoclonal antibodies have revolutionized our approaches to understanding the mechanisms regulating gene action and cell structure and function. Moreover, in these fields the lag between new findings and their application is almost nonexistent. Processes for the mass production of hitherto scarce products—hormones and the like—are emerging swiftly. A gene has been injected into a defective living cell, curing that cell's fatal genetic flaw; new forms of organisms are being "tailor-made" for special purposes or for life in special environments.

Today's successes point us toward tomorrow's goals. Yet, opportunities for new departures must be increased by improving the intellectual environment, bearing in mind that the subtle forces that shape the course of research flourish in a climate of freedom and flexibility.

Nowhere is the crucial importance of pioneering independence exemplified better than in the role of the Carnegie Institution's Department of Embryology in the emergence of developmental biology as an exact science.

Over the past twenty-five years the field of developmental biology has been reshaped. When the Department of Embryology was founded there were three parallel streams of inquiry in the field: *first*, what one might call an anatomy-physiology stream led by Franklin P. Mall, George Streeter, and later Carl Hartman of the Department; *second*, a mechanics of development stream, emphasizing cell and tissue interactions, led by Hans Spemann, Ross Harrison, and Carnegie's Warren and Margaret Lewis; and *third*, an almost completely separate "biochemical embryology" stream for which Joseph Needham became the principal spokesman. There were few points of contact between the three streams.

During the same period the field of genetics was exploding. How remarkable it is that these two fields, embryology and genetics, which are today almost one, went their separate ways. A few pioneers like Goldschmidt, Wright, and Brachet, Caspersson and Schultz, attempted a synthesis. Even Morgan's career was "compartmentalized": embryology—genetics—embryology. In the late 1940's a student might study embryology with only casual reference to genetics, if any. Slowly, however, the remarkable achievements, first of Ephrussi and Beadle, and then of Beadle and Tatum, in elucidating the one gene-one enzyme hypothesis became a part of the literature of embryology as well as of genetics. Students of development began to perceive that it should be possible ultimately to understand the factors regulating the synthesis of specific products and their assembly during the shaping of a new organism.

By the mid-1950's the field was ripe for change. The Department of Embryology played a leading role in those changes. However before discussing them let us examine the success of the Department in fulfilling its original objectives.

When the Carnegie Institution of Washington made, in 1913, the original grant which led to the organization of the Department of Embryology, it did so in response to a statement by Franklin P. Mall about the possibility of making great advances in the knowledge of human development if a concerted attack could be organized. Mall's "Plea for an Institute of Human Embryology," published in the Journal of the American Medical Association in 1913, is thus the scientific charter of the Department of Embryology (5).

In that document Mall wrote, "With a very large collection, a competent staff and the very best material equipment the institute would naturally take up problems which bear on anatomy, physical anthropology, comparative embryology, physiology of gestation, pathology and teratology." He listed certain large questions which at once presented themselves: (1) Curve of growth. (2) Anatomy of various stages; modeling, dissection, *etc.* of internal structure. (3) Morphology of the brain. (4) Histogenesis; the differentiation of the various tissues. (5) The causes of spontaneous abortion. (6) Study of monsters. (7) Study of moles (proliferative malformations of the placenta and membranes). (8) Comparative and experimental embryology.

Questions like these were, of course, merely some of those which Mall considered to represent attainable aspects of the whole task of understanding human reproduction. They formed part of a systematic program, in which, first, the development of the human embryo must be observed. The norm of growth and of external form must be established. Embryos must be arranged in stages. Abnormal and pathologic embryos should be classified, with a careful study of the uterus and Fallopian tubes in which they were found. Mall went on to say that the program might seem to emphasize unduly the making of a collection, but that was not the intention, for he considered that the chief function of an institute of human embryology should be the formulation and solution of problems.

Mall himself lived only long enough to get the enterprise started, and his plans had to be developed by his successors, George L. Streeter and George W. Corner (6). Some of his objectives are only now being realized in continuing studies of the "very large collection" of human embryos Mall envisaged; by far the largest in the world, it is safely housed at the University of California, Davis, and kept usable by good records and ample indexes, under the direction of Ronan O'Rahilly, who has undertaken to complete the unfinished work of Streeter, Heuser and others. The first part of this superb work was published in 1973 (7); a revision of Streeter's catalogue of later stages is now in progress.

What of Mall's final objective, "Comparative and experimental embryology"?

In 1955, Corner (6) wrote "Because the original emphasis of Mall and Streeter was on human embryology, experimental embryology did not at first receive much attention in the Department. Experimental studies on embryonic differentiation have been well supported at the marine biological laboratories and in various university departments of biology, and such work could be appropriately left to those institutions." The Department took for its own field the embryology of the non-human primates. It became a pioneering center of the study of primate reproduction and development. Its monkey colony was the forerunner of today's primate centers. It is interesting to contemplate how a "sideline" assumed major importance. Streeter wanted a monkey colony to produce embryos of known age. However out of this program came Hartman's monumental analysis of the reproductive cycle, including especially the phenomenon of menstruation and of cyclic changes in the uterus and ovaries. It yielded great contributions to the understanding of the human cycle and influenced endocrinology, gynecology, and obstetrics, and led to highly productive research on the anatomy and physiology (and to a limited extent, biochemistry) of the primate uterus and placenta by Bartelmez, Markee, Ramsey, Reynolds, and Csapo.

In addition, during the years 1940–1955 R.K. Burns pursued his now-classic studies of sex reversal in the opossum. Experimental investigations were made on the biochemistry of the male reproductive tract. And, finally, an investigation of the still obscure physicochemical phenomena of embryonic implantation was begun.

In summary, the Department was, for five decades, the world's leading center for the study of the human embryo. It pioneered in the development of primates for research, having one of the early and most successful breeding monkey colonies.

Using these animals, large strides were made toward understanding menstruation and cyclic changes in the ovaries and uterus, laying much of the groundwork for today's advances in population control. Earlier studies, culminating in Ramsey's classic works, set the standard for work on the primate placenta, and Burns' studies of sex transformation similarly provided a standard in the field of sex differentiation.

But the achievements of the past two decades have been no less impressive.

In the late 1950's it was our conviction that the immediate future of developmental biology might lie in two principal directions. We believed *first* that it should be possible eventually to isolate genes, to study their arrangement and organization within the chromosome, and to study mechanisms regulating their action. We had no "timetable" but we were surprised when this goal was achieved as early as 1970. We were convinced, however, that we should begin by attempting to define gene products, the proteins, and by using them as markers to move increasingly closer to the genes. It was a field in which we had already contributed. However, the technology for studying direct gene products, the ribonucleic acids —especially the ribosomal RNAs—developed more rapidly than might have been anticipated, thus making today's successful attack on the genes possible.

The *second* principle upon which we agreed was the need to find ways of studying the manner in which cells impinge upon and influence each other. Cell and tissue interactions are the hallmark of development. How are the inner controls of one cell regulated by contact with another? To understand these problems, it would first be necessary to perfect techniques for analyzing cell differentiation in clonal cultures thereby making it possible to study the genetics of somatic cells. Ultimately attention would have to be paid to the genesis, structure, and functioning of cell membranes themselves.

How successfully have these objectives been met?

Donald Brown and more than a score of colleagues have pioneered in the isolation and characterization of animal genes and their immediate products, the ribonucleic acids; these studies have, in turn, made possible major advances in understanding the mechanisms of gene regulation. His study with Carnegie Fellow, John Gurdon, of an anucleolate mutant strain of the South African clawed toad, *Xenopus*, constitutes a modern landmark, offering decisive proof of the role of the nucleolus in the synthesis of ribosomal RNAs. By further exploitation of the anucleolate mutant, Brown was able to map the genes encoding these RNAs. He and his coworkers then proved that during oogenesis the genes for ribosomal RNAs are amplified into about nine hundred copies. This research then evolved to the point where it became possible to isolate the genes coding for the 28S and 18S ribosomal RNA fragments—the first genes to be isolated from any eucaryotic cell; and the further extension of these techniques culminated in the isolation and characterization of the genes coding for 5S ribosomal RNA.

Brown and his coworkers then turned to the question whether the amplification of specific genes can be demonstrated in specialized cell types. Together with Yoshiaki Suzuki, now at the National Institute of Basic Biology (Okazaki, Japan), he isolated the messenger RNA coding for the silk protein, fibroin, in the larval silkworm, *Bombyx mori*. Using this mRNA as a probe, they were able to show that the fibroin gene is not amplified in the cells of the silkglands as compared to other cells. Suzuki's further studies have provided what may be the most complete account thus far available of the waxing and waning of gene action during development.

The concept of gene amplification continues to be a prime focus of research in the Department, where Allan Spradling studies the genes that encode a group of eggshell (chorion) proteins in the fruit fly, *Drosophila melanogaster*. These genes are under developmental control. The power of using *Drosophila* lies in its suitability for traditional genetics that can now be coupled with the newer methods of gene isolation, characterization, and function. Recently, Spradling has shown that genes for proteins are amplified in follicle cells just before they are expressed—the first reported example of amplification for protein genes that is programmed into normal development. Mutants have already been found that affect this amplification (8).

In the most recent, and perhaps most promising developments, the genes for 5S ribosomal RNA in *Xenopus* have been exploited in analyzing gene function. A protein has been isolated from ovarian tissue that interacts specifically with a control region that is located in the center of the 5S RNA gene. This protein has the remarkable property of binding the gene product, 5S RNA, and forming a stable particle with it. The synthesis of this protein almost certainly accounts for the activation of 5S RNA genes in early oocytes. This is the first discovery of a control protein that is specific for one eukaryotic gene and at least partly responsible for the differential expression of that gene in development (9–11).

As we have already observed, cell and tissue interactions are the hallmark of development. The differentiation of most cells requires, at one or more points in their life histories, interactions with their neighbors, and failures in close range tissue interactions may result in alteration in the normal patterns of development. The question has been posed in a general way as follows. How does the micro-environment make its impact on cell function *via* the cell membrane? Genes direct the synthesis of the components of cell membranes, and must, in turn, be regulated by signals from the cell membrane.

Several of the Department's contributions have had a profound impact on our understanding of cellular differentiation, and work currently in progress offers similar promise. In this account I shall refer only briefly to "past and already cold events," however significant, preferring to concentrate on recent findings.

Of the Department's achievements in the 1960's one stands out; Konigsberg's clonal analysis of myogenesis. His was the first clonal study of any differentiating cell. Puck first developed the "feeder layer" technique, enabling the cloning of established cell lines. It was Konigsberg, however, who modified the technique and applied it significantly. It deserves special mention, too, that two other members of the Department, Coon and Kaighn, followed Konigsberg's initial success with uncommonly important clonal analyses of chondrocytes and liver, and that a former Carnegie Fellow, Tokindo Okada (now at Kyoto University) has used his experience in the Department to great advantage in his studies of "transdifferentiation" of pigmented retinal to lens cells (12).

Several years ago Douglas Fambrough and his colleagues developed methods for measuring the metabolism of a membrane protein, the acetylcholine (ACh) receptor of muscle cells. They studied the rates of synthesis and degradation, and the route by which these molecules went from their sites of synthesis to their final destination in the cell surface. These were the first detailed biochemical analyses of a membrane component with a known function. The essential key to these studies was the availability of a specific assay system that could detect just the ACh receptor and nothing else. Fambrough recognized that the new and powerful method of monoclonal antibody production was a source of specific probes that would enable him to extend his observations to other membrane proteins. Now he and his associates have characterized several of the monoclonal antibodies that they have made against cell surface components. They have diverse, and in some cases surprising, characteristics. One reacts with and kills fibroblasts without affecting muscle cells, solving in an original way the age old problem of getting pure muscle cultures free of fibroblasts (13).

The problems of cell differentiation stand at the leading edge of biology. A great frame of investigation has been built steadily and the frontier is more vital than it has ever been.

Some of the immediate future of the field and of the Department can now be visualized. Even more direct attacks on gene regulation will become possible, arising out of studies of genes in chromatin, that is genes in their "natural environment." The injection of genes and their incorporation in the DNA of their hosts will become commonplace. And it should be possible to faithfully reconstruct gene action *in vitro*. We will continue to amass fundamental knowledge of the synthesis, degradation, and functional role of phospholipids in cell membranes. As Pagano and his colleagues have recognized, this will require new techniques—ways of introducing tagged molecules into membranes, localizing them and following their fate with time. Molecules inserted into cell membranes can be traced by fluorescence energy transfer and the use of haptenated lipids that can be detected with antibodies. These methods can then be applied to problems of membrane "capping," the metabolism of the inner and outer leaflets of the membrane, and the flow of lipids relative to fixed structures in the lipid bilayer (14).

Just as it has been necessary to isolate and characterize specific genes, so it will be essential to continue isolating specific functional membrane components, for example the specific "receptor sites" at which other cells, or molecules, interact with a given cell type. The field is a continuum in which we must take account of the biogenesis and turnover of membrane macromolecules; membrane genetics and the differentiation of functional components in specialized cell membranes; the molecular basis of membrane changes in response to hormones; and finally, the basis of changes related to cell differentiation and growth, both normal and abnormal.

Nothing short of the isolation and characterization of a wide variety of receptor sites and a demonstration of their distribution on the cell surface will begin to answer the questions that now confront us. Considering the enormous range of specificities ascribed to putative receptor sites, there may yet be surprises in store for us.

Looking further ahead, our ultimate goal is a better understanding of the pericellular environment—which provides conditions under which the division of labor of cell specialization occurs—and its regulation.

Cell-cell interaction is a necessary condition for the formation of the cellular architecture of organs as well as for their organization and interconnection. Several large questions remain unanswered: what is the role of cell-cell interactions in the regulation of morphogenetic movements? Do these interactions require the exchange of interactants *via* cytoplasmic bridges, specialized junctions, or other mechanisms, or are signals generated at the membrane, transmitted *via* specific receptor sites and intracellular mediators like calcium and cyclic AMP to specific genes? What is the role of ions and trans-cellular currents in the regulation of differentiation and growth, and in the initiation of regeneration (*15*)?

These, of course, represent but a few of the problems that will be occupying developmental biologists. As Caryl Haskins has observed, the very fact that these problems can be visualized brands them as among the more conventional, the less potentially novel. Sir Peter Medawar emphasized that the truly great new developments of tomorrow by that very fact cannot possibly be conceived even in rough outline today.

REFERENCES

1. Morgan, T.H.: 1976. *Thomas Hunt Morgan*, The University of Kentucky Press, Lexington, Ky.
2. Schmeck, H.M., Jr.: 1980. *The New York Times* 129, No. 44,695, September 3, A1 and A16.
3. Ebert, J.D.: 1980. *BioScience* 30, 503.
4. Ebert, J.D.: 1980. *Carnegie Institution of Washington Year Book* 78, 3–43.
5. Mall, F.P.: 1913. *J. Am. Med. Assoc.*, 60, 1599–1601.
6. Corner, G.W.: 1955. *Carnegie Institution of Washington Year Book* 54, 189–192.
7. O'Rahilly, R.: 1973. *Developmental Stages in Human Embryos. Part A: Embryos of the first three weeks (Stages 1 to 9)*. Carnegie Institution of Washington Publication 631, 167 pp., Washington, D.C.
8. Spradling, A.: *Carnegie Institution of Washington Year Book* 79, 73–78.
9. Sakonju, S., Bogenhagen, D.F., and Brown, D.D.: 1980. *Cell* 19, 13–25.
10. Bogenhagen, D.F., Sakonju, S., and Brown, D.D.: 1980. *Cell* 19, 27–35.
11. Pelham, H.R.B. and Brown, D.D.: 1980. *Proc. Natl. Acad. Sci. USA* 77, 4170–4174.
12. Eguchi, G.: 1979. *Mechanisms of Cell Change*, pp. 273–292, John Wiley and Sons, New York.
13. Fambrough, D.M.: 1981. *Carnegie Institution of Washington Year Book* 79, 19–28.
14. Pagano, R.: 1981. *Carnegie Institution of Washington Year Book* 79, 28–37.
15. Leffert, H.L. (ed.): 1980. *Ann. N.Y. Acad. Sci.* 339, 1–335.

The Development of the Shanghai Institute of Biochemistry

YING-LAI WANG

Y.-L. WANG was born in Jin-men (Quemoy), Fujian (Fukien), China in November 1907. He obtained his B.Sc. degree from the Chemistry Department of the University of Nanking in 1929 and Ph.D. degree in biochemistry from the University of Cambridge in 1941. He had been an assistant, lecturer in chemistry at the University of Nanking, research professor in biochemistry at the National Central University Medical School and senior member of the Medical Research Institute (Preparatory) of the Academia Sinica before the Liberation of China in 1949. After that he had served as a senior member and deputy director of the Institute of Physiology and Biochemistry, Academia Sinica. Since 1958, he has been senior member and Director of the Institute of Biochemistry. He is a member of the Science Council (Biology Division) of the Academia Sinica, President of the Chinese Biochemical Society, and President, Shanghai Branch of the Academia Sinica.

Biochemical research in China began in the early twenties when Wu Hsien and his coworkers in Peking Union Medical College opened up a number of important areas of work in biochemistry, such as protein denaturation, immuno-chemistry, clinical analysis, and nutrition, but for more than ten years before the liberation of the country, biochemical research was almost at a complete standstill, until in 1950 when the first institute for biochemical research was established within the Academia Sinica under the name of the Institute of Physiology and Biochemistry. The Institute started with a rather small staff of about twenty persons, including a secretary and two janitors. It consisted of three sections: the physiology section headed by Professor T.P. Feng who at the same time bore the responsibility of directorship of the Institute; the biochemistry section led by the author who also acted as the deputy-director; and the organic-chemistry section with Professor

Yu Wang as its chief; the last section was amalgamated into the Institute of Organic Chemistry in 1952. The biochemistry section incorporated Professor S.C. Shen, a Toronto trained biochemist from the Institute of Physical Chemistry of the Academia Sinica in 1950. In 1951 and 1952 successively we were strongly reinforced by Drs. C.L. Tsou and T.C. Tsao, both from Cambridge, England; the former specialized in the cytochromes while the latter is muscle proteins. Therefore in 1953 the biochemistry section was subdivided into three groups, namely proteins, enzymes, and metabolism. In 1954, Dr. Y.D. Zhang, a vitaminologist returning from Cambridge joined our group on metabolism followed in 1955 by Dr. T.P. Wang, who had acquired wide research experience in nucleotide-metabolizing enzymes in the United States of America. Dr. C.Y. Niu who had worked in the United States under R.J. Williams and H. Fraenkel-Conrat joined our protein group in 1956 and Dr. G.Y. Zhou, a microbiologist trained in Louvain, moved to our Institute from Peking in 1957. By that time, the Institute had grown to such a size that it was considered necessary to have separate institutes of biochemistry and physiology. Thus, the Institute of Biochemistry was established in 1958 simultaneously with the Institute of Physiology; the former is now known as the Shanghai Institute of Biochemistry of the Academia Sinica. In 1961, the research work of the Institute was reorganized to include the following five divisions and an independent group. They were the divisions of proteins, enzymes, nucleic acids, metabolism and radiobiology, and the group of theoretical biology. The protein division was chiefly concerned with the study of muscle proteins and the chemical synthesis of insulin and also work on plant viruses. The work of the division of enzymes involved studies on succinate dehydrogenase and other components of the respiratory chain of mitochondria, proteolytic enzymes, phosphoesterases and other enzymes with emphasis on their nature and mechanism of action. The division of nucleic acids was mainly engaged in the isolation, purification and analysis of tRNA and the effect of various factors on the stability of the nucleic acid toward enzymatic hydrolysis. In the division of metabolism studies were made on the control of amino acid metabolism in microorganisms, on the effect of protein depletion on liver enzymes, the relation between certain vitamins and amino acid metabolism, and the metabolism of lipids. In radiobiology, studies had been made on effect of radiation on nucleic acid stability and on the oxidative phosphorylation system of animal mitochondria. In 1963, molecular biology was selected as a major area of fundamental research and the Institute of Biochemistry was assigned the chief task of developing this aspect of biology. With the clear emphasis on basic research, the work of the Institute proceeded smoothly in spite of an interruption of three years between 1958 and 1961, during which all traditional research projects were suspended. In the fields of muscle proteins, enzymes, and nucleic acids high-level research turned out rapidly. In collaboration with sister institutions, the total synthesis of bovine insulin with full biological activity and identical crystalline form with the natural insulin was accomplished in 1965. Unfortunately, with the advent of the cultural revolution, owing to the interference by Lin Piao and the "Gang of Four," all basic researches in biochemistry and molecular biology were condemned to be bourgeois trends and were suspended for over ten years. With the downfall of the "Gang of Four," it has been possible to reorient our research work to the best interests of our country. In 1977, molecular biology has again been selecetd as the major line of biological research in the Academia Sinica with special emphasis on the structure, function, and synthesis of biomacromolecules, especially proteins and nucleic acids, molecular genetics including recombinant-DNA research, and biomembranes. Since then the work of the Shanghai Institute of Biochemistry has been reorganized to include the following main divisions: (1) the relationship between the structure and function of proteins and peptide hormones such as insulin, glucagon, luteinizing hormone, releasing hormone, and their analogues. Special attention has been given to the development of more effective methods for the synthesis of proteins and peptides. (2) The synthesis of yeast tRNA^{ala}. This is being carried out in collaboration with other institutes by a combination of chemical and enzymatic methods, and the object of the synthesis is to employ it as a means for the study of structure function relationship of tRNA. New restriction enzymes are being screened and methods for the sequencing of both RNA and DNA are under way. (3) Enzymes and biomembranes—the enzymes under study are the glycolytic enzymes aldolase, fructose bisphosphatase, and D-glyceraldehyde phosphate dehydrogenase, their subunit interaction, allosteric effect, enzyme-enzyme interaction, and also comparative studies of the enzymes from different species of animals; immobilized enzymes, the mode of linking of enzymes on the inner membrane of mitochondria, Na^+, K^+ -ATPase and the interaction of certain plant lectins and cell membrane. (4) Cell recognition and cellular regulation. (5) Mechanism of action of steroid hormones especially the oestrogens in the cell. (6) Protein structure and assembly—plant viruses that infect the cereal crops and fruit trees; animal virus, especially the replication of the double stranded RNA of cytoplasmic polyhedrosis virus; molecular basis of the hybridization of widely separated species of cereal plants, such as rice and sorghum, rice and maize; a toxic protein from pinellia which causes abortion in early pregnancy of animals; snake venom toxins; trypsin inhibitors from plant sources; and muscle proteins. (7) Molecular genetics—the work under way consists of the cloning of ribosomal RNA gene and also of chick globin gene into *Escherichia coli*. In addition to the above 7 main divisions, there are two independent groups: one working on theoretical biology such as irreversible thermodynamics and enzyme kinetics, the other on the biochemistry of hepatoma, including the control of the expression of the gene of α-foetal protein, the effect of α-foetal protein on cell immunity and the relation between hepatitis B and hepatoma. The institute has at present approximately 180 research workers and about 120 laboratory technicians. It operates a biochemical reagent factory with a personnel of about 100. This factory produces over 500 different biochemical reagents to supply the needs of the Institute and other research institutes of the Academia Sinica. The Institute has also a small machine shop with over 70 skillful workers working in collaboration with our instrumentation division on the making and repair of instruments used in biochemical research.

With the new orientation of our research and re-emphasis on molecular biology, the scope of our work has been considerably widened, but it will take some years before our research will get into full swing. We have lost fully ten precious years during which world research in this area of science has made the most spectacular progress in the whole history of science. But we are confident that with stability and unity within our country and under the able leadership of the Chinese communist party, we shall be able to make up for the time loss in the not too distant future.

Organisation and Administration of Governmental Research in Israel

EPHRAIM KATCHALSKI-KATZIR

E. KATCHALSKI-KATZIR was born in Kiev, Ukraine on May 16, 1916, and arrived in Israel with his parents in 1922. He studied chemistry, botany, zoology, and bacteriology at the Hebrew University, Jerusalem, and received his Ph.D. degree in 1941. In 1949 he joined the staff of the Weizmann Institute of Science, Rehovot, and headed its Department of Biophysics from 1951 until his election as 4th President of the State of Israel on April 11, 1973. Upon completion of his term of office in 1978, he returned to scientific research at the Weizmann Institute of Science and at the Tel-Aviv University, where he heads the newly established center for biotechnology. He is a member of numerous learned bodies in Israel and abroad, including the Royal Society, London, and the National Academy of Sciences, USA. He has received numerous awards including the Weizmann, Rothschild and Israel Prizes in Natural Sciences, the Linderstrøm Lang Gold Medal, the Hans Krebs Medal and the Alpha Omega Achievement Medal.

The history of scientific research in Israel is an integral part of the story of the return of the Jewish people to their homeland. Theodor Herzl, the first of the modern visionaries to articulate the concept of a Jewish State, dreamed of a land which would flourish not only as a physical centre for the Jewish people, but also as a great spiritual and scientific centre. Two of the country's major scientific institutions, the Hebrew University of Jerusalem and the Technion-Israel Institute of Technology, Haifa, were established almost a quarter of a century before the state itself came into being. The Daniel Sieff Institute, where Chaim Weizmann had his laboratory, was established in Rehovot in 1934. In 1944, to commemorate Weizmann's seventieth birthday, his friends embarked on the expansion of this

Institute into the Weizmann Institute of Science, which was formally opened in 1949.

The beginnings of agricultural research in the country date back to the end of the 19th century, with the establishment of the Mikvah Israel Agricultural High School. The first experimental agricultural research stations were founded in the early 1900's.

Thus, unlike in other fledgling independent states, when the State of Israel was established in 1948 some important elements in the scientific and technological infrastructure were already in operation. Other institutes of higher learning followed: Bar-Ilan University (1955), Tel-Aviv University (1956), the Ben Gurion University of the Negev (1969), and Haifa University (1963).

In the early years, after the establishment of the State of Israel, agriculture and industry continued to expand, accompanied by a massive though undirected growth in research and development (R and D) activities. During the 1950's it became apparent that while the achievements of basic research were already impressive, industrial R and D was decidedly inadequate. The government, sensing the need for a national policy of applied research and the desirability of involving itself directly in all ongoing R and D in the various government institutions and the institutes of higher learning, set out to crystallise its plans for the future. It began by establishing the National Council for Research and Development (NCRD) in 1959. This body was charged with formulating a national policy for directed scientific research, and coordinating the R and D activities of various government agencies within the framework of that policy.

The NCRD concentrated its efforts on encouraging industrial research. It founded and directly administered the Fermentation Unit (1960), the Center for Scientific and Technological Information (1961), the Oceanographic and Limnological Research Company (1967), and others. In addition, it carried out some important groundwork in examining and evaluating the research and technological needs of various industrial enterprises, most of which were then in their infancy.

By the mid-1960's it was evident, however, that this was not enough, and that a re-examination of the national policy of scientific research was essential. The NCRD could not adequately fulfil the requirement of promoting research activities on a national level, and applied research was still the Cinderella of the system. There was no shortage of sophisticated and highly trained manpower in the country, but only a very small part of it was engaged in applied research. True to the Jewish tradition of reverence for knowledge and study, and the desire to maintain centers of wisdom and learning in Israel, most of the country's scientists were involved in basic research and teaching. Within the limited sphere of applied science, agricultural research then occupied the chief place, and its outstanding contributions to the expansion, diversification and improvement of agricultural output are well known. But Israel's land and water resources are limited, and it was a matter of urgency to seek new vistas for economic expansion. Clearly the road lay in industry, and would involve devising a framework for Israel's future industrial development which would suit her unique circumstances. Rich though she is in human material, Israel is poor in natural resources. We have no oil, no coal, no metals like gold, iron, or aluminium; our only sizeable mineral resources are the phosphate deposits in the southern parts of the country, and the potassium, bromide, and magnesium in the Dead Sea.

It was thus evident that the challenge ahead lay in exploiting our human resources, in channeling the intelligence, imagination and skill of our scientists into the development of a highly sophisticated industry—a science-based industry, which would enable us to export our products and compete in the world market.

By the mid-1960's, therefore, it was clear that the national policy for R and D must accept as its first priority the stimulation and promotion of industrial development. In view of the small scale of the country's industrial concerns, any programme of applied research in the industrial sector would clearly need adequate government backing in order to share the financial risks involved.

It was at this juncture that the late Prime Minister, Mr. Levi Eshkol, appointed a Committee in May 1966 to enquire into the organisation and administration of governmental research. He invited me to chair the Committee and for the next two years I found myself deeply involved in clarifying the problems and in formulating recommendations.

In his letter of appointment the Prime Minister instructed the members of the Committee to survey the research being carried out in the various governmental research institutions, to determine the most suitable ways for the different ministries to carry out both short-term and long-term working programmes of research, and to recommend a pattern for the enhancement of applied research in the various institutes of higher learning.

The main recommendations of the Committee were duly formulated, as follows:

a)   To accelerate the shift from basic research to applied research with special emphasis on industrial innovation.

b)   To reorganise governmental research and the institutions in which it is carried out in such a way that the policy-making and coordinating function is clearly defined, and that actual research activity is made the administrative and budgetary responsibility of the ministries directly concerned.

c)   To encourage all institutions actually or potentially involved in research to expand their technological and applied science facilities; moreover, to request institutions of higher learning to emphasize these aspects in their educational programmes.

d)   To reorganise government-directed R and D as follows:

i)   By broadening the scope of activities of the NCRD so that it will be able to formulate and subsequently implement national R and D policy.

ii)   By creating in each of the ministries involved in R and D an office of

Chief Scientist. The Chief Scientist in each ministry will assist the Minister in defining specific R and D needs and formulating R and D policy within his ministry, in allocating priorities and in stimulating R and D activities in all areas within his sphere of responsibility. He will, moreover, ensure the provision of an appropriate budget, supervise the work done in his ministerial research organisation, and make full use of the available scientific potential.

   iii)   By uniting the various governmental institutions engaged in related research into larger research organisations under the supervision of the appropriate Chief Scientist. Three full-scale research organisations were recommended: the Agricultural Research Organisation in the Ministry of Agriculture, the Industrial Research Organisation in the Ministry of Commerce and Industry, and the Earth Sciences Organisation in the Ministry of Energy and Infrastructure. The Agricultural Research Organisation comprises seven research institutes: the Institutes of Soils and Water; Field and Garden Crops; Horticulture; Animal Science; Agricultural Engineering; Plant Protection; Technology and Storage of Agricultural Products. The Industrial Research Organisation comprises the following research institues: Israel Fiber Institute, National Physical Laboratory of Israel, Paint Research Association Ltd., Israel Ceramic and Silicate Institute, Rubber Research Association Ltd., Israel Institute of Metals, Fermentation Unit.

   In 1969 the recommendations of my Committee were adopted by the government. Since that time, and as a result of the adoption of the Report, there has been an unprecedented growth in applied research, which has caught up with and overtaken the level of basic research. In 1966–67, 63% of the gross national expenditure on R and D went to support basic research and only 37% went to applied research; by 1979 this trend had been reversed, with 64% of the R and D expenditure supporting applied research and only 36% going to basic research. Moreover, in 1966–67, most research (other than for defence) was carried out in institutes of higher learning, which were allocated approximately 62% of the total sum available for R and D; governmental and public research institutes received 27% and industry only 11%. However, the picture had changed drastically by 1978–79; the share in the research budget for the institutes of higher learning had dropped to 45% and that within the industrial sector had increased to 43%. The total sum spent on R and D today exceeds 2% of the gross national product, half of it going to the Ministry of Defence and the other half to civilian projects. The government continues to supply most of the money for applied research, although increasingly large sums of money for R and D are now supplied by the more prosperous industrial companies themselves.

   The increasing participation of the government in R and D has been facilitated by the Chief Scientists in the Ministries of Commerce and Industry, Agriculture, Communications, Transport, Health, and Energy Infrastructure. The Chief Scientist of the Ministry of Commerce and Industry, for example, has a special fund at his disposal for supporting industrially promising R and D projects. To date, some 600 projects have been supported to an extent of 50% in areas such as metallurgy, electronics, and chemistry. R and D projects of national importance, such as those concerned with desalination, solar energy, and other schemes promising a significant increase in exports, have been almost totally financed by the Chief Scientist's fund.

One of the functions of the Chief Scientist is to encourage contact between scientists in Israel and those working in related fields in other countries. Joint R and D projects of mutual interest are also encouraged, and a considerable number of such projects have been endorsed by the Chief Scientists in the various ministries.

A United States-Israel Binational Foundation for encouraging industrial R and D was recently established, and is financing various projects of mutual interest to Israel and the United States. A second U.S.A.-Israel Binational Foundation supports agricultural R and D, and a third supports basic research. The research projects supported by these foundations are chiefly those which are carried out in collaboration by scientists working in the United States and in Israel. Each of the foundations is administered by a joint committee of experts representing the appropriate bodies in the two countries.

The contribution of industrial R and D to the Israeli economy is strinkingly illustrated by the fact that Israel's exports of locally developed industrial products and processes based on original R and D increased from $2.5 million in 1967 to almost $1 billion in 1980. The share of such science-based exports in the total value of industrial exports rose from 4.5% to 31% in the same time period.

Consistent with the pattern of expansion following the government's adoption of the Report, there has been a gradual increase in the total number of science-based industries in the areas of electronics, optics, medical equipment, agricultural implements, innovations in industrial and water technology, mini-computers, pharmaceuticals, fine chemicals, new branches of biotechnology, and medical engineering. Many of these industries are sited in the industrial parks surrounding institutes of higher learning such as The Weizmann Institute of Science in Rehovot, the Technion in Haifa, Tel-Aviv University, and the Ben Gurion University of the Negev in Beer Sheba. All of these industrial companies engage highly qualified personnel and in some cases develop remarkable novel processes and products. The Ormat Company, for example, has successfully designed and built small power stations consuming different low-level energy sources in isolated areas, including the energy accumulating in solar ponds. Elscint Ltd. has developed tomographs competing in price and reliability with those produced by large, well-established companies abroad. Microcomputers in control units for single line irrigation systems have been developed and are being produced at Ein-Tal Industries. Glycerol, β-carotene, and proteins are extracted from marine algae in the south of the country. Computer-based intensive care and cardiac diagnostic monitoring systems for hospitals have been devised and are being produced by MG Electronics. These are only a few examples of the many successful new industrial ventures over the last few years.

I belong to the generation of Israeli scientists who were the first to receive all their professional training in this country. While studying at the Hebrew University on Mount Scopus I had the good fortune to be taught by outstanding scholars in the natural sciences, and it is to them that I owe my own dedication to the study of living organisms and life processes. I continue to be fascinated by the challenges and achievements of modern biochemistry, biophysics, and molecular biology. When still a young assistant at the Hebrew University I was keenly aware of my duty, as a member of the scientific community in an undeveloped but rapidly advancing country, not only to tackle basic research problems but also to make some meaningful contribution to the country. This I could best do by helping to lay the foundations for an organizational and administrational framework in which basic and applied research could flourish. To this day I still feel conscious of this great responsibility. I am therefore thankful that I have had the opportunity to contribute in some small measure to the promotion of basic and applied research. It has been a source of great satisfaction to me, as one of the founding members of The Weizmann Institute of Science, to follow its growth in size and stature over the years; it is gratifying to see that the Center for Biotechnology at Tel-Aviv University, in which I am currently involved, is making a name for itself; and it is good to reflect that, together with my late brother Aharon, the late Professor Ernst David Bergman, and others, I have been able to advise our Government and Prime Ministers, the various ministries, and some of our institutes of higher learning on how to create suitable institutions and organisations for applied research.

It has always been my belief that in Israel, blessed as we are with highly skilled and dedicated people, it would be possible not only to build up a sophisticated R and D infrastructure, but also to further the advancement of industry. Israel emerged from her agricultural period several years ago, and we now find ourselves living through her industrial age. This too will draw to a close and we will then be able, I hope, to pass rapidly into a post-industrial era, in which knowledge, understanding of systems analysis, and utilisation of the most advanced technologies will perhaps help to create a society in which people can enjoy not only material comforts but also a life enriched by fine moral and spiritual values.

I believe that the part I have played in Israel's development has been somewhat similar to that of Professor Egami in Japan. I have carried out some original research work in my own field. I have helped to raise a new generation of fine students. I have tried to assist my people and my country, both by promoting research and development, and in attempting to harness some of the achievements of science and modern technology for the benefit of man and society.

I always look upon this small country of mine as a pilot plant, in which one can check the impact of scientific and technological achievements on a people in the act of building a nation and creating a socity. The true aim of Zionism has always been, I believe, to build a sovereign state in which the quality of life, the culture, and the spiritual vitality of its people would be a source of pride. Modern science and technology have an important part to play in the achievement of this lofty ideal.

Personality Factors of Major Importance for Creative Efficiency of a Scientist

A. E. BRAUNSTEIN

A. E. BRAUNSTEIN was born in 1902 in Kharkov, Russia. M.D. (Kharkov, 1925); D.B.Sc. (Moscow, 1936); Professor in Biochemistry (1938). Member of USSR Academy of Medical Sciences (1945), and Academician (1964) of USSR Academy of Sciences. National honours: State Prize in medical sciences (1941), USSR Ministerial Council Prize for scientific achievements (1978), Lenin Prize in sciences (1980). Decorated Labor Red Banner (twice), Lenin Order and Hammer-and-Sickle Medal with honorary title "Hero of Socialist Labor" (1972). Foreign distinctions: D.h.c.,h.c. (Universities Brussels, Greifswald, Paris VII); member Academy "Leopoldina" (Germany); member of foreign academies in USA, France, GDR, Hungary *et al.* Staff Positions: Senior investigator (1928–1931) and Head of Laboratories in Research Centers of USSR Acad. Med. Sci. in Moscow: the Institutes of Biochemistry (1931–1934), of Exptl. Medicine (1935–1944), of Med. and Biol. Chem. (1944–1960), and in the Institute of Molecular Biology of USSR Academy of Sciences (since 1960). Main areas of research: glycolysis- and respiration-linked phosphate cycles (1925–1936); intermediary N metabolism and vitamin-B_6-dependent enzymes (since 1936). He discovered transamination; elucidated the structure, metabolic functions, reaction mechanisms of aminotransferases and other pyridoxal-P dependent enzymes.

A branch of humanistic inquiry nowadays in high fashion, as witnessed by the steady growth in number of relevant periodicals and sporadic publications, is the field comprizing philosophic, socio- and psychologic or other approaches to scientific creativity—a field designated by various terms such as "scientology," "research on research," and the like. In the writer's view the contents of these writings, when stripped of sophisticated terminologic camouflage and boiled down to essential residue, are usually highly disappointing. Covering a broad range from unmotivated individualistic phantasy to dogmatic platitude, most "scientologic" statements represent, at best, trifling variations of basic concepts set forth, since more than a century, by the classics of materialistic gnoceology and dialectic logics—Feuerbach, Hegel, Marx, Engels, Plekhanov, and Lenin. More recent emphasis on additional aspects, relating mainly to "environmental" rather than to personal factors greatly potentiating the productivity of modern scientific research—such as automation and computerized control of research operations, planned organization of cooperative interdisciplinary investigations, etc.—is also rapidly sinking into triviality.

Nevertheless, I venture briefly to discuss in this essay a few basic aspects I consider of crucial importance for personal creative effectivity of a researcher in most fields of the exact and life sciences, i.e., in the "Naturwissenschaften." The excuse is my own experience accumulated during more than a halfcentury of active experimental research, of postgraduate tuition, editorial and organizational activities in several areas of biochemistry and molecular biology, of professional and informal contacts with numerous outstanding colleagues. Apologies are due to those readers whom similar personal experience may entitle to qualify my statements, in turn, as rather trivial.

What follows relates to certain preconditions I regard as most basic for superior fitness of a successful investigator, often qualified as "talented scientist," but does not necessarily apply to the exceptional cases of creative "intuition," or out-of-the-range capability for epochal discovery we use to name genius (one of the best examples is Albert Einstein, and D. Mendeleyev is another).

In my belief one of the most essential sources of creative potential is the acquisition, as early as possible, and lifelong accretion by all available means, such as reading, travel, contacts with diverse people, openminded observation, etc., of an outlook of broadest scope on nature, mankind, the phenomena and laws of their existence, interaction, and history. I would prefer the old adage defining a specialist as "one who knows all about something and a little about everything else" to be modified to "...knowing all about something and as much as he can about everything." In point of fact, an encyclopedic thesaurus of knowledge spontaneously entails development of memory, of the habit to systematize and classify information, and of associative and combinatory thinking. Conscious self-training (and competent tuition) of young people striving to become scientists helps to develop these attributes which are of decisive importance for the augmentation of creative potential. A most valuable asset is early initiation to one or more predominant foreign languages, preferably—including conversational fluency.

Given a broad scholarly thesaurus of knowledge, efficacy of its utilization in scientific work will rest on the student's ability for selection of a promising and realistic study aim, for screening out appropriate information, and for steady, arduous effort in surmounting obstacles and achieving the chosen aim. A crucial requirement is personal modesty on the part of the student—strict avoidance of self-indulgence, and unbiased critical evaluation of his own contributions. Essential for noteworthy achievements is skill of the young scientist in manoeuvering between the Scylla of dogmatic adherence to concepts previously canonized by outstanding authorities, on one hand, and the Charibdis of disregard for established facts and of uncontrolled phantasy, aptly designated by the eminent psychiatrist Bleuler as "autistically undisciplined thinking," on the other.

This seems to me an appropriate place for stating that I do not agree with currently fashionable claims as to increasing interpenetration (respectively, complementarity) of scientific, humanitary, and artistic modes of insight into the essence of natural and human problems. There does not exist a firm, consistent basis for generalization of this kind.

However, a scientist must certainly know and obey the basic principles of socio-psychologic, ethic and aesthetic approaches to man and the surrounding world. The facts and laws of natural sciences have no intrinsic ethic connotation. But their development, applications and conceptualization can, depending on the moral principles of an individual scientist or scientific community (or on lack of principles) be either beneficial or gravely harmful to other people and mankind as a whole, to proximate or global natural surroundings. Today, joint efforts aiming at global protection of peace, health, and wellbeing of mankind and the conservation of its natural environment, as well as uncompromising opposition to the adverse activities of aggressive political and national powers, are the supreme moral duty of the international scientific community.

As regards aesthetic education, its value lies in development of the scientist's general culture and his awareness of the features of perfection and elegance, i.e., of *beauty*, in the essence and form of works of science. These features include *novelty* and strategic significance of theoretic concepts, reliability (and reasonable simplicity) of research methodology, precision and clarity of verbal, mathematical and pictorial presentation of the results, the prognostic and practical value of the work.

The late Academician A.N. Nesmeyanov, a renowned organic chemist and former president of the USSR Academy of Sciences, used to emphasize that the most fruitful new branches of science originate at the borderlines or crossings of existent disciplines. This thesis is amply supported by the current trends of scientific progress; it proves most helpful in the selection and planning of promising new research problems and approaches.

Such planning is an immensely difficult task, in view of the nearly boundless scope of information to be considered (and the formidable, unaccountable gaps in our knowledge). The complexity is surmounted, in part, by the current worldwide practice of collective investigations performed jointly by specialists qualified in the different branches of knowledge involved. Yet even this progressive mode of team-work will not be proficient unless the team is headed by a leader whose scholarly outlook is sufficiently broad to define, at least in rough outline, the general aims of the planned project, its strategy and appropriate tactical approaches.

In an interview published several years ago in the Soviet weekly "Literaturnaya Gazeta" (14.II.1968) under the heading: "To foresee the attainable, not to overlook the unexpected," I commented on the unrewarding task of experts asked to make prophesies. It is especially unfeasible to predict reliably the forthcoming advances in science (if this could be done, the very existence of science would be in question). Each discovery is a discovery, *i.e.*, "finding out what 'was there all the time' " (J.R. Vane, *TIBS*, v.3, no. 1, p. N2); it is often triggered by a happy inspiration or propitious occurrence. Progress in science is often rooted in unforeseen observations, in "spotting something that you were not looking for." Walter Cannon designated this route to discovery as "serendipity" (from Serendip, the name of a legendary Indian prince said to have found a treasure of great value accidentally). Yet one should bear in mind L. Pasteur's wellfounded caution that (in science) only a prepared mind can profit from fortuitous findings.

In my own itinerary as a scientist there occurred several crucial advances due to serendipity. Thus, the discovery of enzymatic transamination (1936) resulted from studies primarily aiming at identification of the source of amino groups for the resynthesis (reported shortly before by J.K. Parnas and associates) of adenylate (AMP) from the deamination product, inosinate (IMP), in muscle damaged by contractile overstrain or by trauma. A few years earlier, D. Moyle-Needham and some other authors observed that in muscle and other animal tissues oxidation of L-glutamic acid could proceed without accumulation of ammonia or related deamination products, *e.g.*, urea or amide groups. Following this lead, I attempted, in collaboration with Maria G. Kritzman, to identify the source of nitrogen for regeneration of adenosine phosphates in muscle minces on incubation with glutamate, but we failed to obtain any evidence for reamination of IMP. The real immediate NH_2-group precursor, identified $\sim$20 years later, is aspartic acid in all living cells. Had we happened to use aspartate rather than glutamate as the presumable amino donor in the early attempts to reaminate IMP, this could possibly have been found from the outset.

Instead, our very first experiments provided evidence for a much more general and fundamentally important metabolic process, most ancient in the evolution of life, but new to science. Actually, our results revealed rapid, reversible enzyme-catalysed transfer of the amino group, a proton and an electron pair between L-glutamate and pyruvate (arising from oxidation of endogenous lactate or supplied as such), to produce L-alanine and ketoglutarate. There followed intensive researches, in the author's laboratory and elsewhere, demonstrating the crucial biological significance, in intermediary nitrogen and energy metabolism, of enzymes catalysing the above and a large number of analogous amino transfer reactions, their ubiquity in all living cells, broad substrate range with preference for dicarboxylic NH_2 donors and acceptors, and rough outlines of the catalytic mechanism (1–4).

A prominent landmark in the biochemistry of amino acids was the detection by E.E. Snell (1945) of the chemical interconversion of the new vitamin B_6 species —pyridoxal and pyridoxamine—by exchange of amino groups with amino and oxo acids, respectively. His suggestion that these vitamins might participate in biological transamination was soon confirmed by the identification in several laboratories (including Snell's) of pyridoxal-5-phosphate (PLP) as cofactor of amino acid decarboxylases and of transaminases. Within a few years, a number of new coenzyme functions of PLP in a variety of metabolic reactions of amino acids were detected in my laboratory and elsewhere (5–7).

Our involvement in the study of the roles of vitamin B_6 in amino acid metabolism resulted, in part, from a curious serendipitous event. We regarded glutamic acid as the predominant amino donor in transamination; on the look-out for natural analogues of this substrate we became interested, in 1940, in a new tryptophan metabolite—kynurenine, discovered by Y. Kotake who erroneously described it as a substituted derivative of glutamic acid. The structure suggested by him for L-kynurenine, its correct formula and, alongside, the structure of L-aspartic acid are shown below.

Kynurenine   Aspartic acid

Kotake's formula   Actual structure

We failed to observe transamination of L-kynurenine with pyruvate in muscle tissue (currently known to contain no kynurenine transaminase). But in experiments with extracts from liver of vitamin-B_6-deficient rats we observed marked depression (and restitution on supplementation with PLP) of the activity of Kotake's kynureninase—an enzyme cleaving kynurenine, as we could demonstrate (1949), to anthranilic acid and L-alanine (8, 9). Other aroyl-alanines were later shown to undergo similar enzymic cleavage.

This reaction closely resembles the β-decarboxylation of L-aspartic acid by a bacterial enzyme considered as a PLP-enzyme (S. Mardashev (10)), and is obviously similar to the wellknown chemical reactions of "acid split" of β-diketones and of the analogous β-iminoketones. The latter type of molecules includes the tautomeric ketimine structures in equilibrium with aldimines formed from amino acids and pyridoxal or related electrophilic aldehydes.

This analogy laid the ground for my first outline (1949) of a unified interpretation of reactions catalysed by kynureninase and other PLP-dependent enzymes in terms of modern ideas of physical-organic chemistry (9). These concepts served as a basis for the general theory of PLP-dependent enzymic reactions of amino acid metabolism, elaborated in 1952–1953 by me and M. Shemyakin (3, 11, 12); E. Snell and associates (6, 7) published closely similar concepts in 1954. The theory gained general acceptance and, with minor revision, remains essentially valid.

In later years, fortunate instances of serendipity repeatedly initiated new proficient trails in our studies, for example, in the delineation and characterization of a peculiar subgroup of the PLP-dependent enzymes catalysing important metabolic transformations of amino acids, namely, the exclusively β-replacement-specific lyases participating in the synthesis and metabolism of cysteine and related amino acids. I refer to recent articles (13, 14) for an overview of these researches.

In the concluding section of this essay I would like to consider briefly one special aspect closely related to our topic. It concerns the question—much debated and still controversial—as to the definition of a "scientific school." In my opinion, an ever so large number of disciples, educated by a distinguished investigator and conducting work of high quality but limited to multiplication and continuation of mental and experimental approaches predetermined by their master, does not constitute a scientifical school. Such a situation is typically reflected in the well-known sentence of a German biochemist renowned in the first decennia of our century: "I have brains of my own; what I need are skilled executants." Although the contributions from this research center were highly valued, its head can certainly not be considered as leader of a scientific school.

As I see it, former and present associates constitute a "school" (to which followers from outside can also adhere) only if their own ideas and research methodology contribute, by a peculiar "feedback," to enrichment and remodelling of the leader's concepts, initiate the branching of research topics and generation of novel ones, and open up new approaches to their investigation.

REFERENCES

1. Braunstein, A.E. and Kritzman, M.G.: 1937. *Biokhimiya* 2, 242–262, 859–874 (in Russian).
2. Braunstein, A.E. and Kritzman, M.G.: 1937. *Enzymologia* 2, 129–151 (in German).
3. Braunstein, A.E.: 1960. *The Enzymes*, 2nd ed., Vol. 2 (Boyer, P.D., Lardy, H., and Myrbäck, K., eds.), pp. 113–184, Academic Press, New York.
4. Braunstein, A.E.: 1977. *Mol. Biol.* 6, 1238–1257 (in Russian).
5. Braunstein, A.E.: 1973. *The Enzymes*, 3rd ed., Vol. 9 (Boyer, P.D., ed.) pp. 379–481, Academic Press, New York.
6. Metzler, D.E., Ikawa, M., and Snell, E.E.: 1954. *J. Am. Chem. Soc.* 76, 648–652.
7. Snell, E.E. and Di Mari, S.J.: 1970. *The Enzymes*, 3rd ed., Vol. 2 (Boyer, P.D., ed.), pp. 335–370.

8. Braunstein, A.E., Goryachenkova, E.V., and Paskhina, T.S.: 1949. *Biokhimiya* 14, 163–179 (in Russian).
9. Braunstein, A.E.: 1949. *Dokl. AN SSSR* 65, 715–718 (in Russian).
10. Mardashev, S.R. and Etingof, R.N.: 1948. *Biokhimiya* 13, 402 (in Russian).
11. Braunstein, A.E. and Shemyakin, M.M.: 1952. *Dokl. AN SSSR* 85, 1115–1118 (in Russian).
12. Braunstein, A.N. and Shemyakin, M.M.: 1953. *Biokhimiya* 18, 393–411 (in Russian).
13. Braunstein, A.E. and Goryachenkova, E.V.: 1976. *Biochimie (Paris)* 58, 5–17.
14. Braunstein, A.E., Goryachenkova, E.V., Kazaryan, R.A., Polyakova, L.A., and Tolosa, E.A.: 1980. *Frontiers in Bioorganic Chemistry and Molecular Biology* (Ovchinnikov, Yu.A. and Kolosov, M.N., eds.) pp. 167–189, Elsevier/North-Holland, Amsterdam.

Reflections on the Art and Science of Doing Research

ERWIN CHARGAFF

E. CHARGAFF was born in Austria on August 11, 1905. Ph.D. degree (chemistry) summa cum laude, University of Vienna, June 1928. Milton Campbell Research Fellow in Organic Chemistry, Yale University, 1928–1930. Assistant in charge of chemistry, Department of Bacteriology and Public Health, University of Berlin, 1930–1933. Research associate, Institut Pasteur, Paris, 1933–1934. Since 1935 at Columbia University. Pasteur Medal, Paris 1949. Carl Neuberg Medal, New York, 1958. Soc. Chim. Biol. Medal, Paris, 1961. Gregor Mendel Medal, Halle, 1973. National Medal of Science, Washington, 1975. N.Y. Academy of Medicine Medal, New York, 1980. Charles Leopold Mayer Prize: Académie des Sciences, Paris, 1963. Dr. H.P. Heineken Prize, Royal Netherlands Academy of Sciences, Amsterdam, 1964. Bertner Foundation Award, Houston, Texas, 1965. Member, National Academy of Sciences. Fellow, American Academy of Arts and Sciences. Foreign member, Royal Swedish Physiographic Society, Lund. Member, German Academy of Science, Leopoldina. Member, American Philosophical Society. In his research work he has ranged widely through many fields of chemistry and biochemistry. His foremost contribution has, perhaps, consisted in laying the groundwork for our present understanding of the nucleic acids and their role in genetics.

Scientific research as a mass occupation is a recent phenomenon. When I first entered a chemical research laboratory, more than fifty years ago, the aspect of doing basic research and, even more, the social and economic position of the research scientist were entirely different from what they are now. I have written of these things in a recent book of mine, *Heraclitean Fire* (Rockefeller University Press, New York, 1978), published later in German and Japanese versions, and also in many essays, written in English and in German.

A selection of the German articles is scheduled to appear in book form at the end of 1980, under the title *"Unbegreifliches Geheimnis*—Wissenschaft als Kampf für und gegen die Natur" (*1*). My English essays on those topics are best read in the form in which they appeared in several journals, mostly in *Perspectives in Biology and Medicine, Science* and *Nature*. A selection in book form, published a few years ago, I have been unable to acknowledge as my own, as the publishers mangled the text, refusing to pay heed to the protests of the author.

That adventure, probably not customary among the publications of established scholars, has fortified me in the conviction that it was in my lifetime that scientific truth assumed the character of merchandise. Great scientific discoveries that in the past may have shone with an intellectual brilliance, sufficient to illuminate the entire life of those who made them, are now merely tokens in a power game, subject to the stresses of a continually fluctuating stock exchange. It would have been unthinkable in earlier times for a Harvard professor to address the Investment Analysts of America on the commercial portent and the speculative significance of a biological discovery; nor is it likely to have happened before that a press conference, called by two scientists, served to drive up, overnight, the shares of a company of which they were among the owners. Research as an element of the "futures market," as the producer of commodities, surely is a novel development. Moreover, futures in wheat may be a less questionable item—for there surely will be wheat—than futures in interferon.

It stands to reason that the change in the means of scientific production—a change from an artisanal status, as it were, to one on a large, almost industrial scale—has had enormous economic, and even intellectual, consequences for the individual researcher. Thirty-five or forty years have sufficed to make out of a comparatively rare craftsman a member of a very numerous group of intellectual wage earners; a group that I believe has grown into a new class, that of the knowledge producers, manufacturing so-called scientific facts. The mass production and the planned design of scientific discoveries have long ago swamped and invalidated whatever free market forces may have operated in the beginning. Actually, I doubt that those ever existed, for there is no market for new knowledge; not even in the sense in which one could speak of a market for new musical compositions: a market, incidentally, that is certainly regulated by the number of willing ears. In the case of scientific knowledge, however, the audience consists exclusively of the producers themselves and can, therefore, have no regulatory effect. It is fortunate that the real world is different and that shoes are required not only by shoemakers. (Otherwise, the world would probably suffocate in shoes.)

Before scientific research in a form recognizable by us began, *i.e.*, before the beginning of the last century, the individual scientist owned his instruments and implements; even university professors often had to provide, out of their own pocket, the accessories required by their teaching. That changed, of course, with the times. When I joined the thin ranks of scientific craftsmen, the means of production, in contrast to most other artisans, did not belong to us, but they were few and inexpensive. As I do not consider, in this article, the problem of industrial or goal-directed research, one can say that most of the people of whom I speak were employed by universities or by the few research institutes then in existence.

There was a time when one could study nature by talking with others on the agora. But we have decided to study nature by taking it apart, leaving whatever deeper meditation about nature may be possible to the few survivors of the post-nuclear age. Ours is the way of bringing the smallest available particle into the sharpest attainable focus. No wonder that this has completely blurred our vision of the whole. The overemphasis on accuracy far beyond the decimals of necessity; the substitution of image for substance; the incessant search for indirect ways of handling the invisible, without having to take the trouble of making it visible; the displacement of imagination by apparatus, of experience by information; the urge of making the complicated ever more complex by continually inventing mostly nonexisting controlling systems—receptors that do not receive, agents that do not act—; a rigid and thoughtless creed of a godless age, a form of molecular mono-theism, enthroning DNA on altars where it has no business to sit: all that is only part of the reason why scientific research has become so inordinately expensive and dehumanizing.

The individual research scientist, far from having unrestrained access to his means of production is, hence, forced to seek refuge in one of the huge knowledge factories that begin to monopolize, not only all instruments and other tools of research, but also the admission of the individuals to the literature and even their ability of receiving support from the state. Whoever wants to can see it: the forma-tion of monolithic trusts, the sprouting of monopolistic corporations, the concen-tration of the means of production in ever fewer hands, they have not stopped before the so-called temple of science.

In future, I am afraid, scientists will have to resign themselves to working in ever larger groups, directed and controlled by scientific impresarios whose lode stars—"achievement," "success," "efficiency," "money"— are, in my opinion, not compatible with true research and scholarship. The ultimate goal would seem to be the fully automated knowledge factory, with the scientists limited to lubri-cating and servicing the machinery and perhaps also to interpreting the findings for the next generation of service personnel.

The "doing research" of which my title speaks clearly can have nothing to do with the activities described in the preceding paragraphs. Trend-setters, scien-tific fashion designers, with hordes of underlings filling in the obvious colors in the predictable patterns: can those be the scientists with whom I spent my life? Of course, not. The sciences themselves have been overtaken by their much too sudden growth and their practitioners stunned by forces they no longer understand.

As a matter of fact, I believe that there is not a single human being alive now who understands what is really happening; and I do not think only of the natural sciences, but just as much of politics, religion, philosophy, industry, economics, the arts, and so on. The increasingly unmanageable complexity of everything we are in contact with creates leaders who do not know what they are doing; it creates, necessarily, charlatans. We must thank God for them, for without charlatans our world would already have collapsed. The scientific impresarios whom I have mentioned before may, therefore, also represent a necessary stage in the decay of the Western world.

In any event, it is not that novel breed that I want to consider, but the scientists of my own generation. The title of my article speaks of the art and of the science of doing research, as if there existed two alternatives. They do exist, indeed, but "alternatives" is not the right word, for the choices an individual makes presumably are not free in most instances. It is, therefore, better to interpret my title as indicating that there are two principal types of research men.

In the book mentioned before, *Unbegreifliches Geheimnis*, I have on pages 213 to 216 attempted to give a brief typology of the two types of investigators to whom I refer as "Platonians" and as "Cartesians." I cannot repeat myself here, but I may summarize the distinction metaphorically. The Platonian dreamily floats in an ocean; the Cartesian vigorously crawls in a swimming pool. To quote from my previous text: "The Cartesian wishes to explain, the Platonian hopes to understand. To the Cartesian world and life are a mixture of numerous forces, radiations, and substances that can be sorted, described, and related to each other; to the Platonian they are a beautiful riddle whose mystery forms a large part of its beauty."

The Platonian evidently represents the art of doing research, the Cartesian the science. Although there are, of course, no pure types, the vast majority of scientists active now belong to the Cartesian category. Since people nowadays like to get everything explained, the Cartesians are also more successful: it is so much easier to explain a swimming pool than an ocean. Nevertheless, there also exist a few Platonians. In earlier times, when the sciences were young, there were more of them.

It is notoriously hard to know oneself, but I have the impression that, if I am anything, I am a Platonian. All my students, with perhaps one or two exceptions, were Cartesians.

Since our sciences succumbed to the curse and the blessing of induction, it goes without saying that both the Cartesians and the Platonians perform more or less the same kind of experiments. On any given problem, it is even likely that the Platonians do more experiments than do the Cartesians, because the latter perceive clearly what they ought to find and limit themselves to the minimum sufficient to make their case. The so-called progress of science often rests, in fact, on the not always unavoidable inadvertence of those almost too clear-sighted thinkers. Where the two types of investigators differ, however, most is in the use they make of their results. To the Platonian, his experimental findings appear merely as the shadows of reality; and their performance is only the prelude to the real contemplation of nature. To the Cartesian, his results are reality itself, tiny polished blocks of what his world is made of; and out of that vast minuteness he constructs a simplified compatible universe.

If I am to judge from my own, admittedly not paradigmatic, case, the art of doing research consists primarily in letting oneself float. Who casts the initial die, that shall be left open; but once that is done—I wish to avoid such empty tautologies as chance and necessity—the progression becomes both associative and aleatory.

A reader of *Heraclitean Fire* (2) could find many instances of what I have said here. If he were to glance at the results of the first fifteen years of my independent research activity, at the topics I investigated, at the papers I wrote, this is the line —neither straight nor zigzag—that he could make out. After my doctorate at the University of Vienna I went to Yale University, in order to help R.J. Anderson in his study of the lipids of tubercle bacilli. The topic ordained for me was, hence, *lipids* and *fatty acids*. Going in 1930 to my first independent laboratory at the University of Berlin, I worked on the lipids of diphtheria bacilli and BCG, on the synthesis of branch-chained fatty acids and on their coordination complexes with cholic acid, the choleic acids. Returning to the United States in 1934, this time to Columbia University, I brought the lipids with me; but since my first job called for the investigation of blood coagulation, the topic I chose initially was the activation of clotting by tissue lipids. Then I began to float back and forth. Blood coagulation took me to *heparin*, *cephalin* took me to the *cerebrosides*, and those to *sugars*. ^{32}P becoming available at that time, it was natural to study the *metabolism of the phospholipids*. One of those, it then became known (Jordi Folch), contained a hydroxy-amino acid; and from phospatidylserine it was only a short step, or better float, to the investigation of the enzymic dehydration of *hydroxyamino acids* and, therefore, to the study of the *hydroxy ketoacids*. And from those it is not far to the *inositols* and their oxidation.

Another path led from the clotting activity of phospholipids to the tissue activator of blood coagulation, the *thromboplastic protein*. This again imposed the study of *cellular lipoproteins* in general. Accordingly, lipoproteins and blood coagulation formed the subjects of my first comprehensive review articles. I used to say jokingly that in my first few years I had already studied three of the four main compound classes of the cell: lipids, proteins, polysaccharides. Only the nucleic acids were missing, and they came later.

There is no point in going on. The branch lines diversified and multiplied, but they seldom broke. It is quite possible that in a few cases metamorphosis was reversed, a beautiful butterfly giving rise to an ugly caterpillar; but throughout my life I had the feeling that I was borne by gentle and protective waves. It is, however, likely that at no time could I have stated clearly what I was after, what I hoped to find.

Floating is, of course, not everything. One learns from experience, though probably not as much as is usually claimed. Whether inborn or acquired, I do not know, but a gift for recognizing the significant, a revulsion from the trivial, a certain feeling for the fitness of things, an awe of violating the sacrosanct, a measure of mild inflexibility are also necessary.

I am sure that the objection could be raised that what I say here is useless, that nobody could learn the art of research from these lines. I am afraid, that is correct. I have said it before: "What I can teach, cannot be learned." On the other hand, the beginner ought not to be frightened by the cool and deliberate manner in which scientific papers are written. They all carry a thick Cartesian overcoat; everything is crystal clear, it could not have been otherwise. But these exercises in successful logic conceal more than they reveal. Laboratory work looks differently in the laboratory than it does in the scientific journal. Even science has not abolished the human condition. If the natural sciences appear limitless, that is due to the limitlessness of human error and foolishness.

And, finally, what about the science of doing research? There is not much of a problem here: everyone does as everybody else. Besides, there also exists an excellent primer, Peter Medawar's *Advice to a Young Scientist*.

REFERENCES

1. Chargaff, E.: 1980. *Unbegreifliches Geheimnis; Wissenchaft als Kampf für und gegen die Natur*, Klett-Cotta, Stuttgart.
2. Chargaff, E.: 1978. *Heraclitean Fire*, Rockefeller University Press, New York.

You Carry Out Eukaryote Experiments on Shellfish Selfish DNA: An Essay on the Vulgarization of Molecular Biology

ATUHIRO SIBATANI

A. SIBATANI was born on August 1, 1920 in Sakai, Osaka, graduated from Department of Zoology, Kyoto University 1946, and awarded a D.Sc. degree from Nagoya University in 1953 and D.M.Sc. degree from Yamaguchi Medical School in 1958. Immediately after the War he worked in a pharmaceutical company in Tokyo and then moved to Osaka University. Then he held a professorship in Biology at the Yamaguchi Medical School during 1955–1962 and at Hiroshima University during 1962–1968 in Biochemistry and Biophysics. In 1966 he visited the CSIRO Division of Animal Genetics, Sydney, and since 1968 has been a senior principal research scientist in that institution, a part of which later became the CSIRO Molecular and Cellular Biology Unit. His work included the molecular biology of nucleic acids, but since 1975 he has turned to developmental biology (pattern formation). He is the author of many books published in Japan on those disciplines as well as general biology and the philosophy and critique of science.

Like Gunther Stent (1), who belongs to what one might call the "inner circle" or the "thought-collective" of molecular biology, I have a high appreciation of Horace Judson's *The Eighth Day of Creation* (2), for, contrary to what is generally believed, it has demonstrated that molecular biology is not biology at the molecular level. As Stent (1) says, such biology is actually biochemistry. Judson's book has made it abundantly clear that molecular biology is a working style for the investigation of life that to a large degree does not operate at the molecular level. For example, according to Sydney Brenner:

"Finding out about the machinery without touching the biochemistry. There was a culture—well, a *cult*, almost—that became typical of molecular biology......And of course, the simpler the methods used, the more highly prized. In other words, the rII genetics, the acridine mutants, the nonsense mutants, will always be a classic: why? Because it was just done on bits of paper......I mean, it doesn't need all the bloody tubes and counters and so on. And I think that the cult got founded around these ideas of how to solve the code without ever opening the black box" ((2), p. 488).

I think that this fact, so lucidly depicted in the book, should have been the principal conclusion put in its last chapter. Unfortunately, the last chapter of the book is full of compromise and becomes banal as a result. But the thesis that one finds in the text as a whole is of the greatest interest and importance.

I read the book with vivid memories, which suddenly flooded back to me, about how we in Japan started and lived through the period of emergent molecular biology, almost in the same way as those in the "inner circle" lived it. It is true that Judson did not interview any Japanese scientists for his book, for they contributed rather little to the rise of molecular biology (though I think that the story of the discovery of ribonuclease T_1 by Professor Fujio Egami, at least, should have been mentioned). We were too isolated, having only limited contact with the "inner circle," mainly *via* Stent, and our group was too small to attain the necessary momentum or critical mass for the generation of first-rate ideas and experiments. Nonetheless, during the fifties we were obsessed with the same problems; we were attracted to the same sources of new information (including Oswald Avery and André Boivin who was an old acquaintance of Professor Egami's), and we had the same way of looking at things as did the "inner circle" at the other end of the Pacific and the Eurasian continent. Let me examine the parallelism.

Yes, we were isolated, not only from the prospective molecular biologists in Europe and North America, but also from the main academic body in Japan—at least I myself was feeling that way. But also:

"As elsewhere, [and] even more in France, what came to be molecular biology began small, ill-regarded, pinched for funds and space. As elsewhere, in compensation, the beginning was attended by a zest that was at once playful and rebellious" ((2), p. 349).

In Japan, too, as Mutsuo Sekiguchi (3) has pointed out, we were rather like guerillas scattered here and there in a few university departments and in some private enterprises.

At the beginning we were not concerned about who would make the breakthrough—towards the solution of the secret of life. I recall that in the early fifties we decided—within our own small group of people working in separate laboratories—that anyone might seize upon any good ideas and work on them, no matter who was the first to think of them, and without any concern for questions of priority. We encouraged the kind of astuteness that is expressed in the Japanese sayign *ikiuma no me o nuku*, which means that one is so quick in action that one "extracts an eye out of a live horse." "Eye of a live horse" thus became one of our catchphrases for some time. And indeed, Max Delbrück said:

> "......the spirit was—open......In that the first principle had to be openness. That you tell each other what you are doing and thinking. And that you don't care who has the priority" ((2), p. 61).

Regrettably, however, such a spirit had already been eroded by 1963, when, after the necessary (or unavoidable) expansion, even our group was plagued by strife on matters of priority. And Judson quotes an anonymous scientist saying:

> "[In 1956 at Gordon Conference on proteins], these two groups are rather bitter with each other because of priorities, *etc.*, *etc.*... It is clear that the situation is not quite as relaxed as it was a few years ago" ((2), p. 333).

We quickly responded to the celebrated double helix paper in *Nature* by Watson and Crick (1953), with unequivocal and enthusiastic support, for we had known all the relevant facts except for the antiparallel double-helical nature of the DNA strands which was announced at the same time as the Watson/Crick model. I also remember being interested in Dounce's scheme of RNA in protein synthesis ((2), p. 247), and in 1953 devising a modified scheme of it, the details of which I have now completely forgotten. I wrote about it to Syozo Osawa who had just gone to work with A. Mirsky in the Rockefeller Institute. He promptly wrote back that he had had a few sleepless nights trying to devise a good experimental test for the hypothesis. I also clearly remember the paper by C.E. Dalgriesh in *Nature*, and its text-figure, on the multiple use of a template. Crick also remembered the figure "even a quarter-century later" ((2), p. 249).

Around 1953 I co-authored with Professor Egami an introductory book on nucleic acids, in which I intended to argue that a new *logic* might be needed to understand the role of RNA in protein synthesis, but then I erased it, fearing that it might sound too provocative. But Crick is quoted on the genetic code: "......people thought protein synthesis couldn't be a simple matter of coding from one thing to another......It didn't sound like biochemistry to *them* ((2), p. 233; italic by Judson). Judson himself (p. 333) also says: The paper by Crick "On Protein Synthesis" permanently altered the *logic* of biology" (my emphasis).

I had also been heavily prejudiced, like Jacques Monod and many others ((2), p. 387, 391), by the apparent dynamic state of living matter, including proteins. And also like Monod, I gradually came—around the same time (1953) but quite independently, and in my case only by conjecture and inference—to the idea that the apparent dynamic state of the body proteins in vertebrates must be, at least largely, due to protein secretion and cellular turnover rather than molecular turnover, as had been generally believed since Schoenheimer (4). This idea was reinforced by our own experiments which demonstrated the metabolic stability of DNA in *Escherichia coli* and dividing and non-dividing rat liver cells. This last piece of work brought me, I think, close to Stent, who was then addressing himself to the question of the semi-conservative *versus* dispersive mode of DNA replication, which was to be finally resolved in the hands of Meselson and Stahl a few years later. I visited Monod at the Institut Pasteur in 1956 and discussed with him (quite congenially) the philosophical implications of protein turnover (or rather its absence).

Our experience in Japan demonstrates that the rise of molecular biology reflected the emergence of a style of thought which was more universal than local in nature. A few years ago, I once boldly said, to an official review committee for our laboratory in Australia: "The definition of molecular biology is what molecular biologists do." (Then, if asked for a definition of molecular biologists, I would have answered that they were those who called themselves molecular biologists.) This definition is probably not without a parallel. In a book review A. McLaughlin wrote in reference to Thomas Kuhn (6): "[He] defines reason (or at least scientific rationality) as what the scientific community does." And indeed, Crick wrote, not long ago: "Molecular biology can be defined as anything that interests molecular biologists" ((2), p. 201). How similar my saying was to Crick's and both of them were made at about the same time. Judson (2) goes on:

> "[Molecular biology] is expansionist. I recall Max Perutz remarking once. offhand, about a particular field—it happened to be embryology,...... "Molecular biologists are going into that, but the science itself has not yet gone molecular." In such fields not yet gone molecular, more than a few scientists will admit to being, in a quiet way, antimolecular biologists."

So let us reaffirm Stent's claim that molecular biology is *not* biology at the molecular level—that is biochemistry (1). In fact I wrote some years ago (7, 8) that molecular biology ended around 1963, and that what had been with us since was merely biochemistry, perhaps helped by genetics, massivly funded and immensely successful. If molecular biology continued, its character must have changed. But in what way?

With the rise of molecular biology during the decade 1953–1963 one of the outstanding enigmas of life—the secret of heredity—was solved, perhaps much sooner and more easily than anyone had anticipated, by the proposal of the three grand theories of molecular biology, which were, only later, largely verified by experimentation. These theories are, of course, the double helix of DNA, the central dogma for protein synthesis, and the operon and allostery hypotheses. I felt that these discoveries brought a revolution to biology (9).

I vividly recall that, soon after the announcement of the operon hypothesis and the experimental verification of messenger RNA in 1960, some molecular biologists emphasized: "Any experiment not supported by theory—rather than a theory not supported by experiment—is meaningless." Indeed, Judson ((2), p. 93) quotes Sir Arthur Eddington the physicist, writing in 1934: "It is also a good rule not to put too much confidence in the observational results that are put forward *until they are confirmed by theory*" (emphasis in Judson's original).

Also around the same time, Monod ((2), p. 613) claimed that what was true for *E. coli* would be true for the elephant. And so began the exodus of molecular biologists to various tougher fields, above all developmental neurobiology. They had a conspicuous arrogance which was underpinned by an optimism based upon empirical successes, and the supposition that the mode of thinking that had been so successful in molecular biology might help rapid solutions to outstanding problems in other fields of biology.

However, what has happened since then has run counter to these expectations. This is why I used the expression "end of molecular biology" (*7, 8*). But today one might speak more aptly of the "vulgarization of molecular biology."

First of all, the enormous expansion of molecular biology since 1963 brought about a large number of unexpected discoveries: repair of DNA; fundamental differences between prokaryotes and eukaryotes in molecular machinery; fragmentary synthesis of DNA and its priming with RNA; repeated sequences of DNA; reverse transcription; jumping genes; multiple reading frames of a single DNA sequence; split genes; RNA splicing; somatic DNA switching for immunoglobulin-producing cells; and variation of the genetic code.

All being unexpected at the time of discovery, and with many of them still eluding our real understanding, these development should have been regarded, at least at the outset, as meaningless—according to the traditions of good old molecular biology. Moreover, such surprises revealed a fundamental difference between *E. coli* and the elephant. And the migrant molecular biologists, though able and active in their respective new fields, have not brought any spectacular revolution, any new logic, or light to the deeper understanding of life. To be sure, progress was made both inside and outside the bound of molecular biology by both old and new molecular biologists. But, despite the impressive expansion of our knowledge since 1963, we hardly feel that we have come significantly closer to the real grasp of the secret of life.

I had already written on such matters several times before I read Judson (*7, 8*). Now he writes:

> "Many molecular biologists were confident—certainly most of those who had taken part in building that outline [for prokaryotes] were confident—that the outline could be stretched to take care of higher organisms without great difficulty. They spoke of filling in classical molecular biology. Looking for a tough, interesting problem, a new level of molecular biology, many who had taken part in building that outline moved into neurobiology.
>
> The confidence was premature. Filling in classical molecular biology for higher organisms proved to be inseparable from the problem of differentiation—from the long-standing problem of embryology......The problem was harder, both technically and conceptually, than anyone had foreseen." ((*2*), p. 613).

To me, this suggests that there is still some fundamental lacuna in our understanding of the essential features of life, because of which we have failed, on most occasions, to construct adequate theories about the new and surprising discoveries made since 1963. The trouble is that we do not know what this lacuna is about, and how large and fundamental it is (*7, 8*). Therefore, for the traditional molecular biologist, the essential task now should be to fathom the nature of this lacuna. The work to be done in various fields is, then, not to discover something new which is near at hand, but to work out an appropriate framework which will then lead us to see what to do and how to deal with the lacuna.

This situation has also been touched upon, in part, by some leading molecular biologists. For example, in relation to intervening sequences in the structural gene, Crick (*10*) wrote: "When I came to California in September 1976, I had no idea that a typical gene might be split into several pieces and I doubt if anybody else had." Or again, in his concluding remarks at a workshop on developmental biology held in France in May 1979, François Jacob said (if I understood him correctly) something like this: "In those days of "*la belle époque*" ((*2*), p. 272), the available methods were limited. One used to do simple experiments within a day. But one had good ideas! Today people can use fancy techniques, do complicated and sophisticated experiments, but they are not really solving problems in developmental biology."

In 1950, one was well aware that classical biology was incapable of solving the problem of life. So it was natural to turn to molecular biology. But today it is difficult to feel the same way with DNA studies for instance. There are many things to be done on DNA, and undoubtedly many new findings are being obtained. But they are all "solvables," rather than the "unsolvables" such as were tackled in the emerging days of molecular biology. The current quest may be compared to an expedition to some infrequently visited area, where you might expect to find many new species or even genera. The very success of such an expedition makes it difficult for attention to be diverted elsewhere.

However, Seymour Benzer, for one, said:

"It's a new phase. I feel that, y'know, when I came into molecular biology it was a pioneering science. But when a science becomes a discipline, which is essentially true of molecular biology now, when you can buy a textbook, take a course—there's no question there are many surprises left......but a field to work in, to me personally, when it becomes a discipline, becomes less attractive. I find it more fun to be striking out in something which is more on the amorphous side—which was true of molecular biology when I started," ((*2*), p. 271).

For some, however, there is nothing wrong in the present plethora of advancing frontiers in biochemistry and cell biology (*11, 12, 18*). There seem to be more research projects than one can actually take up. But this is *not* molecular biology as it was at the beginning. For example, there is a lot of DNA sequencing going on now, and it may appear that we are learning a great deal through it *a posteriori*. But it is being done according to a style of thought whereby one would set out to crack the genetic code with methodology: first sequence a lot of proteins and a lot of DNA, then put them side by side, and deduce the code out of them—an orthodox method in biochemistry. But as we know, the genetic code was solved without anything like that, and *that* was molecular biology. The current tide of sequencing DNA gives us a kind of *natural history of DNA*, just like a natural history of animals and plants, with specimens in museums, records in libraries, and DNA sequences in a "genothec."

In 1960 I proposed a classification of biology into two areas: fundamental biology and "enumerative biology" (the then new term of mine), placing both biochemistry and taxonomy in the latter category (*9*). In the former I put molecular biology and the prospective population biology (as a part of ecology). Looking back, I think I was right. Now what we are seeing is the "enumerative biology" of DNA, which will not end until all the varieties of DNA sequences on the earth have been exhaustively recorded.

The attitude that one finds in "enumerative biology" represents a framework of thought—an ideology, if you like—for a particular scientific practice. At a workshop on the molecular and developmental biology of insects held in Japan in August, 1980, an European colleague of mine, a *Drosophila* developmental biologist, observed that the main group attending believed that all the problems in developmental biology could be solved by sequencing DNA. In the same year but on a separate occasion, another leading European developmental geneticist declared that one could not have deduced the genetic code, or the existence of the twenty amino acids, from the sequences of DNA alone. Likewise, one has to know first how genes work, that is, one has to understand the "logic" of gene function, in development. Again, Brenner said:

> "What has finally happened is that *all* of that [finding out about the machinery without touching the biochemistry] has been demolished by the fact that in the last five years we are learning to do molecular genetics *directly* by sequencing the DNA. Sanger, and these people, show how you actually sequence the DNA and find the base change. Whereas the attempts to use genetics could be interpreted as just cheap ways of trying to sequence the DNA" ((*2*), p. 488, Judson's emphasis).

Perhaps it was possible to do it the cheap way while the subject remained rather simple. But the amazing complexity of eukaryote systems may preclude such a cheap way; and, after all, though expensive, sequencing DNA may still represent the cheapest and quickest way of solving the problem. Molecular biology has had to change precisely because of the nature of the latest problem it has taken up. But it *that* all? I wonder.

Looking at the situation in broader historical terms, I feel that reductionist (or particulate) biology started around the turn of the century. Following Koch's identification of discrete pathogens, there came Mendelian genetics with its discrete genes, then biochemistry with its essential amino acids, vitamins, and hormones, and finally molecular biology as a discipline with its cohorts of DNA, messenger RNA, and transfer RNA. Basically using the concept of linear systems, it has scored a tremendous success.

And indeed, again according to Brenner:

"This is really asking whether higher organisms have some unique place of molecular biology *that's unknown to us*. Whether the problem of developmental biology could be solved by one insight like the double helix. On one way, you could say, all the genetic and molecular biological work of the last sixty years would be considered as a long interlude—sixty years of following out Morgan's Deviation into the tractable genetic problems. And now that the program has been completed, we have come full circle—back to the problems that they left behind unsolved" ((*2*), p. 209; my italic).

So we start again now, for the first time in all these years asking really seriously how organisms develop. One of the outstanding problems is the scale-invariance of developmental processes: for example, in sea urchins, a double-sized embryo arising from the fusion of two fertilized eggs, and a single cleavage cell at the 4-cell stage, can both develop into normal individuals differing in size by almost an order of magnitude. This looks *very* different from "morphogenesis" as dealt with in molecular biology—of phages and ribosomes, which are manufactured like technological products on an assembly line, *i.e.*, in a uniform size. Here we seem to be faced with an old problem—though one that is new in molecular biology —which drove poor Hans Driesch to the philosophical *cul de sac* of "entelechy." But if we regard such an entity as equivalent to a principle governing in typically non-linear physical systems, it no longer sounds so alarmingly vitalist, although we cannot yet predict that this new approach will lead to a real solution, or merely to another "deviation." If I borrow the analogy used by Stent (*13*), vegetational succession seeking the same end of climax *is* also scale-invariant. So, these days we may have been calling Driesch's entelechy variously field (*14*), chreod (*15*), or positional information (*16*), of which the last is now being linked to such sophisticated mathematical/physical concepts as diffusion-reaction systems, catastrophe theory, dissipative structures, or bifurcation theory—all typically non-linear.

I have learnt that even in this highly competitive era, the open co-operation and fellowship among developmental biologists working with hydra—a typical scale-invariant system—is somewhat reminiscent of the early days of the phage group, in a real contrast to the rampant secrecy and commercial approach among some of the entrepreneurial "molecular biologists" of the present day (*17, 18*).

But all this raises a considerable problem for our understanding of science in general. We have come to learn that the optimism of molecular biologists during the early sixties was premature and even wrong, because they did not realize that they had failed to sense the existence of a vast area which still remained for future investigation. The "progress" of science since then has revealed their unwitting but inexcusable ignorance. Therefore, as long as we expect progress of science in the future (which must be an acceptable argument to everybody), we should admit if we are frank that what we now believe to be true might also be wrong—and to an unknown degree. We had better keep that in mind, and also the *actual* experience of molecular biology—rather than its vulgarized version—whenever we speak out about science and the scientific aspects of whatever faces our society. This may be a financially devastating attitude, but it is more responsible, truthful, and in the real tradition of our learning and culture.

Acknowledgment

I am grateful to Drs. David R. Oldroyd and Graham A. Rockwell for their help in improving the manuscript.

REFERENCES

1. Stent, G.S.: 1979. *Q. Rev. Biol.* 54, 421–427.
2. Judson, H.F.: 1979. *The Eighth Day of Creation*, Simon and Schuster, New York.
3. Sekiguchi, M.: 1976. *Kakubutsu Chichi (Investigating things and refining knowledge)*. Festschrift for Prof. Watanabe, I., pp. 165–166, Tokyo (in Japanese).
4. Sibatani, A.: 1955. *Seibutsu Kagaku (Biological Sciences)* 7, 11–16 (in Japanese).
5. McLaughlin, A.: 1979. *Telos* 41, 189–200.
6. Kuhn, T.S.: 1962. *The Structure of Scientific Revolution*, Univ. Chicago Press, Chicago.
7. Sibatani, A.: 1978. *Seibutsu Kagaku (Biological Sciences)* 30, 175–176 (in Japanese).
8. Sibatani, A.: 1979. *Trends Biochem. Sci.* 4, N161–162.
9. Sibatani, A.: 1960. *Seibutsugaku no Kakumei (Revolution in Biology)*, Misuzu Shobo, Tokyo (in Japanese).
10. Crick, F.: 1979. *Science* 204, 264–271.
11. Bernardi, G.: 1978. *Trends Biochem. Sci.* 3, N241–242, 252.
12. Bernardi, G.: 1979. *Trends Biochem. Sci.* 4, N162.
13. Stent, G.S.: 1981. *Annu. Rev. Neurosci.* 4, in press.
14. Weiss, P.: 1939. *Principles of Development*, Henry Holt & Co., New York.
15. Waddington, C.H.: 1957. *The Strategy of the Genes*, Allen & Unwin, London.
16. Wolpert, L.: 1969. *J. Theor. Biol.* 25, 1–47.
17. Siekevirz, P.: 1980. *Trends Biochem. Sci.* 5(9), VI, VIII.
18. Bernardi, G.: 1980. *Trends Biochem. Sci.* 5(12), I-II.

The Role of University Scientists in Furthering Understanding between Peoples

P. N. CAMPBELL

P. N. CAMPBELL was born on November 5, 1921 in London and educated at University College London obtaining his Ph.D. degree in biochemistry in 1949 after wartime service. He then collaborated with T.S. Work at the National Institute for Medical Research, Mill Hill, London, first on the factor from agenised flour and then on protein synthesis, particularly of milk proteins. In 1954 he moved to the Courtauld Institute of Biochemistry, the Middlesex Hospital Medical School, London, where he worked on the synthesis of many different proteins in animal cells, serum albumin, cytochrome c, α-lactalbumin. He was also involved in the discovery with I. M. Roitt and D. Doniach of autoimmune thyroiditis. In 1967 he was appointed to the Chair of Biochemistry at the University of Leeds. In 1976 he returned to the Courtauld Institute as Director. Together with R. K. Craig a laboratory of molecular biology has been created with particular emphasis on studies on the mechanism of gene expression in animal cells. He is also involved in the activities of the International Union of Biochemistry and is at present Chairman of its Committee on Education and Publications. He is editor of "Essay in Biochemistry" (with R. D. Marshall) and of "Biology in Profile" and is joint author with A. D. Smith of a new book entitled "Biochemistry Illustrated."

Ever since I was elected Secretary of The Biochemical Society in 1958 I have been concerned with various organizations which aim to encourage the cooperation of biochemists on an international basis. I have certainly enjoyed my part time work and especially the friendly relations that I have with so many biochemists throughout the world. It is therefore a pleasure and an honour to respond to the invitation to participate in the celebration of the 70th birthday of Professor Egami whose Institute I visited briefly in 1977. This was one of the highlights of my many enjoyable visits to Japan.

The present invitation included the suggestion that I write on the relation of science to politics. Such a subject is indeed a challenge but I thought perhaps it might be of interest if I tried to knit together some thoughts on how we, as university scientists, may play a modest role in increasing understanding between scientists and between scientists and politicians in our own countries and internationally.

The first thing we want from our politicians is that they create conditons under which the universities can flourish and fulfil their functions as centres of learning. Politicians throughout the world now realize that students are a political force and to some extent must be provided for. All governments therefore spend money on the erection of university buildings and usually devote considerable sums to the payment of the salaries of university teachers. The total salary bill in a University often accounts for about 85% of the total budget of a University. The politicians seem less interested in getting value for money from the Universities and indeed very often they unwittingly create conditions which lead to trouble. I refer to the erection of campuses far out of town which have no living accommodation for the students and very limited transport services. The entry standards to such universities are often low and sometimes non-existent and the teaching is often bad. It is not surprising if, under these conditions, the students turn to politics so that eventually the pot boils over, the army invades the campus, a student is killed and real trouble erupts.

The kind of situation I describe is extremely serious for any country but especially so if the country has limited resources. Too often one feels that the intellectual seed bed of a country, in the form of its youth, is locked away in unproductive activity. We as scientists should spare no effort to point out to the politicians the necessity for them to limit the number and size of their universities to what can be afforded and then to provide not merely salaries but the running expenses to enable productive work to proceed.

Perhaps almost more important than the immediate impact of a government's policies on the universities is the extent to which people are able to exercise their right of free speech. I realize that even in a truly repressive regime it is possible to effectively educate people to a limited extent in institutions of higher education. This would surely apply to the more technical aspects of education. It would be my view, however, that the creation of truly original ideas of the kind that a nation has a right to expect of its universities will be severely hampered under a repressive regime. I certainly had that impression when visiting Iran in the days of the Shah. There seems little that the scientists in such a country can do to correct the situation except to struggle to maintain their integrity and hope for better days. Under such circumstances it seems to me more important than ever that we, who are in a more fortunate position, should extend the hand of friendship in order to maintain the morale of those in trouble.

What of our interactions with our politicians in a country which has "properly" funded universities and where there is freedom of speech. The University budget is a significant proportion of the GNP of such a country so that the funding of universities is now of political interest especially since the immediate beneficaries are a minority of the voters. Politicians are sustained by votes and voters are thought to have a rather short-term view of life.

As scientists we at least have an entree to politicians on the basis that science is good for the future development of the country. Not many politicians have a scientific background and those that have may have escaped from it because they decided it was not for them. The message is that we scientists should take every opportunity to enlighten our politicians, being careful to moderate our enthusiasm for our subject with reality. We can sometimes make an indirect approach to politicians through the media of radio or T.V. for we might then influence the voters. Biochemists certainly face a difficult challenge in explaining their science in simple terms but at least genetic engineering has raised the level of interest and we should accept the challenge whenever it is offered.

I would like to refer briefly to party politics and the role of the Universities. In the U.K. we are careful to try to exclude political parties from our activities within the Universities and I believe this to be an important matter. In science we try to be careful to either avoid discussing political matters with our staff and students or at least to do so in a rational manner. In most Universities we have a rule that members of staff are not allowed to write to the newspapers on political matters using the address of the University. I have noticed that in countries that have other customs in these matters that serious trouble can arise between the Universities and the State. There have been cases *e.g.* in Chile, where the Universities formerly ran T.V. stations which were politically biased and then complained when the new regime interfered with the Universities. In short, party political activities in Universities should be confined to the students.

I would now like to move to the role of scientists in furthering international understanding. I first became deeply involved in this area when The Biochemical Society was trying to create cooperation among European biochemists. FEBS (The Federation of European Biochemical Societies) emerged from these efforts. The main impetus behind international collaboration in science must be to advance our science by exchanging ideas and avoiding the unnecessary repetition of experiments. There is however an important bonus to those who believe that true understanding between individuals is based on mutual respect. In science, and nowhere I believe is this more true than among biochemists, we have common objectives and our conclusions are based on well tried scientific principles. This means that irrespective of our background, language, colour or religion we have a mutual respect for one another as scientists. Our hope is that this respect may be transferred to the wider fields of human activity and that eventually scientists in many parts of the world may have an influence for good. I sometimes think that this is a naive thesis but every now and again my hopes are reinforced.

In the early discussions on the formation of FEBS which took place in England in 1962–63 we were determined not to be involved in politics. It soon became apparent that we had failed in that respect for there was the problem of the two Germanies. At that time the Federal Republic (GFR) in the West did not recognize the Democratic Republic (DDR) in the East. The problem was eventually resolved by agreement that the new Federation should be one of Societies, and not countries so that it was legitimate for one country to have two biochemical societies and for these to be located in the western and eastern parts of the particular country. (This concept has been useful in the more recent discussions concerning the admission of China to IUB). As a result of these lengthy discussions which were marked by the determination of the biochemists not to be submerged by the political problems FEBS was safely launched in London in March 1964. In April 1965 the second meeting was held in Vienna. I well remember sitting on the steps of the University on that occasion and talking to some young biochemists from the DDR who had travelled to the West for the first time in their lives. I sensed that it had all been worthwhile.

Later there was the problem of the admission of Israel. Was it really part of Europe? for otherwise Israel was not admissible. We agreed that there were some fine biochemists in Israel and that they could not belong to any other regional organization. The Council therefore decided that Israel was part of Europe and so Israel was admitted to FEBS. I was surprised later when visiting the Weizmann Institute in Rehovot to be told by the President, who was a physicist, that the physicists were delighted at the decision of the biochemists. I asked the reason and was told that the biochemists had established the principle that Israel was a part of Europe and all the other scientific disciplines were following.

In more recent times I have been encouraged by the spirit of cooperation among the younger scientists in FAOB (Federation of Asian and Oceanian Biochemists). Japanese biochemists have played an important and enlightened role in encouraging this cooperation. At least there are no old inhibitory traditions to be pushed aside in most of the countries of FAOB. I often hear it said that FEBS of course was easy to launch whereas FAOB is difficult. I certainly take the point about the various difficulties of FAOB but it is simply not true that FEBS was without its difficulties.

A problem that sometimes arises with international cooperation is the role of Governments. If a large congress is to be organized it is often necessary for the local biochemists to seek financial aid or at least sponsorship from their Government. For reasons I have already alluded to this aid will not usually be provided entirely without strings. The Government may well calculate that it will gain prestige by its support of a congress of distinguished scientists. As a result the congress may well be opened by a Minister of the Government. Even if the Minister steers clear of any remark that could be offensive to an audience made up of visitors from all over the world, the scientists who attend the congress must be getting close to the point where they are engaged in politics. For this reason I feel that scientists should curb their enthusiasm to have a direct participation of a Minister of the Government in their activities. We should remember that whereas it may be harmless in some countries, a precedent is being created which will make it harder to object on a future occasion when matters have clearly reached danger point.

All of us who try to take a serious interest in world affairs are bound to be fearful of the political and economic instability that exists to-day in many countries. The position indeed seems to get worse and not better so that the number of countries which are now virtually closed to IUB seems to grow daily. We can only go on naively as biochemists rejoicing in the cooperation that has been built up over the last 20 years. It is surely a tribute to our friendship that many biochemists must feel that if only our countries were run by biochemists the world would be a better place.

As I have mentioned before I am delighted that the Japanese biochemists have been so active in encouraging FAOB. I can only hope that they will continue their efforts realizing that patience is required but that in the end the fruits of cooperation in Asia will be very rewarding.

Ethics and Evolution in Boltzmann's and Einstein's Thought

E. BRODA

E. Broda was born in Vienna in 1910. He studied chemistry at the University of Berlin under M. Bodenstein and P. Günther, and of Vienna under H. Mark. He left Austria for Britain after the occupation by Nazi Germany. There he worked first on the physical chemistry of vision, and later he did radiochemical research for nuclear energy. After his return to Austria in 1947 he joined Vienna University, where he is Professor of Applied Physical Chemistry and Biophysical Chemistry. He now mainly works on bioenergetics, cellular ion transport and prokaryotic evolution. He also serves as official expert on problems of nuclear energy, especially, radiation protection. Further he is actively interested in solar energy, science history, social affairs, and world peace. He is now the Chairman of the Austrian Pugwash Group.

Fujio Egami's extraordinary merits as an evolutionary biologist have now also been recognized through his election as President of the International Society for the Study of the Origin of Life. On the other hand, Egami has also devoted himself to problems of society and of the social responsibility of the scientists. So it is hoped that Egami will welcome the dedication of an essay concerned with the views of two of the greatest thinkers of mankind on the origin of morality: Ludwig Boltzmann and Albert Einstein. Both these giants of the mind at later stages of their lives gave deep thought to important matters far outside their original field, physics.

Boltzmann's (1) main area in physics was atomistics. In his days the concept of the atom was still attacked by eminent physicists and chemists, including Ernst Mach and Wilhelm Ostwald. Boltzmann showed that the Second Law of Thermodynamics is to be derived precisely of the basis of atomistics. Moreover, Boltzmann applied the Second Law to photosynthesis. His majestic words (2) were:

"The general struggle of the organisms is not a struggle for the elements—the elements of all organisms are abundant in air, water, and soil—nor for energy, which in the form of heat, unfortunately unchangeably, is contained in all substances, but a struggle for the entropy (more precisely: negentropy. E.B.), which becomes available in the transition of the energy from the hot Sun to the cold Earth. To exploit this transition as far as possible the plants spread out the immeasurable area of their leaves and force solar energy in a way as yet unexplained, before it sinks down to the temperature level of the Earth, to carry out chemical syntheses of which one has no inkling as yet in our laboratories. The products of this chemical kitchen are the object of the struggles of the animals."

Thus Boltzmann was one of the founders of bioenergetics.

Furthermore, Boltzmann was a champion of evolutionist thought, and called the 19th century that of Darwin (3). Long before Oparin (1924) and Haldane (1928) Boltzmann (4) proposed chemical evolution in 1904:

"Here it does not matter to us whether during the millions of years in the enormous mass of water on Earth the first protoplasm evolved "'by accident' in moist mud, whether egg cells, spores or other germs, as dust or embedded in meteorites, some time arrived on Earth from space. More highly developed individuals will hardly have fallen from the skies. Thus at first there were only quite simple individuals, simple cells or lumps of protoplasm. Constant movement, socalled Brownian movement, is shown, as is well known, by all small lumps; growth through absorption of similar components and subsequent proliferation by division is fully imaginable mechanically. Similarly it can be understood that the rapid movements were influenced and modified by the environment. Lumps where modification took place in the sense that on an average (preferentially) they moved towards substances better suitable for absorption succeeded better in growth, and more often in propagation. Hence they soon overran the others."

We shall see that Boltzmann also attempted to derive the human ethical and esthetical senses on an evolutionist, Darwinian, bases.

In physics, Einstein has been called the natural successor to Boltzmann by one of his closest collaborators (5). Indeed all his early work refers to molecular statistics. Two of the three major papers in his "wonderful year," 1905, deal with atomistics in the spirit of Boltzmann, one with the explanation of Brownian movement (6), and the other with the idea of the light quantum (7). Through the hypothesis that electromagnetic radiation is composed of discrete packages of energy Einstein extended atomistics from substances to fields. Now all matter, in the philosophical sense, is composed of discrete particles.

Life owes its continued existence to Einstein's light quanta, and in this sense Einstein might also be claimed as a biophysicist. Photosynthesis is based on photochemical reactions. Now each photochemical step requires a definite minimum quantum energy of the light. If the energy within a electromagnetic field were increasingly diluted with increasing distance from the light source soon not enough energy would be available locally to excite the molecules. Thus the quality of photochemical reactions would depend on light intensity, and at low enough intensity no reaction could occur at all. However, according to Einstein the energy of the quanta does not depend on source strength or distance.

While young Einstein had been influenced by Ernst Mach's positivist philosophy, he later adopted an epistemology akin to the realism-materialism of Boltzmann (1). This has been explained elsewhere (8–11).

At first Einstein had concentrated on physics, but after the shock of the First World War he devoted himself increasingly to public affairs, to the struggle for peace and democracy (12). Thus he became the target of vicious attacks and had to emigrate from Germany in 1933.

In 1939, Einstein's revulsion from Nazism moved him to rewrite and sign the warning letter of Roosevelt about the danger of nuclear warfare that had been submitted to him by his friend Leo Szilard. While Szilard and Einstein did not propose the construction of nuclear weapons (13), the letter did lead to terrible developments. After in 1945 it had turned out that Germany had not succeeded in building nuclear weapons, Szilard and Einstein made frantic attemps to prevent the use against Japan; but it was too late. Possibly the gain of the USA of some months by the Einstein letter made possible the disasters of Hiroshima and Nagasaki. A moral evaluation of Einstein's action in 1939, when a Nazi nuclear weapon appeared possible, is difficult. Einstein regretted his letter bitterly (12, 14, 15).

What were the roots of Einstein's moral convictions? While he called himself "religious," he did not accept the idea of a personal God (16) or the validity of any established religion. He also dissociated himself from the clergy when talking of his "religion" (17).

Einstein tried to define what for lack of a better expression (17) be called religion. In the 1920's he sent a telegram (18) to a New York rabbi who had worried about the convictions of the now-famous Jew:

"I believe in the God of Spinoza who reveals himself in the lawful harmony of existence, not in a God concerned with the actions and fates of humans."

In his last period, Einstein (19) wrote similarly:

"My idea of God is formed by that conviction of a superior reason, tied to deep feeling, which reveals itself in the recognizable world. In the usual terminology it can be called pantheist. Denominational traditions I can consider only historically and psychologically; I have no other link with them."

Spinoza has often been considered as a philosophical materialist (20), e.g., by F. Engels (21). It may be held that the difference between Boltzmann's and Einstein's views lies largely in terminology and emphasis (22).

Yet no ethical rules can be derived from mere contemplation of the physical laws, however much harmony one sees in them, as Einstein did. The world is the home not only of the good and beautiful. In the same world the Nazis operated the gas chambers, the American military employed the nuclear weapons and the Shah tortured the Persian democrats. How can one conclude from mere contemplation of physical laws that there is anything wrong about all that?

Einstein (16) sought an answer by referring to the teachings of great spiritual leaders, including founders of religions. In 1941 he wrote:

"And if one asks whence derives the authority of such fundamental ends, since they cannot be stated and justified merely by reason, one can only answer: they exist in a healthy society as powerful traditions, which act upon the conduct and aspirations and judgements of the individuals; they are there, that is, as something living, without its being necessary to find justification for their existence. They come into being not through demonstration, but through revelation, through the medium of powerful personalities."

Demokritos, Buddha, and Spinoza are mentioned in the context specifically (16, 23). But here again the question is begged. Why are we to accept the rules laid down by these spiritual leaders, but not rather those, say, of superstitious priests or of Adolf Hitler? By which criterion do we distinguish the right and the false prophets?

The trouble stems from Einstein's static world view. He did not connect morality with the changing fates and needs as conditions developed, often in a contradictory way. Neither animal evolution nor human prehistory or history had a place in Einstein's thought. Of course, Einstein accepted the truth of Darwin's ideas; but he cared little. Einstein did not see the need for the development of morality in the struggle for survival. By the way, this also applied, in a rather striking way, to Einstein's colleague as a physicist-philosopher, and esteemed correspondent Moritz Schlick, the head of the Vienna Circle, who was murdered in 1936 by a right-wing fanatic. Schlick's (24) approach to ethics was socially-minded, but formalistic, and therefore unsatisfactory.

The Darwinian Boltzmann had derived all structure and function of organisms through evolution. This applied also, as stated in 1897, to the brain and its mental functions (25):

"The brain is considered by us as the instrument, the organ for the production of world pictures, which because of their great utility for the preservation of the species according to Darwin's theory developed to particular perfection in man, as in the giraffe the neck, and in the stork the beak developed to unusual length."

Further, in 1904 he said (*26*):

"These laws of thinking have formed according to the same laws of evolution as the optical apparatus of the eye, the acoustical apparatus of the ear, the pumping device of the heart. In the course of the evolution of mankind anything unsuited has been left behind."

Ethics were likewise treated from an evolutionist point of view. Actions qualified as good if they furthered the survival of the species. Another morality could not have maintained itself among animals or men. Thus Boltzmann did not seek the roots of ethics in conscious thought, as do positivists, or in religious tradition. In 1897, Boltzmann (*27*) said:

"The origin of violent feelings of pain and joy is explained by Darwin's theory, as they are needed to obtain the reaction energy required for the survival of the species......Noble and lofty is for our subjective feeling whatever furthers and exalts our species. These ideas have no objective existence......"

And in his anti-Schopenhauer lecture (*28*) in 1905:

"Ethics must ask when the individual may maintain his own will, and when he must subordinate it to that of others, so that the existence of the family, the tribe, of mankind—and therefore of all together—be furthered as much as possible...... If morality of any kind had the effect that the tribe following it decays, it would be refuted thereby. In the last analysis not logic, not philosophy, not metaphysics decides whether something is true or untrue, but action (practice. E.B.)."

Thus values are judged according to their role in respect to evolving life. Similar ideas had been applied to esthetics by Boltzmann (*29*) in 1900:

"The explanation of the wonderful beauty of flowers, of the richness of forms in the insect world,......all this herewith becomes the domain of mechanics (natural science. E.B.). We understand that it has been useful and important to our species that certain sensations flattered us and were sought by us, other repelled us......Thus we can explain mechanically the origin of the concept of beauty as well as of that of truth.

We understand, too, why only individuals could continue to exist who abhorred with all intensity of their nerve power and tried to prevent certain highly pernicious influences, but strove with equal nerve power towards others that were needed for their own preservation or that of the species. Thus we understand how the whole intensity and power of our emotional life evolved, joy and pain, hatred and love, desire and fear, delight and despair."

Boltzmann concluded: "The God, by whose grace kings rule, is the fundamental law of mechanics (*30*)."

Thus it appears that Boltzmann, through his evolutionism, had found a sound basis for ethics. His failure in the application to the concrete reality of human society surely is due to his aloofness from politics. Einstein, on the other hand, during most of his life indefatigably worked for progressive causes. But in his admirable activity he was largely guided by mere feeling, without the benefit of an evolutionist, rational basis.

Yet there is a serious weakness in Boltzmann's view. He never explained the historical character of the human ethical (and esthetical) sense. From his writings one would guess that all mankind subscribes to the same moral values, and has always done so. No account was taken of the obvious differences between the moralities dominant among different kinds of society in history. Let us compare ancient Egypt, India, or Japan, the medieval Moslem world, Aztec Mexico and the modern capitalist society. Clearly the tremendous differences in moral (or esthetic) convictions are not determined genetically or inherited, but they developed in social history.

REFERENCES

1. Broda, E.: 1955. *Ludwig Boltzmann*, Deuticke, Vienna.
2. See Ref. 1, p. 71.
3. See Ref. 1, p. 104.
4. See Ref. 1, p. 122.
5. Lanczos, C.: 1964. *The Einstein Decade*, p. 56, Elek, London.
6. Einstein, A.: 1905. *Ann. Phys.* 17, 549–560.
7. Einstein, A.: 1905. *Ann. Phys.* 17, 132–148.
8. Einstein, A.: 1949. *Albert Einstein, Philosopher-Scientist* (Schilpp, P.A., ed.) p. 21, Open Court, La Salle.
9. Broda, E.: 1973. *The Boltzmann Equation* (Cohen, E.G.D. and Thirring, W., eds.) Springer-Verlag, Vienna.
10. Broda, E.: 1979. *Einstein Centenarium 1979* (Treder, H.J., ed.) Akademie-Verlag, Berlin.
11. Broda, E.: 1980. *Einstein und Österreich, Österr. Akad. Wiss.*, Vienna.
12. Nathan, O. and Norden, H.: 1960. *Einstein on Peace*, Simon and Schuster, New York.
13. Broda, E.: 1980. *Phys. Bl.* 36, 86.
14. Clark, R.W.: 1974. *Einstein, Leben und Werk*, p. 400, Bechtle, Esslingen.
15. Herneck, F.: 1976. *Einstein und die Atombombe*, No. 51, p. 6, Archenhold-Sternwarte, Berlin-Treptow.
16. Einstein, A.: 1950. *Out of My Later Years*, p. 21, Thames and Hudson, London.
17. Einstein, A.: 1956. *Lettres à Maurice Solovine*, pp. 102, 114, Gauthier-Villar, Paris.
18. Hoffmann, B. and Dukas, H.: 1973. *Albert Einstein, Creator and Rebel*, p. 95, Hart-Davis McGibbon, London.
19. Einstein, A.: 1955. *Mein Weltbild*, p. 171, Ullstein, Berlin.
20. De Vries, T.: 1970. *Spinoza*, pp. 113, 145, Rowohlt, Hamburg.

21. Thalheimer, A. and Deborin, A.: 1928. *Spinozas Stellung in der Vorgeschichte des dialektischen Materialismus*, Verlag für Literatur und Politik, Vienna.
22. Broda, E.: 1980. *Comp. Biochem. Physiol.* 67B, 373–378.
23. Einstein, A.: 1955. *Mein Weltbild*, p. 15, Ullstein, Berlin.
24. Schlick, M.: 1930. *Fragen der Ethik*, Springer-Verlag, Vienna.
25. See Ref. 1, p. 106.
26. See Ref. 1, p. 109.
27. See Ref. 1, p. 125.
28. See Ref. 1, p. 121.
29. See Ref. 1, p. 127.
30. See Ref. 1, p. 131.

The Predominance of English and the Potential Use of Esperanto for Abstracts of Scientific Articles

RALPH A. LEWIN and DAVID K. JORDAN

R. LEWIN was born in London. In Cambridge, England, he read natural sciences and obtained B.A. and M.A. degrees. He went to the USA in 1947, and in 1950 obtained a Ph.D. degree at Yale University. In 1973 he was awarded an Sc.D. degree from Cambridge. His scientific interests have centred around experimental phycology and marine microbiology. He has specialized in the genetic and physiological control of flagellar activities in *Chlamydomonas*; gliding bacteria (flexibacteria); and a "new" class of algae, the Prochlorophyta. He wrote a book of poems entitled "The Biology of Algae and Other Verses" (1978). With I.K. Reed he translated a children's book, "Winnie-the-Pooh" by A.A. Milne, into Esperanto: it was published in 1972. His interest in the international language began in his boyhood, when he learned Esperanto from his parents. Some 10 of his scientific articles were published in Esperanto.

The General Conference authorizes the Director General to follow the current evolution in the use of Esperanto in science...(Resolution of the General Conference of UNESCO, Montevideo, 1954)

There are many secondary, personal advantages of writing and publishing scientific papers: payment, pride, promotion, even a measure of immortality. But the most obvious and immediate goal of the enterprise is to let other researchers

---

* This photograph shows the authors of this article, Lewin (centre) and Jordan (right) with Dr. Stephen Zamenhof (left), a molecular biologist and a nephew of Ludwig Zamenhof, who created Esperanto about 100 years ago. (Photo by L. Cheng)

know what one is doing and what one has discovered. Accordingly, as a general rule, the more of a scientist's colleagues who can read and understand his publications, the better. This means he ought to write clearly in a language which both he and they know well. But which language?

English as the Language of Science

L atin and Classical Chinese, the *lingue franche* of the two gigantic intellectual communities of Europe and East Asia, have now passed into disuse. In both areas, if anything has replaced them as an international language it is English, which has become the language of science so universally that it seems without peer in the history of our planet. The reasons have little to do with any inherent merits of English itself. It is not easier to learn or to use than other languages—indeed it may be harder than most—and it is not any more logical or less ambiguous than the others. Its present position seems to derive primarily from two facts. First, it is the language of England, home of the Industrial Revolution and seat of the once extensive British Empire. Second, it is the language of the English-speaking diaspora, and particularly of the United States, a country blessed with natural resources in an era when growth of industry made it easy to turn them into wealth.[*1] Whatever the historical details, whether or not English retains its premier position for very long into the future, and whatever its cost (2), it is evident that today more scientific communication—articles and lectures—takes place in English than in any other language is the world. This point was emphasized when an eminent European biologist, addressing colleagues during a visit to the United States, prefaced his heavily accented lecture by announcing that he would deliver it in "the international language of science: broken English." There is much truth in this modest joke.

By way of example of the anglophone[*2] domination of the scholarly world, Table 1 reproduces the findings of a study developed by Clyde Thogmartin (3) to help American students of the social sciences in selecting languages for study to meet university foreign-language requirements. Thogmartin reports that he used UNESCO's *International Bibliography of the Social Sciences* as his data base, fearing that an American (and hence largely anglophone) abstracting service might tend to misrepresent the real distribution of articles by omitting most non-English items. (He was probably right in this hunch. His survey of 2,000 random articles in *Psychological Abstracts*, an American service, included a total of only 1.7% in languages other than English.) Although the table does suggest which foreign languages would

---

[*1] For a fascinating series of papers on the spread of English in the modern world (see Ref. *1*).

[*2] Etymologically, the word "anglophone" would seem to refer specifically to people who *speak* English, and we ought to coin another word—presumably *anglograph*—to refer to those with the ability to read and write English, which is what we are especially concerned with here. As far as we can tell, however, "anglophone" is already used in English (and French) to include the sense of reading and writing.

TABLE 1. Language of Publication of Titles in Four Social Sciences (Source: 3)

Sociology (1976) Sample: 1,100 of 4,827 Titles (22.75%)			Economics (1976) Sample: 1,100 of 6,851 Titles (16.05%)		
Language	n	%	Language	n	%
English	509	46.3	English	423	38.5
French	157	14.3	French	182	16.6
Russian	124	11.3	Polish	130	11.8
Japanese	78	7.1	German	106	9.6
German	63	5.7	Russian	88	8.0
Hungarian	36	3.3	Spanish	48	4.4
Italian	35	3.2	Italian	34	3.1
Spanish	33	3.0	Others	89	8.1
Political science (1976) Sample: 1,000 of 4,253 Titles (25.86%)			Anthropology (1975) Sample: 1,000 of 7,060 Titles (15.58%)		
Language	n	%	Language	n	%
English	513	51.3	English	469	46.9
French	163	16.3	French	260	26.0
German	124	12.4	German	101	10.1
Spanish	65	6.5	Russian	78	7.8
Italian	56	5.6	Spanish	60	6.0
Russian	28	2.8	Italian	18	1.8
Others	51	5.1	Others	14	1.4

be most useful for American students in the four social sciences listed, the most striking fact that leaps from its columns is that there is no language even remotely as useful as English in the published literature of the social sciences, whether one is an American student or not.

This situation suits English-speaking people very well, including the writers of this article. We were both born and raised in English-speaking countries, and we have little difficulty handling the language effectively. But for non-anglophone scientists, the situation may be very different. Sooner or later in their lives, most of them must take English classes some three to five times a week for five to ten years if they are to achieve a reasonable familiarity with the language and a tolerable facility in its use. Meanwhile, for the same three to five hours a week (and the time it takes to do the homework) over the same five to ten years, we native speakers are free to spend our time studying more science or mathematics, or reading literature, or even playing cricket or baseball. At least in theory, we have a linguistic advantage which can help to keep us ahead in our scientific productivity or enable us to lead fuller cultural lives. This is not fair, but it is so.

Things are worse than that, however. Not only is mastery of the English language time-consuming, it seems also to be almost unattainable for many people. It is certainly true that many Europeans—Scandinavians and Netherlanders particularly come to mind—manage, by a combination of good schooling, intellectual discipline, and the fact that they already speak Germanic languages, to achieve a satisfactory mastery of English, sometimes even at the expense of their own mother tongue (as many Swedes now complain). Some can write scientific articles in good English. But for far more scholars, especially outside Europe (as in Asian nations where native languages are not Indo-European), the burden of learning English is for greater and the problem of communication far more serious. Yet when Asian scientists, say, publish in their own languages, their findings fail to reach the large international audience that English articles so readily reach. For instance, the plant hormone gibberellin was known and studied in Japan for several years before most plant physiologists in the West learned of its existence. The main reason was that early publications on the subject were almost exclusively in Japanese. However, when Japanese scientists have felt the need for including at least an abstract in a European language, the necessary linguistic expertise has often been unavailable.

Indeed, so difficult is our language for many Japanese that some native speakers of English used to make unsympathetic jokes about the inscrutability of the instructions that came with Japanese industrial products. They do so less often now; the linguistic standard used in such texts has recently been much improved. But in the 1950's and 1960's, abstracts of Japanese scientific papers, ostensibly in English, were so full of misused synonyms and bizarre syntax that they were frequently unintelligible to a native speaker of English. Some Japanese journals now have English-language editors, and examples of "Anglonipponic" are disappearing from the scientific literature. (Now it seems to be especially Chinese scientists who experience the intellectual discomfort of this gap between their desire to reach the wider scientific world that reads English and their own ability to express themselves in this language.)

The importance of English in reaching an international scientific audience is so great, however, that there has so far been no alternative but to persist. A few journals issued in Japan (such as *Plant and Cell Physiology*, the official organ of the Japanese Society of Plant Physiologists) publish articles in English (and occasionally also in other Western languages), but do not even accept articles written in Japanese. A similar situation exists in Germany, where some publications formerly published predominantly in German now use English almost exclusively. Thus the *Archiv für Mikrobiologie* has become the *Archives of Microbiology*, and the *Zeitschrift für Induktive Abstammungs- und Vererbungslehre* has turned into *Molecular and General Genetics*. The editorial board of the French journal *Biologie Cellulaire* "recommend that English be used since it is a language which is understood by the whole of the scientific community and thus permits a wide distribution of articles." And for *Folia Biochimica et Biologica Graeca*, which bears a Latin name but is published in Greece, neither Latin nor Greek contributions are acceptable: articles have to be in English, French, or German.

One can only guess how much effort and money it costs such journals to employ anglophone editors, but the expense is apparently considered worthwhile, since articles in English are much more widely appreciated than the same articles would be if published in Japanese or German. Much of the pressure to use English seems in fact to come from the writers of articles themselves. We have been told by both German and Chinese scientists that for their really important publications they use English whenever possible. The editor of one international journal of public policy told us that, in order to retain the multilingual character of his journal, he found it necessary to translate contributions in English back into the languages of the countries from which he had received them. (For example, an Italian scholar would send him an article in English, and, because his journal was nominally multilingual, the editor would request permission to have an Italian version made.) This, he explained, was the only way to prevent the journal from "deteriorating" into an all-English one.

The Problem

It seems obvious that a scholar's scientific efficacy must be gravely affected by his ability to understand the technical literature of his field of specialization. We often have the uneasy feeling that a given piece of research might have been improved if the author had read such-and-such a work in a language we suspect him of not knowing. However, it is difficult to compile objective data. There are many reasons other than a writer's linguistic deficiencies for his not citing a given article. Perhaps he thinks the work is not worthy of attention. Perhaps his library has not yet received a copy of it. Perhaps it was published in an obscure journal, or one where he would not think to look for it. Perhaps (as is often the case in the humanities and social sciences) the article belongs to a tradition of research different from that in which he is working.

Such considerations are probably less prevalent in the natural sciences. Accordingly we have compiled a few figures from the recent periodical literature in phycology to determine the extent to which authors writing for one journal or another may cite the phycological literature differentially. Phycology, the study of algae, was chosen because one of us is a professional phycologist with some experience in the literature on that subject, but the choice is not inappropriate for present purposes. Algae, widely distributed throughout the world, recognize no political frontiers. Freshwater algae, especially those that form drought-resistant spores, can be blown freely from lake to lake and nation to nation, and in addition they may be transported across continents on the feet of migratory waterbirds. Marine algae, almost by their very nature, are international because they occupy the high seas or the seashores along continental margins. So *a priori* one would not expect bibliographies of papers on algae to show a strong linguistic bias apart from the language limitations of the authors or of the technical libraries to which they have access.

In several major journals published recently in different countries and languages, we counted citations in different languages in order to assess how much

TABLE 2.   Distribution of Citations across Languages in Selected Phycology and Marine Biology Journals

Journal and country	Number of articles	Number of citations	Number of articles in various languages							Percent in English
			Eng.	Fre.	Ger.	Rus.	Jap.	Chi.	Latin	
1. USA	77	2,041	1,867	58	101	2	1	0	8	91.5
2. France	22	321	184	94	33	0	1	0	4	57.3
3. Czecho-slovakia	20	514	299	27	140	18	0	0	3	58.2
4. Germany	25	614	321	80	136	4	1	0	24	52.3
5. India	21	412	369	11	24	0	0	0	5	89.6
6. Japan	24	243	129	16	32	0	61	0	4	53.1
7. USSR	26	325	92	7	4	218	0	0	0	28.3
8. China	25	1,030	762	73	54	14	29	59	28	74.0

Small numbers of articles in other languages, not listed in this table, prevent the totals being equal to the sum of the listed citations. Some of these languages are Dutch, Spanish, Italian, Czech, Serbo-Croatian, Polish, and the Scandinavian languages, including Icelandic.
Names of journals: 1. Journal of Phycology, 1979. 2. Revue Algologique, 1979. 3. Algological Studies (= Archiv für Hydrobiologie Supplementband), 1979 (The series is edited at the Czechoslovak Academy of Sciences, but published in a German journal.). 4. Nova Hedwigia, 1979 (Phycology only.). 5. Phykos, 1979. 6. Bulletin of the Japanese Society of Phycology, 1976. 7. Okeanologiya, 1979 (Marine biology only. There is apparently no phycological journal published in the Soviet Union; we have therefore chosen an oceanographic one as being fairly comparable). 8. Studia Marina Sinica, 1978 (There being no strictly phycological journal published in China, we used this as a close substitute; cf. note 7, above.).

(if at all) a phycological author tends to be constrained to limit his review of the literature to papers published or abstracted in his mother tongue or national language.*

That there are linguistic biases is evident from the data shown in Table 2. Articles in the American and Indian journals refer predominantly (90% or more of the citations) to papers or books published in English, whereas in journals published in France, Germany, Japan, and the Soviet Union, the proportion of English citations falls below 60%. The differences are made up largely by articles in the language of the country in which the journal is published: French (29%), German (22%), Japanese (25%), and Russian (67%), respectively. In our samples, items in Chinese are cited exclusively in the Chinese journal, and items in Japanese are almost never cited in European journals.

Although these statistics are of course subject to many kinds of criticism, the pattern that emerges is very clear: in general, we read only what we can read easily, tending to by-pass what we cannot read easily even though it might be relevant to our subject. Scholars who use English natively appear to prefer not to read or refer to articles in other languages, and clearly scientists of other language backgrounds have similar preferences.

The hegemony of English may be unfair. Table 2 suggests something about

---

* Differential availability of journals in different libraries is also a factor. However librarians we have spoken to agree that libraries seek to subscribe to journals that their patrons will use, and that one consideration of this is the language of a given journal. Accordingly language still remains a dominant factor.

how (im)practical it is. Having English in its present premier position appears to be better than having the scientific literature spread yet more widely among modern languages, but worse than it would be if all scientific literature were accessible in a single language well known to everybody. But fair or unfair, good or bad, the domination of science journals by English is a fact of life. In fact, this domination is probably even increasing, and the retitling of German journals, mentioned earlier, can be taken as a sign of the times.

The Future

It is reasonable to consider how long this anglophone hegemony is likely to last. Given the difficulty of learning English, can we expect modern school systems, devoted to universal education, to be very successful at teaching it? Given the new nationalisms that have tended to grow up after the decline of the British Empire, can we expect the new nations that once made up that Empire to give as much attention to teaching English as they have in the past? Given the shift of wealth from the Western nations to the Middle East, can we expect that research and education will continue to be so dominated by the languages of those formerly richer nations? If we are truly seeing the beginnings of a huge and viable scholarly community in China, can we really believe that all of its findings will be published principally in English? Frederick Starr (4) has argued convincingly that many of the social supports for the English *oikoumene* are rapidly eroding away even while anglophone intellectuals rejoice in the position of English as the first planet-wide *lingua franca*. If Starr is right, we may expect changes in the position of English during the next few decades. What alternatives are there? Is it in our power to make a rational choice among them?

One alternative is Esperanto, deliberately created late in the Nineteenth Century to serve as an international auxiliary language. Since one cannot communicate in Esperanto if one has nobody to communicate with, realizing its potential as a medium of communication has required the slow development of a community of speakers. Like the telephone or the metric system, Esperanto was, at its beginning, full of promise but so little known that it was more toy than tool. This is still true to some extent. Yet the telephone has become increasingly important to modern life as it has become more prevalent, and in the course of its more than two centuries of history the metric system has become dominant as more and more people have learned to use it. In both cases the utility of the invention at any given time has been partly a function of the number of people using it. Similarly, Esperanto has been quietly evolving into a medium of common use among members of an ever growing language community that spans nearly eighty nations throughout both the communist and the non-communist worlds. A reasonable guess is that about a million people can now use Esperanto in international contacts. (For a description of this movement, see Ref. *5*. For a guide to more general information, see Ref. *6*.

There have been many proposals for its wider use in the sciences. See, for example, Ref. 7.)

One of the most eminent scientists to use the international language is Fujio Egami. Professor Egami, among his other scientific and cultural achievements, not only is a speaker and writer of Esperanto—he learned the language in 1928 —but has been since 1967 the president of the Japanese Esperanto Institute. The Esperanto translation of Egami's elegant little book *The Search for Life* (8) provides abundant evidence that science and Esperanto can be very much at home with each other. (It is ironic that, at a nuclear level, genes of bacteria, plants, and animals all employ the same language, whereas we closely related bipedal primates have grown up to use thousands of mutually unintelligible communication codes.)

What is important for our present concerns is the possibility that Esperanto could or should evolve into a language for scientific publications, or at least abstracting, in the years ahead. It might at least be used for formal taxonomic descriptions of algae and other plants, which now have still to be published in Latin (9). Its peculiar advantages are that it is easy to learn and that it is politically neutral. These things are difficult to measure, but it seems a common experience that the international language can be learnt with roughly ten to twenty percent of the effort necessary to learn a "natural" language. Because its vocabulary (including the scientific vocabulary) is derived from Western European languages, it presents fewer difficulties for students already familiar with at least one of these (such as English) than for others, but it is easy for everybody. The morphology and syntax, on the other hand, are much less European, and in our experience speakers of European background enjoy no particular advantage in their ability to create grammatical and elegant sentences.

For native speakers of English it is a source of complacent satisfaction that so many scientific articles must now be published and abstracted in English. For others it is a regrettable necessity. A more rational decision could perhaps have been made long ago to elect an easier language for such purposes, so that we all—Americans, Chinese, Japanese, Russians, Thais—could get on with the business of scholarship instead of having to be so involved in the business of language. But things have not worked out that way. Nevertheless, if Starr is right in his intimations that the end of the English-speaking era may lie just ahead, we should not let the opportunity pass us by another time, but rather should consider the possibilities of using Esperanto, at least for abstracts, as soon as possible (10).

We propose continuing on the path Egami has already explored, and making increasing use of Esperanto for abstracts, if not for full texts, of scientific and technical articles. To begin with, Esperanto abstracts could be added to English articles, and both Esperanto and English abstracts could be provided for articles in other languages. This is already being done in a few journals, among the latest of which is the Chinese publication *Entomotaxonomia*. The practice could and, we think, should be extended and adopted generally. Like the pioneers who first used telephones or metric measures, we would benefit little from the Esperanto abstracts at first. But their prevalence would contribute to people's motivations to learn the language, and this would rapidly increase its value.

Language is a subject laden with emotion. Learning languages is slow, difficult, and often unrewarding work. And yet, outside the confines of the Twentieth-Century English-speaking world, learning other languages is one of the heaviest burdens a scientist must bear. As we consider the efforts of our scholarly fellowship in the decades and centuries to come, both compassion and reason dictate that we should make every effort to alleviate that burden. We believe that Esperanto can help us to accomplish this great scientific and humanitarian goal.

Resumo en Esperanto

La angla jam fariĝis la precipa lingvo uzata por eldonado kaj resumado sciencaj. Tamen povo uzi la anglan estas malfacile akirebla, kaj la malpovo de sciencistoj legi kaj verki en nepropraj lingvoj kaŭzas tion, ke esplorantoj ankoraŭ nekonas verkojn kiuj plibonigus ilian esploradon. La hegemonio de la angla kredeble jam atingis sian zeniton. Se tio ja okazis, la problemoj de la scienca resumado povos fariĝi eĉ pli malfacilaj. Oni diskutas Esperanton kiel eblan internacian helplingvon por la sciencoj.

Acknowledgment

We are grateful to Lanna Cheng and G. Michael Phillips for their kind advice and assistance in the preparation of this article.

REFERENCES

1. Fishman, J.A., Cooper, R.L., and Conrad, A.W.: 1977. *The Spread of English: The Sociology of English as an Additional Language*, Newbury House Publ., Rowley, MA.
2. Piron, C. and Tonkin, H.: 1979. *Translation in International Organizations. Esperanto Documents*, n.s. 20, Universal Esperanto Association, Rotterdam.
3. Thogmartin, C.: 1980. *Anthropol. Newslett.* 21 (2), 6.
4. Starr, S.F.: 1978. *Change* 10(5), 26–31.
5. Foster, P.G.: 1971. *Pensiero e Linguaggio in Operazioni* 2, 201–215.
6. Tonkin, H.: 1977. *Esperanto and International Language Problems: A Research Bibliography*, 4th ed., Esperanto Studies Foundation, Washington.
7. Moon, P. and Spencer, D.E.: 1948. *J. Franklin Inst.* 246, 1–12.
8. Egami, F.: 1977. *La Serĉado de la Vivo* (translated from the Japanese by K. Matsuba), Japan Esperanto-Instituto, Tokyo.
9. Lewin, R.A.: 1970. *Taxon* 19, 347–348.
10. Tonkin, H.: 1979. *J. Commun.* 29, 124–133.

Message of Thanks

First, let me extend my heartful thanks to my senior associates and close friends in the academic world for their willing contributions of essays to this project at the request of the members of the editorial committee, who were once my colleagues. Upon reading through all the essays, I was filled with emotion and could not help allowing myself to exult in a sense of happiness at having such friends who share the same domain of science.

I wish to add that the contributors hereto include those who are far more senior than I, not to speak of the many who are around my age and have already retired from their public posts in academia, or are about to retire, and those who are presently active in the forefront of the scientific world. I am deeply touched by the warmheartedness of these senior contributors.

The relations that connect me with them are various. The essays compiled in this volume are from associates whom I became acquainted with through my research activities since 1930's up till now in such fields as sulfate esters and sulfatases, glycoconjugates and related enzymes, inorganic nitrogen metabolism, biological oxidations, bacterial toxins, nucleic acids and nucleases, chemical and early biological evolution, etc.; good old friends from the days when I studied in France; friends I made through many international meetings, including those held in Japan, such as the International Symposium on Enzyme Chemistry (1957), the 7th International Congress of Biochemistry (1967), the 5th International Conference on the Origin of Life (1977) and the 26th International Congress of Pure and Applied Chemistry (1977); friends who came to visit me during those days when I was with the laboratories of the Nagoya and Tokyo Universities and the Mitsubishi-Kasei Institute of Life Sciences, respectively, or conversely, friends at the laboratories abroad who welcomed my visits to them; in addition, those who became known to each other through the Esperanto movement, which is more my second life work than a mere hobby, and Japanese friends who are currently engaged in activities overseas. These many ties of friendship, I believe, have made this publication particularly full of variety.

Looking back upon my research life spanning nearly 50 years, I feel myself most happy, not only because I was able to make the acquaintance of many scientists of prominence as such, including the contributors hereto, but also because I was blessed with qualified co-researchers all the way through. On this occasion of my 70th birthday, a group of people who represent all of my past co-researchers have formed an Editorial Committee and laid down the cornerstone for the project of publishing this volume. I wish to present herewith my warmest thanks to them all, and, in particular, although I cannot find the appropriate words, to thank the Executive Editors, who shared a great proportion of their precious research time to undertake the heavy work of editing for eventual publication.

Fujio EGAMI

Postscript

In celebration of the 70th birthday of our respected teacher, Professor Fujio Egami, we asked each of his acquaintances and friends spread across the world to contribute an essay related to science that could be a kind of review of ongoing or past researches, an outlook on the future, or on science administration, research funds or awards or even on the relations of science with such other areas as human society, industry, medicine, arts, language, *etc.*

In response to our request, the 56 essays which found their places in this volume came in with an astonishing swiftness and smoothness. Not only that, almost all of those who received our request letters gave us their warm responses; some apologizing for unavoidable reasons that prevented them from favoring us with their essays; others making contributions from their recent books or work in commemoration of the occasion instead of writing an essay anew; or others writing their congratulations on Prof. Egami's 70th birthday. All of us who were engaged in compiling this volume were struck by a deep sense of gratitude toward these writers, and at the same time we were deeply impressed again by the breadth and depth of Prof. Egami's acquaintanceship, and above all, we felt it was telling evidence of his fascinating personality.

Professor Fujio Egami, upon finishing his chemistry course at the Faculty of Science, the University of Tokyo in 1933, started his own research work on sulfate esters and sulfatase under Professor T. Soda (1932–1942). During this period, he went to France to study and performed research on methanol oxidation in mice at the Institut de Chimie biologique de la Faculté de Médecine de l'Université de

446

Strasbourg under Professor M. Nicloux (1934–1935), which was followed by research on the microbial deamination of alanine at the Institut de Biologie Physico-chemique under Professor E. Aubel (1935–1936). From the former he learned of the importance of analysis in biochemistry, and from the latter he discovered that NO_3^- is utilized in organisms in lieu of O_2, which provided him with a clue for his exemplary work in later years at the Nagoya University, research on inorganic nitrogen metabolism centering upon nitrate reductase.

In 1942, Professor Egami moved to the Nagoya University and began his energetic research and lecturing in biochemistry. He became a professor of organic chemistry in the Department of Chemistry, the Faculty of Science in 1943 and was responsible for guiding many promising scientists into the field of the organic chemistry of natural products in Japan. In those days, there was no professorial position in biochemistry on the Science Faculty and only in 1953 did he become a professor of biochemistry when a lectureship was established there.

Just before moving to the Nagoya University, he spent about six months at the Institute of Infectious Disease of the University of Tokyo (presently the Institute of Medical Sciences) to learn about bacterial toxins under Professor S. Hosoya. This later became one of his major themes paralleling inorganic nitrogen metabolism, as mentioned above, at the Nagoya University. Thus, he proceeded with his biochemical study of a series of toxins such as *Shigella* exotoxin, streptolysin S and endotoxins of Gram-negative bacteria.

When scientific interchange activities with overseas laboratories became brisk in the 1950's, his research achievements on these two major themes received high evaluation around the world. His laboratory was always full of vigor and attracted many young students.

Professor Egami participated in the General Assembly for inaugurating the International Union of Biochemistry as the Japanese representative in 1955, and made great contributions in inviting the International Symposium on Enzyme Chemistry to Japan in 1957. Thereafter, for the nine years after 1961, he was very active as a member of the Council of the International Union of Biochemistry.

At the beginning of the 1950's, he began to get interested in nucleic acid from the viewpoint of the biological implications of nucleic acid, which enhances the formation of streptolysin S. After grasping the massive potential development in the biochemistry of nucleic acid from the research information pouring in from abroad after World War II, his assertion of the importance of the study of nucleic acid created a swirling enthusiasm for that study in the Japanese biochemical world, and he himself took up the study of nucleic acids and nucleic acid degrading enzymes. This provided him with the basis for making his "Discovery of RNase T_1" in 1957, his last year at Nagoya University.

In 1958, he was invited to come and assume a professorship in the Science Faculty of the University of Tokyo when the Department of Biophysics and Biochemistry was launched. In 1960, he published in *Nature* his prediction that the sequencing of RNA would be possible by the use of several species of ribonuclease having different specificities, which turned out to be the driving force that directed the course of world biochemistry in the 1960's toward "Structural Analysis of RNA."

At the same time, that Professor Egami was discovering and purifying a number of RNases to supply to the world's noted nucleic acid researchers, he also continued work in the protein chemistry of RNase T_1 and performed detailed studies from every angle on this protein discovered in Japan by forming a research team for furthering the study on the "Structure and Functions of RNase T_1."

His research work on bacterial toxins while he was at the Nagoya University was succeeded by study of the biosynthesis of streptolysin S and biochemical and genetic study of pyocin, the bacteriocin produced by *Pseudomonas aeruginosa*, which he went on to develop further. His study of polysaccharide sulfate originated from the study of chondroitin sulfate, and charonin sulfate (natural cellulose polysulfate) was succeeded by the study of the glycosidases of marine animals, in which many enzymes were discovered and made available for analyzing the structure of heteropolysaccharides and glycoproteins. Work falling in the sphere of inorganic nitrogen metabolism had not much time to develop at the University of Tokyo, but an elegant research work was accomplished on the biosynthesis of aromatic nitro groups.

Through these studies, many talented scientists spanning a wide spectrum were trained, and, making the most of their own characteristic talents, they are staffing the forefront of today's biochemical circles in Japan with some of them interested in the energy metabolism of organisms or immunochemistry, and others in the biochemistry of glycoconjugates, enzyme chemistry or nucleic acid chemistry.

In March 1971, Professor Egami retired as a university professor and terminated his teaching activities in biochemistry. In the course of his professorial services at university, he devoted himself uninterruptedly to Japan's science administration as a member of the Science Council of Japan, and he fulfilled his duties as President of the Science Council of Japan from 1969 through 1972. It should be added that he was President of the Japanese Biochemical Society (1968) and President of the Chemical Society of Japan (1977).

In June 1971, he took part in founding the Mitsubishi-Kasei Institute of Life Sciences and after incorporation he presided over the Institute as the first Director.

The Institute was and still is aimed at not only biochemical research but also at an integrated development of life sciences under the all-out support rendered by Mitsubishi Chemical Industries Limited, one of the leading industries of Japan today. While it is financially supported by a private business, the Institute does not pursue immediate profit but it was established as an institute in quest of development of the basic sciences, in other words, it is a realization of what is not possible within the framework of the individual Japanese university.

At this institute, Professor Egami, having dreamt of some day explaining the mechanisms of the human brain and the process of the genesis of organisms in the language of physical science, installed therein laboratories of neurophysiology, physiological psychology and developmental biology, and developed new scientific fields such as social life science, a phrase coined by him, and biogeochemistry, with the concept of the possible future impact of the development of life sciences that might wield adverse effects over human society and the earth. He also led a small research group to reopen the research on sulfate, which is his oldest and simultaneously his newest theme, and in addition he launched new research on chemical and early biological evolution. Based on his additional convincing evidence that "life originated from the sea," he discovered the formation of peptide bonds and protocell-like structures from simple compounds in modified sea media.

On November 21, 1980, his 70th birthday, Professor Egami retired as the Director of this novel and unique Institute. However, he has not retired from his own research life yet. As Honorary Director of the Institute and as President of the International Society for the Study of the Origin of Life, he still remains active, and is devoting himself to the study of chemical and early biological evolution with an ever-growing scholarly interest. Quite recently, he has come to advocate a working hypothesis on the origin of the genetic code. This is also the reason why those essays related to evolution occupy a relatively greater proportion in this volume.

Undeniably Prof. Egami's contribution toward bringing up Japan's biochemistry to today's level was great through the years of rapid growth after World War II, and helped to attract global attention to it. We all on the Editorial Committee shall be most pleased if this volume should turn out to be a monument commemorating Prof. Egami's varied and broad activities throughout the past five decades.

With our best wishes for Prof. Egami's best health and everlasting success.

Executive Editors
Makoto KAGEYAMA
Keiko NAKAMURA
Tairo OSHIMA
Tsuneko UCHIDA

Name Index

450